T0228100

NONLINEAR
WAVE
EQUATIONS

PURE AND APPLIED MATHEMATICS

A Program of Monographs, Textbooks, and Lecture Notes

EXECUTIVE EDITORS

Earl J. Taft
Rutgers University
New Brunswick, New Jersey

Zuhair Nashed
University of Delaware
Newark, Delaware

EDITORIAL BOARD

M. S. Baouendi
University of California,
San Diego

Jane Cronin
Rutgers University

Jack K. Hale
Georgia Institute of Technology

S. Kobayashi
University of California,
Berkeley

Marvin Marcus
University of California,
Santa Barbara

W. S. Massey
Yale University

Anil Nerode
Cornell University

Donald Passman
University of Wisconsin,
Madison

Fred S. Roberts
Rutgers University

Gian-Carlo Rota
Massachusetts Institute of
Technology

David L. Russell
Virginia Polytechnic Institute
and State University

Walter Schempp
Universität Siegen

Mark Teply
University of Wisconsin,
Milwaukee

MONOGRAPHS AND TEXTBOOKS IN
PURE AND APPLIED MATHEMATICS

1. *K. Yano,* Integral Formulas in Riemannian Geometry (1970)
2. *S. Kobayashi,* Hyperbolic Manifolds and Holomorphic Mappings (1970)
3. *V. S. Vladimirov,* Equations of Mathematical Physics (A. Jeffrey, ed.; A. Littlewood, trans.) (1970)
4. *B. N. Pshenichnyi,* Necessary Conditions for an Extremum (L. Neustadt, translation ed.; K. Makowski, trans.) (1971)
5. *L. Narici et al.,* Functional Analysis and Valuation Theory (1971)
6. *S. S. Passman,* Infinite Group Rings (1971)
7. *L. Dornhoff,* Group Representation Theory. Part A: Ordinary Representation Theory. Part B: Modular Representation Theory (1971, 1972)
8. *W. Boothby and G. L. Weiss, eds.,* Symmetric Spaces (1972)
9. *Y. Matsushima,* Differentiable Manifolds (E. T. Kobayashi, trans.) (1972)
10. *L. E. Ward, Jr.,* Topology (1972)
11. *A. Babakhanian,* Cohomological Methods in Group Theory (1972)
12. *R. Gilmer,* Multiplicative Ideal Theory (1972)
13. *J. Yeh,* Stochastic Processes and the Wiener Integral (1973)
14. *J. Barros-Neto,* Introduction to the Theory of Distributions (1973)
15. *R. Larsen,* Functional Analysis (1973)
16. *K. Yano and S. Ishihara,* Tangent and Cotangent Bundles (1973)
17. *C. Procesi,* Rings with Polynomial Identities (1973)
18. *R. Hermann,* Geometry, Physics, and Systems (1973)
19. *N. R. Wallach,* Harmonic Analysis on Homogeneous Spaces (1973)
20. *J. Dieudonné,* Introduction to the Theory of Formal Groups (1973)
21. *I. Vaisman,* Cohomology and Differential Forms (1973)
22. *B.-Y. Chen,* Geometry of Submanifolds (1973)
23. *M. Marcus,* Finite Dimensional Multilinear Algebra (in two parts) (1973, 1975)
24. *R. Larsen,* Banach Algebras (1973)
25. *R. O. Kujala and A. L. Vitter, eds.,* Value Distribution Theory: Part A; Part B: Deficit and Bezout Estimates by Wilhelm Stoll (1973)
26. *K. B. Stolarsky,* Algebraic Numbers and Diophantine Approximation (1974)
27. *A. R. Magid,* The Separable Galois Theory of Commutative Rings (1974)
28. *B. R. McDonald,* Finite Rings with Identity (1974)
29. *J. Satake,* Linear Algebra (S. Koh et al., trans.) (1975)
30. *J. S. Golan,* Localization of Noncommutative Rings (1975)
31. *G. Klambauer,* Mathematical Analysis (1975)
32. *M. K. Agoston,* Algebraic Topology (1976)
33. *K. R. Goodearl,* Ring Theory (1976)
34. *L. E. Mansfield,* Linear Algebra with Geometric Applications (1976)
35. *N. J. Pullman,* Matrix Theory and Its Applications (1976)
36. *B. R. McDonald,* Geometric Algebra Over Local Rings (1976)
37. *C. W. Groetsch,* Generalized Inverses of Linear Operators (1977)
38. *J. E. Kuczkowski and J. L. Gersting,* Abstract Algebra (1977)
39. *C. O. Christenson and W. L. Voxman,* Aspects of Topology (1977)
40. *M. Nagata,* Field Theory (1977)
41. *R. L. Long,* Algebraic Number Theory (1977)
42. *W. F. Pfeffer,* Integrals and Measures (1977)
43. *R. L. Wheeden and A. Zygmund,* Measure and Integral (1977)
44. *J. H. Curtiss,* Introduction to Functions of a Complex Variable (1978)
45. *K. Hrbacek and T. Jech,* Introduction to Set Theory (1978)
46. *W. S. Massey,* Homology and Cohomology Theory (1978)
47. *M. Marcus,* Introduction to Modern Algebra (1978)
48. *E. C. Young,* Vector and Tensor Analysis (1978)
49. *S. B. Nadler, Jr.,* Hyperspaces of Sets (1978)
50. *S. K. Segal,* Topics in Group Kings (1978)
51. *A. C. M. van Rooij,* Non-Archimedean Functional Analysis (1978)
52. *L. Corwin and R. Szczarba,* Calculus in Vector Spaces (1979)

Additional Volumes in Preparation

NONLINEAR WAVE EQUATIONS

Satyanad Kichenassamy

University of Minnesota
Minneapolis, Minnesota

Marcel Dekker, Inc.   New York•Basel•Hong Kong

Library of Congress Cataloging-in-Publication Data

Kichenassamy, Satyanad
    Nonlinear wave equations / Satyanad Kichenassamy.
      p. cm. — (Monographs and textbooks in pure and applied mathematics;
  194)
    Includes bibliographical references (p. -) and indexes.
    ISBN 13: 978-0-8247-9328-9
    1. Nonlinear wave equations. I. Title. II. Series.
  QA927.K43 1996
  531'.1133'01515355—dc20

                                                 95-24573
                                                     CIP

The publisher offers discounts on this book when ordered in bulk quantities. For more information, write to Special Sales/Professional Marketing at the address below.

This book is printed on acid-free paper.

Copyright © 1996 by MARCEL DEKKER, INC. All Rights Reserved.

Neither this book nor any part may be reproduced or transmitted in any form or by any means, electronic or mechanical, including photocopying, micro-filming, and recording, or by any information storage and retrieval system, without permission in writing from the publisher.

MARCEL DEKKER, INC.
270 Madison Avenue, New York, New York 10016

Current printing (last digit):
10 9 8 7 6 5 4 3 2 1

PRINTED IN THE UNITED STATES OF AMERICA

To my parents

Preface

This volume is devoted to the mathematical aspects of nonlinear wave propagation, with special emphasis on nonlinear hyperbolic problems. Its aim is to introduce the reader to those tools which proved most effective in recent years for the problems of *global existence, singularity formation and large time behavior of solutions,* and in the study of *perturbation methods.*

Wave equations describe the propagation of physical quantities. While the most important of these equations are hyperbolic, many of their *approximations* are not, and very different types are encountered in the literature.

Basic issues are singularity formation and asymptotic behavior. One may distinguish between "weak" singularities, in which the top order derivatives are more singular than the other terms, and "strong" ("blow-up") singularities where lower order terms compensate higher order derivatives exactly. The complementary issue is global existence. If singularities do form, one would like to find an expansion of the solution near its singularities, and if possible continue the solution after blow-up. If global existence has been proved, one would like to have information about the *large time behavior* of the solution, and only two ideas seem to be available to this end: approximate the equation by a linear one ("nonlinear scattering"), or by a simple nonlinear model, generally one possessing *solitons.*

A detailed description of wave propagation may then be made in terms of the evolution of: the unknown itself; the modulation of its rapid oscillations; the discontinuities, or other singularities, of the unknown or its derivatives. These paradigms lead to: particle-like solutions (such as solitons); envelope equations such as the nonlinear Schrödinger equation; propagation of singularities and microlocal analysis.

General issues of existence and global regularity are nowadays best settled by the use of abstract iteration or fixed-point techniques, and recent, sophisticated estimates for linear equations have proved useful to this end. On the other hand, a detailed description of singularity formation and asymptotic behavior is based on formal asymptotic representations. *Perturbation theory* is concerned with the construction and evaluation of such formal models, in the hope that their first few terms represent accurately the solution under study. The recent revival of the study of analytic solutions of nonlinear wave equations has proved helpful in this direction.

Such are the main themes treated here. They will be seen to provide a convenient way to introduce the main directions of current research in nonlinear wave equations. We now briefly describe the contents of this monograph.

Chapter 1 is introductory, and recalls a few ideas on linear wave propagation with some background material.

Chapter 2 develops current results on local and global existence, for holomorphic or H^s solutions, emphasizing recent developments. The choice of topics roughly follows the historical evolution of the subject, which seems helpful to motivate the direction of current research.

Chapter 3 is devoted to the problem of singularity formation, with emphasis on perturbations of the wave equations. It is shown in particular that suitable changes of variables enable one to "unfold" the process of singularity formation in many cases, thereby providing a detailed description of the blow-up mechanism. A brief account of the very useful theory of shock waves for hyperbolic conservation laws is included. Recent results on microlocal methods for the propagation and interaction of singularities in high-order derivatives are also described. Notions on weakly nonlinear geometrical optics are included in Chapter 5.

Chapter 4 is devoted to the theory of solitons. Their importance as universal models for intrinsically nonlinear waves justifies an emphasis on the rather limited class of equations which possess solitons in the strict sense. This chapter gives a mathematical account of this multi-faceted theory. Many aspects of the "Inverse Scattering Transform," which provides a mathematical description of the main properties of solitons, still await justification, and some rigorous tools are given to this end.

Chapter 5 deals with perturbation methods. The introduction of formal solutions and the study of their validity have formed by far the most effective way of obtaining useful detailed information on nonlinear waves. Reduction to an implicit function theorem, possibly of Nash-Moser type, is shown to justify a large class of expansions. Among very common methods for nonlinear waves, soliton perturbation, Whitham's method and high-frequency asymptotics are presented. Although rigorous results in this area are relatively few, especially for the former method, its importance in applications justifies their inclusion. A dynamical systems approach which was successful in resolving long-standing questions for water waves is included.

Chapter 6 deals with General Relativity, which has of late attracted many mathematicians. In view of the greater amount of differential geometry involved in this subject, it was included as a separate chapter, although General Relativity represents in many ways the culmination of the theory of nonlinear wave propagation. Treatment is limited to the special difficulties in the problems of existence and singularity formation for Einstein's equations.

The construction of numerical schemes, which is a separate issue, has not been discussed. In fact, while there is a sizable, very useful, numerical literature on shock waves and solitons, there are relatively few papers on semilinear hyperbolic equations. Also, the mathematical status of this literature has not been established in many cases.

Parts of this book were taught as a Special Topics course in PDE, and within a first graduate course in PDE at the University of Minnesota, essentially at the pace of one chapter per quarter. Exercises are included with every chapter and often summarize the content of a research article. A good background in PDE at the beginning graduate level should be enough to read the volume with profit, although a few sections are meant for more experienced readers. Conscious attempt has been made to avoid duplicating material available in standard monographs. For this reason, the reader is referred to the notes in each chapter for sources and for proofs which have been outlined or omitted. Despite its size, the bibliography is not meant to be complete; it should nevertheless enable the reader to reconstruct the literature from recent references. Each chapter has a separate bibliography.

A subject index and a notation index are included; since the table of contents is fairly detailed, its entries have not been reproduced in the index.

Satyanad Kichenassamy

Contents

Contents

Chapter 1

Linear Wave Propagation

This chapter presents some material on linear PDE with constant coefficients which will be needed in later chapters. Four basic examples will be used:

(a) The *wave equation* $\Box u := (\partial_t^2 - \Delta)u = 0$, and the closely related *Klein-Gordon* (KG) equation $(\Box + m^2)u = 0$. Here Δ denotes the Laplace operator in $\mathbf{R}^n$, and m is a positive number;

(b) The *Schrödinger equation* (without potential): $iu_t - \Delta u = 0$;

(c) The *Airy equation*: $\partial_t u + \partial_x^3 u = 0 \quad x \in \mathbf{R}$.

After reviewing in §1.1 a few properties of the Fourier Transformation, we discuss the Sobolev inequality and related estimates in §1.2.

We then turn to the initial-value problem in §1.3, which is readily handled by Fourier transformation in the space variables. Solvability of the initial-value problem in spaces based on L^2 requires a condition on the "symbol" of the equation. Section 1.3 also includes a brief discussion of hyperbolicity, and gives representation formulæ for the solution of the wave and Klein-Gordon equations with sufficiently regular data.

Precise information on solutions with less regular data can be obtained, by density, from estimates on smooth solutions (§1.4). These results actually take the form of pointwise *decay estimates* as $t \to \infty$, and therefore provide information on asymptotic behavior as well. They are proved here for the wave and Klein-Gordon equations; they play a role comparable to that of L^p estimates for the Laplacian.

This study of overall regularity is followed, in §1.5, by a few results on the propagation of singularities: a solution which is, say, locally square-summable in space and time may in fact be of class C^∞ except on a small set which "evolves in time." This situation has been considered from two perspectives:

1

1. By postulating that the singularity occurs precisely across a prescribed hypersurface, where some derivatives of the solution have a jump;

2. By relating the normals to possible singularity surfaces to directions of rapid decay of the Fourier transform of "localizations" of the solution.

The second view-point can in fact be used to describe much more general singularities, which need not be localized on hypersurfaces. A closely related and classical construction, the "method of geometrical optics," is also described.

Finally, §1.6 recalls and motivates the usual definition of a "wave" as the propagation of a disturbance.

A few more specialized results that were omitted in the text are included among the problems (§1.6). Some familiarity with the basic tools of PDE is assumed; for background information, the reader is referred to the texts by Hörmander (1983) (vol. 1), Courant-Hilbert (1962), John (1989), Garabedian (1964), Trèves (1975); see also Stein (1970, 1993) and Stein-Weiss (1971) for a more thorough coverage in harmonic analysis.

The notation is more or less standard, and is summarized at the end of the volume; we occasionally use $\|u; X\|$ for the norm of u in the space X.

1.1   THE FOURIER TRANSFORM

> After defining the Fourier transform on smooth, rapidly decreasing functions, we briefly recall the inversion theorem, the Parseval relation and the Paley-Wiener(-Schwartz) theorem. The Fourier transform of Gaussians is reviewed, and elementary applications to the wave equation are given.

Generalities

The Fourier transform of $u \in L^1(\mathbf{R}^n)$ is defined by

$$\hat{u}(\xi) := \mathcal{F}u(\xi) := \int_{\mathbf{R}^n} e^{-i(x,\xi)} u(x) \, dx.$$

One can extend $\mathcal{F}$ to much more singular distributions by finding a space of very smooth functions which is invariant under $\mathcal{F}$, and by then appealing to a duality argument:

Let us say that u belongs to the Schwartz space $\mathcal{S}$ if $u \in C^\infty$ and

$$\forall \alpha, \beta, \quad \sup_{x \in \mathbf{R}} |x^\alpha D^\beta u(x)| \leq C_{\alpha,\beta},$$

where $x^\alpha = x_1^{\alpha_1} \ldots x_n^{\alpha_n}$ and $D^\beta = D_1^{\beta_1} \ldots D_n^{\beta_n}$, with $D_j = -i\partial_j = -i\partial/\partial x_j$.

Integration by parts shows that $\mathcal{F}$ maps $\mathcal{S}$ to itself.

The Fourier transform is its own tranpose with respect to the usual L^2 pairing:

LEMMA   *If $u, v \in S$, then*

$$\int v \mathcal{F} u \, dx = \int u \mathcal{F} v \, dx = \int\int u(x)v(\xi)e^{-i(x,\xi)} \, dx \, d\xi.$$

The proof is immediate.

Let φ be a linear functional on S. We say that it is continuous, and write $\varphi \in S'$, if there are constants C, k and l such that for any $u \in S$,

$$|\varphi(u)| \leq C \sup_x (1 + |x|)^k \sum_{|\alpha| \leq l} |D^\alpha u(x)|.$$

Such functionals are called temperate distributions. We often write (φ, u) for $\varphi(u)$. Any compactly supported distribution is in S'. The smallest possible value of l is called the *order* of φ. One defines the Fourier Transformation on S' by

$$(1.1) \qquad\qquad (\mathcal{F}\varphi)(u) = \varphi(\mathcal{F}u).$$

One checks that (1.1) does define an element of S'. If we identify a function $u \in S$ with the functional defined by $\varphi(v) = \int vu \, dx$, the above lemma ensures that $\mathcal{F}\varphi(v) = \int (\mathcal{F}u)v \, dx$. This definition therefore agrees with the previous one on S.

Basic properties

By integration by parts, one derives

$$(1.2) \qquad\qquad \mathcal{F}(D_j u) = \xi_j \mathcal{F}u \quad \text{and} \quad \mathcal{F}(x_j u) = -D_j \mathcal{F}u,$$

which is valid in S and, by duality, in S'.

The *Fourier Inversion Theorem* enables one to recover u from $\mathcal{F}u$.

THEOREM 1.1 *For any $u \in S'$, one has $\mathcal{F}(\mathcal{F}u) = (2\pi)^n Ru$, where $Ru(x) := u(-x)$.*

Proof: It suffices to prove the result in S. Let $G = R\mathcal{F}^2$. By (1.2), G commutes with D_j and with multiplication by x_j. Let us show that these properties imply that G is a constant multiple of the identity. The constant will be determined from a special case.

Let therefore $u \in S$, with $u(0) = 0$. We claim $Gu(0) = 0$ as well. To see this, let us write $1 = \varphi_1(x) + \varphi_2(x)$, where $\varphi_1, \varphi_2 \in C^\infty$, $\varphi_1 = 0$ for $|x| > 2$, and $\varphi_2 = 0$ for $|x| < 1$. One can then write $u = \sum_j x_j w_j$, where

$$w_j = \varphi_1 \int_0^1 \partial_j u(tx) \, dt + \varphi_2 u x_j / |x|^2.$$

By inspection, the functions w_j are in S, and therefore $Gu = \sum_j x_j G w_j$, which vanishes for $x = 0$.

More generally, let $u, v \in S$. If $u - v$ vanishes at some point x_0, so does $Gu - Gv = G(u - v)$. Now take $v(x) = u(x_0)w(x)$, with $w(x) = \exp(-(|x|^2 - |x_0|^2)/2)$. We find

$$Gu(x_0) = u(x_0)Gw(x_0).$$

Since we prove in the lemma below that $Gw(x_0) = (2\pi)^n$, the proof is complete.

One thus recovers, for $u \in S$, the familiar formula

$$u(x) = (2\pi)^{-n} \int_{\mathbf{R}^n} \mathcal{F}u(\xi) e^{i(x,\xi)} \, d\xi.$$

We now compute, as announced, the Fourier transform of an exponential:

LEMMA   *For any non-singular symmetric matrix A with a positive definite real part,*

$$(1.3) \qquad \mathcal{F}[\exp(-(Ax, x)/2)] = \frac{(2\pi)^{n/2}}{\sqrt{\det A}} \exp(-(A^{-1}\xi, \xi)/2).$$

Remark 1: The determination of the square root is the one which gives 1 when $A = I$; it is well-defined, by analytic continuation, since the set of matrices under consideration is convex.

Remark 2: Using the continuity of the Fourier transform, we may extend this formula to the case when $A = -iA_0$, A_0 real. The result is

$$(1.4) \quad \mathcal{F}[\exp((iA_0 x, x)/2)] = \frac{(2\pi)^{n/2}}{\sqrt{|\det A_0|}} \exp(-(A^{-1}\xi, \xi)/2 - i\pi \mathrm{sgn}(A_0)/4),$$

where $\mathrm{sgn} A_0$ is the number of negative eigenvalues of A_0.

Proof: Let $u = \exp(-(Ax, x)/2)$. Since

$$(D_j - i(Ax)_j)u = 0,$$

we have

$$(\xi_j + i(AD)_j)\hat{u} = 0.$$

Multiplying by iB, where B is the inverse of A, we find $(i(B\xi)_j - D_j)\hat{u} = 0$, so that $\hat{u}\exp((B\xi, \xi)/2) = C$ (constant). If A is real, we may write $A = {}^tP\Lambda P$ with P orthogonal and $\Lambda = \text{diag}(\lambda_1, \ldots, \lambda_n)$. Therefore

$$\hat{u}(0) = \int e^{-(\Lambda y, y)/2}\,dy = \prod_1^n (2\pi/\lambda_j)^{1/2} = (2\pi)^{n/2}/\sqrt{\det A}.$$

This proves the result for real A. The general case follows by analytic continuation. This completes the proof.

A classical consequence of the Fourier inversion theorem is

THEOREM 1.2 *For u and v in $\mathcal{S}$,*
  (i) $\int u\bar{v}\,dx = (2\pi)^{-n}\int \mathcal{F}u\overline{\mathcal{F}v}\,d\xi$;
  (ii) $\mathcal{F}(u * v) = (\mathcal{F}u)(\mathcal{F}v)$, and $\mathcal{F}(uv) = (2\pi)^{-n}(\mathcal{F}u) * (\mathcal{F}v)$. In these formulae, bars indicate complex conjugate.

Part (i) is the Parseval relation; when $u = v$, it shows that $\mathcal{F}$ is, upto a constant factor, an isometry in L^2 (Plancherel); to prove (i), use (1.2). The first claim in (ii) is straightforward; the second follows by Fourier transforming the first.

The Fourier transform of a compactly supported distribution u has an analytic extension as an entire function. Its rate of growth in the imaginary directions actually characterizes the size of the support of u. More precisely we have the Paley-Wiener-Schwartz theorem:

THEOREM 1.3 *A temperate distribution is supported in the ball of radius R and has order N or less if and only if its Fourier transform is entire and satisfies, for some constant C,*

$$|\hat{u}(\zeta)| \le C(1 + |\zeta|)^N e^{R|\text{Im}\zeta|}.$$

Remark: A related, very useful, result is that $u \in L^2(\mathbf{R})$ is supported by $\{x \le 0\}$ if and only if $\hat{u}$ has an analytic extension to $\text{Im}\,\zeta > 0$ and its L^2 norm on the sets $\{\text{Im}\,\zeta = \text{const.}\}$ are uniformly bounded. Equivalently, u belongs to the Hardy space H^2. All these results are based on the same basic idea of pushing the contour of integration in the inversion formula into the complex domain.

First applications

The Fourier transform gives immediately the solution to the initial-value problem with data on $(t = 0)$ for our four basic examples, if the data are very well-behaved. Take for instance the Cauchy problem for the Klein-Gordon equation:

$$(1.5) \qquad \begin{cases} \Box u + m^2 u = 0; \\ u(x,0) = f(x), \quad u_t(x,0) = g(x). \end{cases}$$

The functions f and g are called the Cauchy data. The Cauchy problem with data on a more general surface Σ can be reduced to an initial-value problem on $\{t = 0\}$ for a variable-coefficient equation, after straightening the initial surface by a change of coordinates. If Σ is limited to the past, it can also be dealt with by extending the solution by zero to the past of Σ; this extension now solves an inhomogeneous equation with vanishing data on $\{t = -a\}$ if a is very large and positive.

Let $\hat{u}(t, \xi)$ be the Fourier transform of u with respect to the space variables. We are led to an ODE which is readily solved; the net result is that *if f and g are in $\mathcal{S}$, there is a solution which is in $\mathcal{S}$ for every fixed t.* It is given by

$$(1.6) \qquad \hat{u}(t, \xi) = \hat{f} \cos(t\sqrt{m^2 + |\xi|^2}) + \hat{g} \frac{\sin(t\sqrt{m^2 + |\xi|^2})}{\sqrt{m^2 + |\xi|^2}}.$$

One verifies directly the *energy identity*:

$$(1.7) \qquad \int (m^2 |u|^2 + |\nabla u|^2 + |u_t|^2)\, dx \quad \text{is independent of } t.$$

This is a special case of Noether's theorem, which gives a systematic procedure for deriving such identities for equations having a variational formulation (see the references in the Notes). The case of the wave equation ($m = 0$) is settled similarly.

For both equations, note that if u_g (resp. (u_f)) is the solution with data $(0, g)$ (resp. $(0, f)$), we have

$$u = u_g + \frac{d}{dt} u_f.$$

Applying the Paley-Wiener-Schwartz theorem, we find

THEOREM 1.4 *If $\Box u + m^2 u = 0$ and u and u_t vanish for $|x| > R$, then $u \equiv 0$ for $|x| > R + |t|$.*

Remarks: This means that the support travels "with speed 1 at most." One can show by examples that the speed is indeed one. Another proof of this follows from the representation formulae derived in §1.3. It will show in addition a stronger property of the wave equation in an *odd* number of space dimensions: if u does not vanish at (x, t), then there is a point x_0 on the *sphere* of radius t about x at which the Cauchy data do not vanish. This latter property, known as the (strong) Huygens principle, is very seldom valid; a related property holds also for the equations of three-dimensional linear elasticity. In other words, data generally have an effect throughout their sphere of influence and not only at its boundary. However, it is sometimes possible to show that the influence in any given region is very small for large t (local energy decay, see Lax-Phillips (1989), and also Friedlander's formula below).

The uniqueness of the solutions follows quite generally from Holmgren's theorem. Another proof follows from the energy identity (1.7). We next need to investigate the regularity of solutions corresponding to less regular data; this is accomplished in §1.4.

For the inhomogeneous equation $\Box u + m^2 u = h(x,t)$, we must add to (1.6)

$$\int_0^t \frac{\sin((t-s)\sqrt{m^2 + |\xi|^2})}{\sqrt{m^2 + |\xi|^2}} \hat{h}(\xi, s) \, ds,$$

which follows from the formula of variation of parameters. In this context, this formula is known as *Duhamel's principle*, and is interpreted as giving the solution of the inhomogeneous problem by superposition of solutions of homogeneous problems with data of the form $(0, g)$. One frequently sees it written in the form

$$u(.,t) = \int_0^t R(.,t-s) * h(.,s) \, ds,$$

and $R(x,t)$ is often called the Riemann function; it may be a distribution.

1.2 THE SOBOLEV INEQUALITY

A few versions of the Sobolev inequality are reviewed. In addition to its standard versions, regarding the spaces $W^{m,p}$ and L_s^p defined below, we mention the global Sobolev inequality, where derivatives are replaced by certain vector fields with linear coefficients. The latter is closely related to the decay properties of the wave equation. We also state a form of the Gagliardo-Nirenberg inequalities, with an application to the action of nonlinear functions in Sobolev spaces.

The standard Sobolev inequality

This inequality relates the integrability of the gradient of a function to its integrability or continuity. We state the results for smooth functions; they can be extended by density to less regular functions, as usual. We use (throughout this volume) $\|u\|_p$ to denote the L^p norm of u.

THEOREM 1.5 *If u is smooth and compactly supported in $\mathbf{R}^n$, then*

(i) *If $1 \le p < n$, $\|u\|_{p^*} \le C(p,n)\|\nabla u\|_p$, where $1/p^* = 1/p - 1/n$.*

(ii) *If $p = n \ge 2$, $\| \exp(c|u|/\|\nabla u\|_n)^{n/(n-1)} \|_1 \le C(n) meas(supp(u))$, for suitable constants c and C.*

(iii) *If $p > n$, $|u(x) - u(y)| \le C(\|\nabla u\|_p)|x - y|^{1-n/p}$.*

Remark 1: This theorem is loosely referred to as the Sobolev inequality. The case $p = n$ is due to Trudinger and Moser, and the case $p > n$ goes back to Morrey. The case $p = 1$, from which (i) follows, essentially goes back to Gagliardo (1958) and Nirenberg (1959). See problem 11(c) for Sobolev's proof.

Remark 2: The completion of C_0^∞ for the norm $\sum_{|\alpha| \leq m} \|\nabla u\|_p$ is, if $p < \infty$, the Sobolev space $W^{m,p}(\mathbf{R}^n)$. Its norm will be written $\|u\|_{m,p}$. The space $W^{m,\infty}$ consists of functions with Lipschitzian $(m-1)$th derivatives. The space $W^{m,2}$ is also denoted by $H^m(\mathbf{R}^n)$. More generally, one defines, for s real, the Sobolev space H^s as the space of locally square-summable functions such that the H^s norm

$$|u|_s := \left(\int (1 + |\xi|^2)^s |\hat{u}|^2 \, d\xi \right)^{1/2}$$

is finite. It is a Hilbert space. There is a natural identification of H^{-s} with the dual of H^s.

Remark 3: The space $W^{m,p}$ is embedded in L^q for $1/q = 1/p - m/n$, if $mp < n$, and in C^0 if $mp > n$. This follows easily from theorem 1.5. Elements of $W^{n/p,p}$ are not continuous in general, if $p > 1$. However, if $p = 1$: $W^{n,1} \subset C^0$. This can be proved directly, but we will obtain it as a consequence of L^1-L^∞ estimates for the wave equation.

Idea of Proof: (for more details, see Nirenberg (1959) or Adams (1975)) We give the proof of (i), and very briefly comment on the other two inequalities. We first assume $p = 1$. If $u \in C_0^\infty$, we have $u(x) = \int_{-\infty}^{x_i} |D_i u| \, dx_i$, and therefore,

$$|u(x)|^{n/(n-1)} \leq \left(\prod_{i=1}^{n} \int_{-\infty}^{x_i} |D_i u| \, dx_i \right)^{1/(n-1)}.$$

We now integrate with respect to each variable, one at a time. The key observation is that at each stage, there are exactly $n-1$ factors in the product which undergo the integration; this allows the use of Hölder's inequality in the form

$$\|f_1 \dots f_{n-1}\|_1 \leq \prod_j \|f_j\|_{n-1}.$$

We find successively

$$\int |u|^{n/(n-1)} \, dx_1 \dots dx_n$$
$$\leq \int dx_2 \dots dx_n \left(\int_{\mathbf{R}} |D_1 u| \, dx_1 \right)^{1/(n-1)} \left(\prod_{j>1} \iint |D_j u| \, dx_1 \, dx_j \right)^{1/(n-1)}$$
$$\leq \int dx_3 \dots dx_n \left(\iint_{\mathbf{R}^2} |D_2 u| \, dx_1 \, dx_2 \right)^{1/(n-1)} \left(\iint |D_1 u| \, dx_1 \, dx_2 \right)^{1/(n-1)} \times$$

$$\times \quad \left(\prod_{j>2} \iiint |D_j u|\, dx_1\, dx_2\, dx_j\right)^{1/(n-1)}$$

$$\leq \int dx_{l+1}\ldots dx_n \left(\prod_{k\leq l} \int\cdots\int |D_k u|\, dx_1\ldots dx_l\right)^{1/(n-1)}$$

$$\times \left(\prod_{k>l} \int\cdots\int |D_k u|\, dx_1\ldots dx_l\, dx_k\right)^{1/(n-1)}$$

$$\leq \left(\prod_{j=1}^{n} \int |D_j u|\, dx\right)^{1/(n-1)} \leq C\|\nabla u\|_1^{n/(n-1)},$$

which proves the result for $p = 1$. The result for $1 < p < n$ follows by applying this result to $u|u|^{\gamma-1}$, with $\gamma = p(n-1)/(n-p)$.

Case (ii) ($p = n$) can be proved by writing the exponential as the sum of its Taylor series, or by using a suitable form of the Hardy-Littlewood-Sobolev inequality. More generally, if $n = mp$, $p > 1$,

$$\int \exp\left[\left(\frac{|u|}{c_{n,m}\|u\|_{m,p}}\right)^{p/(p-1)}\right]\, dx \leq C_{n,m}|\text{supp}\,(u)|.$$

The modulus of continuity for $p > n$ can be obtained from a version of Poincaré's inequality; the following intermediate result may be noted:

$$|u - u_B| \leq C R^{1-n/p}\|\nabla u\|_p,$$

where B is a ball, and u_B the average of u over B.

Let us now define a closely related family of spaces: We say that $u \in L^p_s$ if $u \in L^p$ and $\mathcal{F}^{-1}(1 + |\xi|^2)^{s/2}|\hat{u}| \in L^p$. We write its norm as

$$|u|_{s,p} := \|\mathcal{F}^{-1}((1 + |\xi|^2)^{s/2}\hat{u})\|_p.$$

This space coincides with H^s for $p = 2$, and with $W^{s,p}$ if s is an integer. It also satisfies the Sobolev inequality for $1 < p < \infty$ (exercise 11).

The next result shows how to estimate kth derivatives in terms of mth derivatives if $m > k$. It is known as the *Gagliardo-Nirenberg inequality*.

THEOREM 1.6 *One has* $\|D^k u\|_q \leq C(n,p,q,k,m)\|u\|_r^{\theta}\|D^m u\|_s^{1-\theta}$, *provided that*

$$1/q - k/n = \theta/r + (1-\theta)(1/s - m/n),$$

and $1 \leq r, s \leq \infty$, $k < m$, $q \geq 0$, $k/m \leq 1 - \theta \leq 1$. *The result is not valid if* $\theta = 1$, *when* $s = n/(m-k) \neq 1$.

The case $1 - \theta = k/m$ is particularly useful in practice.

As an application, let us prove an estimate on the action of nonlinear functions on Sobolev spaces.

THEOREM 1.7 *Let $F \in C^l(\mathbf{R})$ with $l \geq 1$ and $F(0) = 0$. Then, if $|u| \leq M$, $l \geq m$ and $mp > n$, one has*

$$\|F(u)\|_{m,p} \leq C(M, F, m, p, n)\|u\|_{m,p}.$$

Remark: The point is that the r.h.s. grows linearly with the norm of u even though derivatives of $F(u)$ involve products of derivatives of u. A similar statement is valid for vector-valued functions u; this yields an estimate for the product of two functions. The easy proof of this generalization is left to the reader.

Proof: We must estimate terms of the form

$$H(u)D_{a_1}^{k_1}u \ldots D_{a_l}^{k_l}u,$$

where $\sum_j k_j \leq m$. The Gagliardo-Nirenberg inequality gives

$$\|D_{a_j}^{k_j}u\|_p \leq C\|u\|_\infty^{1-\theta_j}\|D^m u\|_m^{\theta_j},$$

where $\theta_j = (n - k_j p)/(n - mp)$. Since u is bounded, we estimate $H(u)$ by a constant and find, for the desired term, the estimate

$$CM^{l-\theta}\|D^m u\|_p^\theta,$$

where $\theta = \sum_j \theta_j \leq 1$. We now use the Sobolev inequality to estimate $M^{l-\theta}$ by $CM^{l-1}\|D^m u\|_p^{1-\theta}$. This ends the proof.

There is a variant of the Sobolev inequality which rapidly yields decay results for the wave and Klein-Gordon equations. It is the "invariant," or "global" Sobolev inequality. To state it, we must introduce a few Lie algebras related to the Lorentz group.

The Lorentz and conformal groups

Note: We will, in this paragraph (as in chapter 6), write indices in their proper position: up for coordinates and contravariant objects, down for forms and other covariant quantities. In other words, we have so far been raising and lowering indices with respect to the Kronecker delta, but this is not consistent (nor convenient) in Minkowski space.

Recall that the Lorentz group is the group of linear transformations of Minkowski space $\mathbf{R}^{n+1}$ which leave invariant the "line element"

$$(1.8) \qquad\qquad ds^2 := \eta_{ab}\, dx^a\, dx^b, \quad (0 \leq a, b \leq n),$$

where $x^0 = t$ and $\eta_{ab} = \delta_{ab} - 2\delta_a^0\delta_b^0$. The usual summation convention on repeated indices in different positions (one up, one down) is used. The Lorentz group is isomorphic to the group O(3,1). It is generated by spatial rotations and "boosts" (*i.e.* special Lorentz transformations). It is a Lie group, and its Lie algebra is generated by vector fields of the form

$$X^c\partial_c = \omega^{ab}M_{ab} := \omega^{ab}(x_a\partial_b - x_b\partial_a),$$

where $\omega^{ab} = -\omega^{ba}$ are constants; $M_{0\alpha}$, $\alpha > 0$ correspond to boosts, and the others to spatial rotations.

The Poincaré group, also called the inhomogeneous Lorentz group, is generated by the elements of the Lorentz group and translations, which contribute the fields $P_a = \partial_a$ to the Lie algebra.

The conformal group is the group of all transformations which transform the line element into $\Lambda(x^a)\,ds^2$. It is a non-trivial exercise to show that it is generated, if $n \geq 2$, by: the elements of the Poincaré group, dilations ($x^a \mapsto \lambda x^a$ with λ constant), and a four-parameter family of nonlinear transformations:

$$(x^a) \mapsto \frac{x^a - A^a x^c x_c}{1 - 2A^c x_c + (A^b A_b)(x^c x_c)}.$$

These latter transformations can also be written

$$g_v(x) := \mathcal{I}_0 T_v \mathcal{I}_0,$$

where $T_v(x) := x + v$, and $\mathcal{I}_{x_0}(x) := (x - x_0)/[(x - x_0)^a(x - x_0)_a]$ are the "Lorentz inversions." One may therefore recover all conformal transformations as products of Lorentz transformations, dilatations, translations, and Lorentz inversions, even though $\mathcal{I}_0$ does not belong to the connected component of the identity in this group.

It is easy to check that the dilatations are generated by $S = x^a\partial_a$; using the above formula, we find, corresponding to the nonlinear conformal transformations, the generators

$$K_a = (2x^c x_a - x^b x_b \delta_a^c)\partial_c.$$

The Lie algebra of the conformal group is generated by the fields P_a, M_{ab}, S and K_a. The commutation relations are

$$
\begin{aligned}
[M_{ab}, M_{cd}] &= \eta_{ad}M_{bc} - \eta_{ac}M_{bd} + \eta_{bc}M_{ad} - \eta_{bd}M_{ac}; \\
[M_{ab}, P_c] &= \eta_{bc}P_a - \eta_{ac}P_b; \\
[P_a, P_b] &= 0; \quad [M_{ab}, S] = 0; \quad [K_a, K_b] = 0; \\
[P_a, S] &= P_a; \\
[K_a, S] &= -K_a; \\
[M_{ab}, K_c] &= \eta_{bc}K_a - \eta_{ac}K_b.
\end{aligned}
$$

A weighted Sobolev inequality.

After these preliminaries, let us introduce Sobolev-type norms associated with the above Lie algebras, and give an embedding theorem for the corresponding spaces.

We let

$$(1.9) \qquad [u]_{m,p} = \sum_{l \le m} \|X_1 \ldots X_l u\|_p,$$

where the sum extends over all possible products of m vector fields chosen among the generators of the Lorentz group. If the scaling operator is included, we denote the resulting norm by $[u]'_{m,p}$. We also consider functions $u(x,t)$ of $n+1$ variables. The result is

THEOREM 1.8 *For any smooth function with compact support, we have*

1. $|u(x,t)| \le C(1 + t + |x|)^{-(n-1)/p} \sup_{t \ge 0}([u]_{m,p} + \|u\|_{m,p}),$

2. $|u(x,t)| \le C(1 + t + |x|)^{-(n-1)/p}(1 + |t - |x||)^{-1/p} \sup_{t \ge 0}([u]'_{m,p} + \|u\|_{m,p}),$

provided that $1 \le p < \infty$ and $m > n/p$.

Proof: See Hörmander (1985) and Klainerman (1987) and their references.

1.3 THE INITIAL-VALUE PROBLEM

An evolution problem, linear or not, is said to be well-posed in a space X if solutions with data in X at time $t = 0$ (i) exist for later times; (ii) are uniquely determined by their data; and (iii) depend continuously on these data in an appropriate topology. For linear problems, this means that the solution operator is bounded from X to itself, or to some other space Y. We focus here on this issue for hyperbolic equations.

The Fourier transformation gives easy first results for contant-coefficient systems, and helps motivate the notion of hyperbolicity. The non-persistence of L^p-based regularity conditions for hyperbolic equations explains the central role of H^s spaces in the theory, as compared with the case of parabolic equations. Definitions of hyperbolicity are given next.

For the case of the wave and Klein-Gordon equations, a few representation formulæ for the solution of the Cauchy problem, in preparation for the proof of decay estimates.

Two types of questions can be raised about an evolution equation:

(a) (Cauchy problem) If f has support in a half-space ($t < 0$), is there a solution with the same property? If smooth, this solution would vanish, together with some of its t derivatives, for $t = 0$.

(b) (Initial-value problem) Is there a solution defined for $t > 0$ which has a prescribed (possibly singular) behavior as $t \to 0+$?

The first question essentially leads to hyperbolicity, and amounts to the propagation of disturbances which are given at $t = 0$. The second question may have a positive answer for non-hyperbolic equations as well, such as the Schrödinger and heat equations; it also arises in the construction of singular solutions.

Auxiliary issues are:

(c) (Well-posedness) Is the solution unique? Is it continuous in f in a given function space?

(d) (Regularity) If there is a solution, does it have m derivatives in L^2, or less?

These issues are related since we are dealing with linear problems. In terms of H^s spaces, hyperbolic equations are expected to display a gain of $m - 1$ derivatives, as opposed to m for elliptic equations, but the situation is more delicate with other measures of regularity, see §1.4.

Elementary criteria for well-posedness.

Let us consider the problem

$$(1.10) \qquad \begin{cases} u_t = P(D)u, \\ u(0) = u_0(x), \end{cases}$$

where $x \in \mathbf{R}^n$ and $P(D)$ is a differential operator; u may be a vector. Fourier transforming in space, we find:

PROPOSITION   Problem (1.10) is well-posed in L^2 if and only if

$$\| \exp(tP(\xi)) \| \leq c(t).$$

This very simple criterion applies for example to the Airy equation. A similar result can be given for equations of order higher that one in time, by finding an appropriate reduction to a first-order system. There may, however, be a simpler approach. Thus, from (1.6), it easily follows that the Cauchy problem for the wave or Klein-Gordon equations is well-posed in $H^{s+1} \times H^s$ for any $s \geq 0$.

Remark 1: It is possible to obtain criteria for higher order systems by converting them into first-order systems in one of the usual ways. One checks easily that $c(t)$ can be assumed to have exponential growth.

Remark 2: A lower-order perturbation of a well-posed problem need not be well-posed in the same space. Thus, $u_{tt} = \varepsilon u_x$ is not well-posed in spaces based on L^2 for $\varepsilon \neq 0$.

Another simple situation is the case of the constant-coefficient system:

$$\begin{aligned} Qu_t &= \sum_{j=1}^{n} A^j \partial_j u \\ u(0) &= u_0(x), \end{aligned}$$

where Q is invertible. Let $A(\xi) = \sum_j i A^j \xi_j$. Fourier transformation leads to the solution

$$u(x,t) = \mathcal{F}^{-1}[\exp(tQ^{-1}A(\xi))\hat{u}(0)].$$

It follows that the initial-value problem is well-posed in $L^2(\mathbf{R}^n)$ if the exponential is uniformly bounded in ξ. It is necessary for this that for every ξ, the equation

$$p(\tau, \xi) := \det(i\tau Q - A(\xi)) = 0$$

have only real roots. It is sufficient that these roots be in addition *distinct*. The polynomial p is the *symbol* associated with the differential operator $Q\partial_t - \sum_j A^j \partial_j$, and these two conditions are called hyperbolicity and strict hyperbolicity respectively. For a detailed discussion, see Kreiss and Lorenz (1989).

Non-persistence results.

A Cauchy problem may fail to be well-posed in spaces of continuous functions, or in spaces based on L^p, $p \neq 2$. The first statement is perhaps to be expected, since boundary-value problems for elliptic equations are not "correctly set" in spaces of continuous functions, and require the introduction of Hölder spaces. The second is however more surprising. Let us illustrate this on the wave equation.

Consider first the function

$$(f(t + |x|) - f(t - |x|))/|x|,$$

with $f(r) := (1 - r^2)^{5/2} H(1 - r)$, where H is Heaviside's function, and $x \in \mathbf{R}^3$. This is a solution of the wave equation, and, by inspection, u is of class C^2 for $0 \leq t < 1$. Nevertheless, since $u(0, t) = 2f'(t)$, we see that $u_{tt} \to \infty$ as $t \to 1-$. This example shows incidentally that classical solutions cannot be continued as such for all time, even though they may be *bona fide* distributional solutions in $H^1(\mathbf{R}^4)$. Thus, the introduction of weak solutions is necessary in the simplest of cases.

Next, let us turn to L^p spaces. Motivated by the energy identity, one might surmise that for solutions of the wave equation, the "L^p energy"

$$\int_{\mathbf{R}^n} (|u_t|^p + |\nabla u|^p)\, dx$$

is finite for $t > 0$, and any $p \in (1, \infty)$, if it is for $t = 0$. This is false in space dimension 2 or higher unless $p = 2$:

THEOREM 1.9 *Let u solve the wave equation in n space dimensions. Let $e(x,t) = |u_t|^2 + |\nabla u|^2$. Then for any $p \in (1, \infty)$, $p \neq 2$, and any $T, A > 0$, there is a choice of u such that*

$$\|e(., T)\|_{p/2} \geq A \|e(., 0)\|_{p/2},$$

provided that $n \geq 2$.

Remark: This easily implies that the operator which multiplies the Fourier transform of u by $\exp(it|\xi|)$ is bounded on L^p precisely when $p = 2$.

Proof: We prove this result for $n = 3$, which implies the result for $n \geq 3$ as well, using finite speed of propagation; indeed, one can view this counter-example as a function in $\mathbf{R}^n$, and cut it off or large values of its argument, without changing the solution in a large ball. The result in two dimensions is similar, but more technical, see the Notes.

We consider a solution of the form $u = f(|x| - t)/|x|$. The strategy is to choose $\varepsilon > 0$ and f so that u is supported in $\{\varepsilon \leq |x| - t \leq 2\varepsilon\}$. This set has volume $O(\varepsilon)$ for $t = T$, but only $O(\varepsilon^n)$ for $t = 0$. This discrepancy will make the L^p energy much larger at time T than at time 0, which will prove the claim for $p > 2$; a similar argument will apply for $p < 2$. The details follow.

Define

$$G_\pm(f, \xi) = \frac{1}{2}\left\{ \frac{f}{\xi + T} \pm \sqrt{2(\xi + T)^2 - f^2/(\xi + T)^2} \right\}.$$

Let us define $f_\pm$ by

$$f'_+ = G_+(f_+(\xi), \xi) \quad \text{for} \quad \xi \geq \varepsilon, \quad f_+(\varepsilon) = 0,$$

and

$$f'_- = G_-(f_-(\xi), \xi) \quad \text{for} \quad \xi \leq 2\varepsilon, \quad f_-(2\varepsilon) = 0.$$

One sees easily that f_+ and f_- are both defined on $[\varepsilon, 2\varepsilon]$ if ε is small enough. One can also assume that $f_+(2\varepsilon)$ and $f_-(\varepsilon)$ are both positive. There is therefore a number $\eta \in (\varepsilon, 2\varepsilon)$ at which they coincide. Define f by patching these two functions at $\xi = \eta$, and by extending the result by 0 outside $[\varepsilon, 2\varepsilon]$. Let now $u = f(t - r)/r$, with $r = |x|$. One checks directly that

(i) supp $u \cap (t \geq 0) = \{\varepsilon \leq r - t \leq 2\varepsilon\}$;

(ii) $e(r, T) = 1$ if $\varepsilon \leq r - T \leq 2\varepsilon$, and 0 otherwise.

To complete the proof, we let $v(t) = \text{vol}(\text{supp}(u(., t)))$, and $\phi(t) = \|e(., t)\|_{p/2}$. Since $e(r, t) = e(r, t)^{p/2} = 1$ on the support of u, and since $\|e(., t)\|_1$ is independent of t (by the usual energy identity), we find that if $p > 2$,

$$
\begin{aligned}
\phi(T) &= (\int_{T+\varepsilon \leq |x| \leq T+2\varepsilon} e\, dx)^{1+(2/p-1)} \\
&= \|e(.,0)\|_1 v(T)^{2/p-1} \\
&\leq \phi(0)v(0)^{1-2/p}v(T)^{2/p-1}
\end{aligned}
$$

using Hölder's inequality. Now $v(0) = \omega_n(2^n - 1)\varepsilon^n$, while $v(T) \sim n\omega_n \varepsilon T^{n-1}$. Thus $v(0)/v(T)$ can be made arbitrarily small, and the result follows in the case $p > 2$. A smooth counter-example can be found by regularization.

If $p < 2$, we write $\int e^{p/2}\, dx \leq (\int e\, dx)^{p/2}(\int_{u \neq 0} 1\, dx)^{1-p/2}$, so that $\phi(0) \leq \|e(.,0)\|_1 v(0)^{2/p-1}$. On the other hand, we still have $\phi(T) = \|e(.,0)\|_1 v(T)^{2/p-1}$. We therefore find this time

$$
\phi(T) \geq \phi(0)(v(T)/v(0))^{2/p-1}
$$

which, using time-reversal, proves the claim for $p < 2$.

Hyperbolicity.

We recall here, without proof, a few facts about hyperbolic polynomials and hyperbolicity of single equations; the case of systems is dealt with in §2.2.

DEFINITION $P(\zeta)$ is a hyperbolic polynomial with respect to the direction N if there is a number τ_0 such that $P(\xi + i\tau N) \neq 0$ for $\xi \in \mathbf{R}^n$ and $\tau < \tau_0$.

DEFINITION   The Cauchy problem for equation $P(D)u = f$ with respect to the plane $(x, N) = 0$ consists in finding a solution such that $u - \phi = O((x, N)^m)$, where $m = \deg P$ and ϕ is a prescribed function.

If $N = (0, \dots, 0, 1)$, and $x^n = t$, this means that $u, \partial_t u, \dots, \partial_t^{m-1}u$ are prescribed for $t = 0$.

Hyperbolicity of a single equation is necessary and sufficient for the Cauchy problem to be solvable in C^∞. A hyperbolic polynomial has a hyperbolic principal part, but the converse is false. However, if the principal part is strictly hyperbolic as defined below, then hyperbolicity cannot be destroyed by lower-order terms:

THEOREM 1.10 *Any operator with principal part P_m of degree m is hyperbolic (w.r.t. N) provided P_m satisfies one of the following equivalent conditions:*

(i) P_m is hyperbolic with respect to N and of principal type (i.e., $\nabla P_m(\zeta)$ doesn't vanish if $P_m(\zeta) = 0$ and $\zeta \neq 0$).

(ii) $P_m(\xi + \tau N) = 0$ has m real, distinct roots if ξ is not parallel to N (strict hyperbolicity).

The role of these conditions will become apparent in the development of the energy method in §2.3.

Representation formulae

a. Spherical means. Let us define, for any function of x and t, its *spherical mean* about the point x by

$$(1.11) \qquad I_u(r,x,t) := \frac{1}{n\omega_n} \int_{|\xi|=1} u(x+r\xi,t)\, dS(\xi),$$

for $r > 0$, where $\omega_n = \pi^{n/2}/\Gamma(n/2+1)$ is the volume of the unit ball in $\mathbf{R}^n$, and $dS(\xi)$ is the surface measure on the $(n-1)$-sphere. The subscript u is omitted in this paragraph. One can extend I by reflection to negative values of r: $I(-r,t) = I(r,t)$. This turns out to yield a function of class C^2 if u is of class C^3. Since $I(x,0+) = u(x,t)$, the knowledge of the spherical means suffices to determine u.

Assume now that $\Box u = 0$. The Cauchy data $u(x,0)$ and $u_t(x,0)$ give rise to

$$F(r) := I(r,0) \quad \text{and} \quad G(r) := I_t(r,0).$$

Differentiating the definition of I and integrating by parts, we find

$$(1.12) \qquad I_{tt} = I_{rr} + \frac{n-1}{r}I_r.$$

This equation can be solved in closed form for all dimensions, but the result is particularly simple and important in three space dimensions. Indeed, if $n = 3$, we find $(\partial_{tt} - \partial_{rr})(rI) = 0$. It follows that

$$2rI(r,t) = (r+t)F(r+t) + (r-t)F(r-t) + \int_{r-t}^{r+t} sG(s)\, ds.$$

Letting $r \to 0$, we find

$$(1.13) \qquad u(x,t) = \frac{d}{dt}(tF(t)) + tG(t).$$

As already mentioned, if u_g is the solution for $f \equiv 0$, the solution in the general case is $u_g + \partial u_f/\partial t$. More generally, one has *Kirchhoff's formula*: if $\Box u = f(x,t)$ in three space dimensions, and V is any smoothly bounded volume enclosing a point x_0,

$$
\begin{aligned}
4\pi u(x_0,t) \;=\; & \int_V \frac{f(x,t-r)}{r}\, dV - \int_{\partial V} \Big[u(x,t-r)\frac{\partial}{\partial n}\Big(\frac{1}{r}\Big) \\
& - \frac{1}{r}\frac{\partial u}{\partial n}(x,t-r) - \frac{1}{r}\frac{\partial r}{\partial n}\frac{\partial u}{\partial t}(x,t-r) \Big]\, dS
\end{aligned}
$$

where $r = |x - x_0|$.

For the solution of (1.12) in the general case, see exercise 18. The net result is that for odd n the solution is given by a relation of the form

$$(1.14) \qquad u = \sum_{k=0}^{(n-3)/2} a_k t^{k+1} \left(\frac{d}{dt}\right)^k I_g(x,t).$$

The case when n is even is treated by descent, *i.e.* by treating u as a function of $n+1$ space variables which is constant with respect to the last variable x_{n+1}.

b. Comparison principles. While hyperbolic equations do not, as a rule, satisfy maximum principles similar to those valid for elliptic equations, the wave equation in low space dimensions does satisfy a limited comparison principle:

THEOREM 1.11 *If $n \leq 3$ and $\Box u_j = h_j$ for $j = 1$ or 2. Assume that $h_1 \leq h_2$ for all x, t, and $u_1(x,0) \leq u_2(x,0)$, $\partial_t u_1(x,0) \equiv 0$, $\partial_t u_2(x,0) \equiv 0$. Then $u_1 \leq u_2$ for all t.*

This result follows immediately from (1.13).

Such results have recently been used skillfully to obtain results on singularity formation (see Ch. 3).

Remark: In one space dimension, it suffices to assume $\partial_t u_2(x,0) \leq \partial_t u_1(x,0)$ instead of $\partial_t u_1(x,0) \equiv \partial_t u_2(x,0) \equiv 0$.

c. Fundamental solutions. Let $P(D)$ be a constant coefficient operator (i.e., a polynomial in $D_1, \ldots, D_n$). A fundamental solution for $P(D)$ is a distribution E such that

$$P(D)E = \delta.$$

It enables one produce a solution of $P(D)u = f$ for $f \in C_0^\infty$ in the form $u = E * f$. But since there are several solutions to $P(D)u = 0$, , there are several possible choices for E. If P is hyperbolic with respect to the direction N, one can write down a distinguished fundamental solution as follows:

$$(1.15) \qquad \forall \phi \in C_0^\infty, \ (E, \check\phi) = (2\pi)^{-n} \int \frac{\hat\phi(\xi + i\tau N)}{P(\xi + i\tau N)} \, d\xi,$$

where $\tau < \tau_0$ and $\check\phi(x) = \phi(-x)$.

It turns out that this fundamental solution can be characterized:

THEOREM 1.12 *There is exactly one fundamental solution which is supported by the half-plane $\{(x,N) \geq 0\}$; it is given by (1.15).*

In fact, conversely, if there is a fundamental solution supported in a cone K with vertex at the origin, such that $K \cap \{(x,N) \geq 0\} = \{0\}$, then $P(D)$ must be hyperbolic. We refer to Hörmander (1983) for the proof of these two results.

We next give the explicit form of a few fundamental solutions for the wave and Klein-Gordon equation.

Let us first define a family of distributions. We let $\chi_+^a = x_+^a/\Gamma(a+1)$. This is locally integrable only for Re $a > -1$, but can be analytically continued to all values of $a \in C$ thanks to the relation $(\chi_+^a)' = \chi_+^{a-1}$. Since $\chi_+^0 = H$, the Heaviside function, we have $\chi_+^{-k} = \delta_0^{(k-1)}$ for $k \geq 1$. We also use freely the notion of composition of distributions by smooth maps.

Now let A be a non-singular symmetric matrix, and let $A(x)$ be the associated quadratic form. Define $B = A^{-1}$. Let (n_+, n_-) be the signature of $A(x)$. We then have

THEOREM 1.13 $B(\partial)(A(x) \pm i0)^{(2-n)/2} = (2-n)n\omega_n|\det A|^{-1/2}e^{\pm i\pi n_-}$.

Idea of Proof: The result is first proved for the "elliptic" case where A is real and positive definite, by a change of variables which reduces the operator $B(\partial)$ to the Laplacian. By continuation, this remains true if ReA is positive definite. One then considers $A_\epsilon = -i(A(x) + i\varepsilon|x|^2)$, which tends to $-iA$ as $\varepsilon \to 0$.

For the wave equation, this produces, after a short calculation, the fundamental solution

$$E = \frac{1}{4}\pi^{(1-n)/2}\chi_+^{(1-n)/2}(t^2 - |x|^2)$$

for $n \geq 1$. It is supported in the inside of the double light cone $t^2 \geq |x|^2$. We may write $E = E_+ + E_-$, where $E_+ = 2E$ for $t \geq 0$, 0 otherwise. E_+ and E_- are still fundamental solutions, and are called respectively the retarded and advanced fundamental solutions. Thus, in three space dimensions, there is one solution of $\square u = h$ which vanishes for t large and negative, if h is compactly supported, namely

$$E_+ * h(x, t) = \int \frac{h(x - y, t - |x - y|)}{4\pi|x - y|}\, dy.$$

This clearly shows that E_+ has the property that the solution at time t only depends on the sources at earlier times, justifying its name.

Remark: Observe that the fundamental solutions for hyperbolic equations are nothing but analytic continuations of fundamental solutions for elliptic equations. It is because of the particular properties of the distribution χ_+^a for negative values of a that they look so different from their elliptic counterparts; thus, the wave equation in 3+1 dimensions has a fundamental solution which appears to be homogeneous of degree -1 inside the light cone, while the Laplacian in 4 variables has a fundamental solution homogeneous of degree -2. Furthermore, as the above proof suggests, solutions with particular properties are obtained by a careful choice of the contour of integration in the inversion theorem; thus, $1/(x \pm i0)$ are both fundamental solutions of $-id/dx$, but they correspond to different conditions at infinity.

The uniqueness of the retarded fundamental solution is easy to prove for the wave equation. Any other solution with support in the half-plane $t \geq 0$ would have to have the form $E_+ + v$ where $\square v = 0$ and $v = 0$ for $t < 0$. But then the convolution $E_+ * v$ is well-defined and we may also use the associativity of the convolution product to obtain

$$v = \delta * v = (\square E_+) * v = E_+ * (\square v) = 0,$$

QED.

1.4 DECAY ESTIMATES

We discuss a few decay estimates for linear wave equations. The issue is to estimate the solution of the initial-value problem in terms of Sobolev norms of the data. By fixing the value of t, one may also view these as regularity estimates on the solution.

The results can be classified by their method of proof: stationary phase; representation formulae; use of Sobolev inequalities; complex interpolation. In addition, we derive closely related local smoothing estimates which have recently played an important role in some global existence questions, and show how the estimates on the initial-value problem translate into estimates for inhomogeneous problems.

The method of stationary phase.

In its simplest form, this method estimates the decay, as $t \to \infty$, of integrals of the form

$$I(f, \varphi, t) := \int_{\mathbf{R}^n} \hat{f}(\xi) e^{it\varphi(\xi)} \, d\xi.$$

Assume $\hat{f}$ is compactly supported. We prove that $I = O(t^{-k})$ for every k if $\nabla \varphi$ never vanishes, and that $I = O(t^{-n/2})$ if φ has a finite number of non-degenerate critical points. A precise statement is as follows:

THEOREM 1.14 *Let φ be as above. The integral I satisfies*

$$|I(f, \varphi, t)| \leq \|\hat{f}\|_{C^l} \|\varphi\|_{C^m} t^{-n/2}$$

if l and m are large enough.

The case of the Klein-Gordon equation in n space dimensions corresponds to $\varphi = \pm\sqrt{m^2 + |\xi|^2}$. The Schrödinger equation leads to a similar result ($\varphi = |\xi|^2$). The Airy case is however not covered by the Theorem (see exercise 15).

Proof: Reduce the exponent to a linear form by a change of variables near any non-critical point, and use integration by parts; near critical points, replace the exponent, using the Morse lemma, by a quadratic form. The assertion is then clear.

Remark: In case $D^2\varphi$ has rank $\rho < n$ at finitely many points, the decay rate becomes $t^{-\rho/2}$; see Littman (1963), and Pecher (1976) for an application.

Use of representation formulae.

This approach yields rather precise information with relatively simple technical tools. We prove the following:

THEOREM 1.15 *Let u be the solution of the wave equation with data (f, g). One has*

$$|u(x,t)| \leq C(\|f\|_{n,1} + \|g\|_{n-1,1})(1 + |t|)^{-(n-1)/2}.$$

Moreover, for $t > 1$, we have

$$|u(x,t)| \leq C(\|f\|_{[n/2]+1,1} + \|g\|_{[n/2],1})t^{-(n-1)/2}.$$

We also prove

THEOREM 1.16 *Let u be the solution of the Klein-Gordon equation with data (f, g). One has*

$$|u(x,t)| \leq C(\|f\|_{n+1,1} + \|g\|_{n,1})(1 + |t|)^{-n/2}.$$

Moreover, for $t > 1$, we have

$$|u(x,t)| \leq C(\|f\|_{[n/2]+2,1} + \|g\|_{[n/2]+1,1})t^{-n/2}.$$

These are known as "L^1-L^∞ estimates." The energy estimates can be thought of as L^2-L^2 estimates. Interpolating between the two, one can obtain some simple L^p-L^q estimates (exercise). Other estimates of u in L^∞ can be obtained from the Sobolev inequality and the energy identities. These are the L^2-L^∞ estimates such as

(1.16) $$|u(x,t)| \leq (|f|_{[n/2]+1} + |g|_{[n/2]}).$$

for the wave equation if $n \geq 3$. Their easy proof is left to the reader. Note that Th. 1.10 implies that $W^{n,1}$ embeds into L^∞, and, therefore, into C^0. As will be clear from the proofs, one can in fact give more precise bounds; they will not be needed, however; see the notes for more results.

Proof: In view of the representation formulae for the wave equation, we have to estimate the spherical means I_g and its time derivatives. The Klein-Gordon equation can always be treated by descent, since $ue^{imx_{n+1}}$ solves the wave equation in $n+1$ space dimensions if u solves the Klein-Gordon equation. We also take $m=1$.

STEP 1: PRELIMINARIES.

Recall that $I_g(x,t) = (n\omega_n)^{-1} \int_{|\xi|=1} g(x+t\xi)\, d\xi$. If j is an integer greater than k, we have

$$\left(\frac{d}{dt}\right)^k I_g(x,t) = -\int_t^\infty \left(\frac{d}{ds}\right)^{k+1} I_g(x,s)\, ds$$
$$= \frac{(-1)^{k-j}}{(j-k-1)!} \int_t^\infty (s-t)^{j-k-1} \left(\frac{d}{ds}\right)^j I_g(x,s)\, ds.$$

Therefore,

(1.17)
$$\left| t^{k+1} \left(\frac{d}{dt}\right)^k I_g(x,t)\, dt \right| \leq C \int_t^\infty t^{k+1}(s-t)^{j-k-1} \left| \int_{|\xi|=1} \left(\frac{d}{ds}\right)^j g(x+s\xi)\, d\xi \right| ds.$$

Since the volume element dx equals $s^{n-1} ds\, d\xi$, we require that m be such that

(1.18)
$$t^{k+1}(s-t)^{j-k-1} \leq C s^{n-1} t^m$$

for $s \geq t$. This amounts to $m+n = j+1$ and $j \leq n+k$. Since $j \geq k+1$, we finally have

(1.19)
$$t^{k+1} |I_g^{(k)}(x,t)| \leq C t^{1-n+j} \int_{|y|>t} \sum_{|\alpha|=j} |D^\alpha g|(x+y)\, dy.$$

if $1+k \leq j \leq n+k$.

One can also transform the spherical mean into an integral on $|y| < t$, by using the divergence theorem: for any $q = 1, \ldots, n$,

$$\int_{|\xi|=1} f(t\xi)\, \xi_q\, dS(\xi) = t^{1-n} \int_{|y| \leq t} \partial_q f(y)\, dy.$$

Thus, if $I_g^{(k)} = (d/dt)^k I_g(x,t)$,

$$(n\omega_n) I_g^{(k)} = \int_{|\xi|=1} \sum_{|\alpha|=k} t^{-|\alpha|-1} \partial^\alpha g(x+t\xi)(t\xi)^\alpha \xi_q(t\xi^q)\, dS(\xi)$$
$$= t^{-n-k} \int_{|y| \leq t} \sum_{|\alpha| \leq k} \partial_q \{ \partial^\alpha g(x+y) \cdot y^\alpha y_q \}\, dy.$$

Estimating $|y|$ by t, we find

(1.20)
$$(n\omega_n) I_g^{(k)} \leq t^{-n-k} \int_{|y| \leq t} \sum_{|\alpha|=k \text{ or } k+1} t^{|\alpha|} |\partial^\alpha g(x+y)|\, dy.$$

In applying the method of descent, we will also need to know that if $\tilde{\xi} = (r\xi/t, \xi_{n+1}) \in S^n$, with $0 \leq r \leq t$, then

$$dS(\tilde{\xi}) = (r/t)^{n-1} \, dr \, dS(\xi)/\sqrt{t^2 - r^2}.$$

Or course, a given pair (r, ξ) corresponds to two points on S^n.

STEP 2: WAVE EQUATION, n ODD.

We start from the representation formula for $f = 0$:

$$u_g = \sum_{k=0}^{(n-3)/2} a_k t^{k+1} \left(\frac{d}{dt} \right)^k I_g(x, t).$$

We wish to use (1.19). To use the same value of j for all the terms in this sum, we must take $(n-1)/2 \leq j \leq n$, since k goes from 0 to $(n-3)/2$. For $j = (n-1)/2$, we find

$$|u(x, t)| \leq C t^{(1-n)/2} \int_{|y|>t} \sum_{|\alpha| \leq (n-1)/2} |D^\alpha g|(x + y) \, dy,$$

and for $j = n - 1$, we find

$$|u(x, t)| \leq C \int_{|y|>t} \sum_{|\alpha| \leq n-1} |D^\alpha g|(x + y) \, dy.$$

This proves Th. 1.15 for $f = 0$.

We must now estimate $t^k I_f^{(k)}$ for $0 \leq k \leq (n-1)/2$ to conclude the proof. To this end, it suffices to use (1.19) with $j = (n+1)/2$ and $j = n$ as is readily verified.

The proof of Th. 1.15 is therefore complete for n odd.

STEP 3: WAVE EQUATION, n EVEN.

By descent, we have

$$u = \sum_{k=0}^{(n-2)/2} a_k t^{k+1} J_g^{(k)},$$

where

$$J_g^{(k)} = \int_0^t \int_{|\xi|=1} t^{1-n} \frac{r^{n-1} \, dr \, dS(\xi)}{\sqrt{t^2 - r^2}} \sum_{|\alpha|=k} \partial^\alpha g(x + r\xi)(\frac{r\xi}{t})^\alpha.$$

To prove the estimate for $t > 1$, we split the integral as $\int_0^{t-1/2} + \int_{t-1/2}^t$. For the first part, we estimate $1/\sqrt{t^2 - r^2}$ by $(t^2 - (t - 1/2)^2)^{-1/2}$ which is $O(1/\sqrt{t})$, while the other terms are estimated by

$$t^{1-n} \int_{|y| \leq t-1/2} \sum_{|\alpha|=k} |\partial^\alpha g(x + y)| \, dy.$$

For the second part, we write it as

$$\int_{t-1/2}^{t} \frac{r\,dr}{\sqrt{t^2 - r^2}}(r^{n-2}I_g^{(k)}),$$

which, using (1.19) for $j = k+1$, is $O(t^{1/2-1})$. Putting both pieces together, we find

$$|t^{k+1}J_g^{(k)}| \leq Ct^{1-n+[(n-2)/2+1]-1/2}\|g\|_{n/2},$$

for $t \geq 1$, as desired.

For small t, we argue as follows:

$$
\begin{aligned}
t^{k+1}J_g^{(k)} &\leq Ct^{1-n}\int_0^t \frac{r^{n-1}I_g^{(k)}(x,r)\,dr}{\sqrt{t^2 - r^2}} \\
&\leq C\int_0^t (t/r)^{k+2-n}\frac{dr}{\sqrt{t^2 - r^2}}r^{k+1}I_g^{(k)}(x,r) \\
&\leq C\|g\|_{n-1,1}
\end{aligned}
$$

as desired (we used (1.19) with $j = n-1$).

This proves the result for $f = 0$. The small time estimate follows by a similar argument in the general case. For the large time estimate, we may argue as follows: The solution we must estimate is $u_g + \partial u_f/\partial t$, and we need only consider the second term. Now, let $v = Su_f$, where S is the scaling operator $t\partial_t + \sum_i x^i\partial_i$. One checks that v solves the wave equation with data $(0, (1 + x^i\partial_i)f)$. One now chooses a smooth function η which equals 1 on $(0, T)$, where T is fixed, and 0 outside $(0, 2T)$. We now solve the wave equation with data $(0, \eta(|x|/T)v_t(0))$. The solution coincides with v at $(0, T)$, by a domain of dependence argument, and the results we just obtained therefore give the estimate:

$$|v(0,T)| \leq C\|\eta(|x|/T)v_t(0)\|_{[n/2],1} \leq C(1+T)\|f\|_{[n/2]+1},$$

and since $v(0,T) = Tu_t(0,T)$, we have the required estimate for $T \geq 1$. Note that the above argument applies irrespective of the parity of n.

STEP 4: KLEIN-GORDON EQUATION, n EVEN.

Any solution of the Klein-Gordon equation in n space dimensions generates a solution of the wave equation with $n + 1$ space dimensions via the formula $v(x, x_{n+1}) = u(x)\exp(ix_{n+1})$.

We must now estimate, instead of $J_g^{(k)}$, an expression of the form

$$\int_0^t \frac{(r/t)^{n-1}}{\sqrt{t^2 - r^2}} \sum_{|\alpha| \leq k} \int_{|\xi|=1} \partial^\alpha g(x + r\xi) \cdot \xi^\alpha e^{i\xi_{n+1}}(i\xi_{n+1})^{k-|\alpha|}\,dS(\xi)\,dr,$$

where $\xi_{n+1} = (\pm)\sqrt{1 - r^2/t^2}$. The new terms do not contribute more than numerical factors. One then uses (1.19) again on each term of the sum, with $j = n - 1$, to obtain the small time estimate.

Remark: A simpler, slightly less precise argument, consists in applying directly the estimate for the wave equation to $u(x)e^{ix_{n+1}}\psi(x_{n+1})$, for a suitable cut-off function ψ. If we are only interested in small t, we may choose the support of ψ to be a fixed interval about 0, and use a domain of dependence argument to prove that the cut-off doesn't alter the solution for small t.

For large times, we must convert the integral on S^n in the representation formula into a volume integral on a ball of radius t about x, as in Step 1. Upto a power of t, we are led to expressions of the form

$$\int_{|y|\leq t} \sum_{|\alpha|\leq l} \partial^\alpha[g(x+y)]e^{iy_{n+1}}(y_{n+1}/t)\,dy.$$

Expanding the derivative, we observe that one may integrate first with respect to y_{n+1}; this produces a bounded factor. One then sorts out the exponents and checks that the decay rate is indeed $t^{-n/2}$.

If $f \not\equiv 0$, we may obtain the small time estimate as above, and the large time estimate by, for instance, the argument of Step 3; the verification is left to the reader.

Step 5: Klein-Gordon equation, n odd.

This case combines the difficulties of the previous two. One can handle the small time estimate using for instance the above remark (although one could be more precise). One must now estimate integrals of the form

$$\int_0^t \frac{r^n\,dr}{\sqrt{t^2-r^2}} \int_{\tilde{\xi}\in S^n} \partial^\alpha \tilde{g}(x+r\tilde{\xi}) \cdot \xi^\beta\,dS(\tilde{\xi}).$$

Here, $\tilde{g}(x,x_{n+1}) = g(x)e^{ix_{n+1}}$. One must transform the surface integral into a volume integral, and again separate $0 \leq r \leq t-1/2$ from $t-1/2 \leq r \leq t$, and integrate with respect to the $(n+1)$st variable, which does not contribute a spurious factor of t. The somewhat lengthy details are omitted.

Use of global Sobolev inequalities.

Granted the weighted Sobolev inequality in Minkowski space, this method is one of the fastest ways to prove decay estimates: Observe that the global Sobolev inequality involves norms that are constant in time if $p = 2$. They are indeed essentially energy norms of the functions $X_1 \ldots X_l u$ which, in case $\Box u + u = 0$, will all solve the Klein-Gordon equation as well. The supremum can therefore be determined from the initial data alone, and gives immediately an L^2-L^∞ estimate for the Klein-Gordon equation.

An advantage of this method is that it can be adapted to nonlinear situations.

A very simple (and loosely related) proof will follow from the conformal method in Ch. 2.

Complex interpolation.

These methods involve two generalizations of the Riesz-Thorin interpolation theorem. The first consists in working with an analytic family of operators depending on a parameter in a strip of the complex plane, and allowing the operator norms to grow exponentially at infinity in the imaginary directions (instead of being bounded as in the usual set-up). The second consists in observing that BMO and the Hardy space H^1 may replace L^∞ and L^1 in certain cases.

To illustrate the first idea, let us prove

THEOREM 1.17 *The solution of the Klein-Gordon equation in n space dimensions satisfies*

$$\|u; L^q(\mathbf{R}^{n+1})\| \leq C(|u(0)|_{1/2} + |u_t(0)|_{-1/2}),$$

where $2(n+1)/(n+3) \leq p \leq 2(n+2)/(n+4)$ for $n \geq 2$, and $1 < p \leq 6/5$ for $n = 1$.

STEP 1. The first point is to recognize that what we need is an estimate on the Fourier transform of a measure carried by the two sheets of a hyperboloid (resp. a cone, if $m = 0$). Indeed, let $B = \sqrt{m^2 - \Delta}$ and $\phi_\pm = (1/2)\mathcal{F}(Bu(0) \pm iu_t(0))$. Then u is the sum of the two integrals

$$\int e^{-i((x,\xi) \pm \sqrt{m^2 + |\xi|^2} t)} \frac{\phi_\pm(\xi)}{\sqrt{m^2 + |\xi|^2}} \, d\xi.$$

We therefore must estimate u, given the measure $F d\mu = \frac{F}{\sqrt{m^2 + |\xi|^2}} \, d\xi$ on a two-sheeted hyperboloid, where $F = \phi_\pm$. Note that

$$\|F; L^2(d\mu)\| = |u(0)|_{1/2} + |u_t(0)|_{-1/2}.$$

We now denote this hyperboloid (resp. cone, if $m = 0$) by S.

STEP 2. Let $p \in [1, 2)$. The statement

$$\|\mathcal{F}(F \, d\mu)\|_q \leq C\|F\|_{L^2(d\mu)}$$

for all F, where $q = p'$ is equivalent to

$$\int_S |\hat{f}|^2 \, d\mu \leq C\|f\|_p^2$$

for all f. But this latter statement follows from

(1.21) $\|(d\mu)\check{\ } * g\|_q \leq C\|g\|_p.$

To see this, it suffices to write

$$\int \overline{\hat{f}} \hat{f} \, d\mu = \int \bar{f} \mathcal{F}^{-1}(\hat{f} \, d\mu) \, dx = \int \bar{f}((d\mu)\check{} * f) \, dx \le \|f\|_p \|(d\mu)\check{} * f\|_q.$$

STEP 3. Let us introduce the analytic family of operators

$$T_z : g \mapsto \mathcal{F}^{-1}(G_z \mathcal{F} g),$$

where $G_z = \gamma(z)(R(x) - r)^z_+$. The point is that if γ has a simple zero at $z = -1$, G_z reduces to integration against $d\mu$ if $z = -1$. If therefore we can establish the boundedness of T_z for $\operatorname{Re} z = 0$ or $-\lambda$, for some $\lambda > 1$, we will obtain the desired result by Stein's interpolation theorem with the value $p = 2\lambda/(\lambda + 1)$, provided that γ grows exponentially at most in the imaginary directions.

STEP 4. An explicit calculation of Fourier transforms gives $\lambda = (n + 1)/2$ and $n/2$ for the Klein-Gordon and wave equations respectively. These give the values $p = 2(n + 1)/(n + 3)$ and $p = 2n/(n + 2)$.

As an example of the use of BMO estimates, we mention

THEOREM 1.18 *For solutions of the KGE,*

$$\|u_g\|_{p'} \le C t^{-(n-1)/2} \|g\|_p$$

where $1/p = 1/2 + 1/(n + 1)$ and $1/p + 1/p' = 1$.

The idea of the proof consists in the "Fourier multiplication operator" T^α associated with the function

$$(1 + |\xi|^2)^{-[\alpha + (n-1)/2]/2} \sin[t(1 + |\xi|^2)^{1/2}],$$

and showing that its kernel is bounded for $\alpha = 3/2$, and BMO for $\alpha = 1$. One then interpolates between $\operatorname{Re} \alpha = 1$ and $\operatorname{Re} \alpha = (1 - n)/2$.

Local smoothing.

Even though a given equation may define a group of isometries in all H^s spaces, it may have smoothing properties in spaces based on L^p. The additional smoothness in question is valid either for almost every t, or sometimes for $t > 0$, so that no contradiction ensues. We present two such results, one for the Airy equation, and the other for the Schrödinger equation.

Consider a solution of

$$u_t + u_{xxx} = 0.$$

We denote by $L^q(L^p)$ the space defined by the completion of smooth functions with compact support for the norm

$$\left(\int_{-\infty}^{+\infty} [\int_{-\infty}^{+\infty} |u(x,t)|^p \, dx]^{q/p} \, dt\right)^{1/q}.$$

We also denote by D^μ the Fourier multiplier associated with $|\xi|^\mu$; thus, $\mathcal{F}D^\mu u = |\xi|^\mu \mathcal{F}u$. The result is the following:

THEOREM 1.19 *If $(\theta, \alpha) \in [0.1] \times [0, 1/2]$,*

$$\|D^{\theta\alpha/2}u; L^q(L^p)\| \leq C\|u(0)\|_{L_2},$$

where $p = 1/(1 - \theta)$ and $q = 6/[\theta(\alpha + 1)]$.

Remark: For $\theta = 0$, this gives a Strichartz-type estimate: $\|u; L^q(L^p)\| \leq C\|u(0)\|_{L^2}$, if $2/q = 1/3 - 2/(3p)$ and $2 \leq p \leq \infty$.

Proof: We must estimate $u(t) = (3t)^{-1/3} \, \text{Ai}(x/(3t)^{1/3}) * u(0)$, where

$$Ai(x) := \int_{\mathbb{R}} e^{i(\xi^3/3 + x\xi)} \, d\xi$$

is the Airy function. Let $s(t) = (3t)^{-1/3} \, \text{Ai}(x/(3t)^{1/3})$. We proceed in four steps.

STEP 1. The van der Corput theorem (exercise 15) shows that

$$\sup_x |D^\alpha s(t)| \leq C|t|^{-(1+\alpha)/3}$$

if $0 \leq \alpha \leq 1/2$.

STEP 2. It follows that

$$\|D^{\alpha+i\beta}s(t) * u(0)\|_\infty \leq C(1 + |\beta|)|t|^{-(1+\alpha)/3}\|u(0)\|_1$$

for $0 \leq \alpha \leq 1/2$ and β real. On the other hand, it is immediately seen that

$$\|D^{i\beta}s(t) * u(0)\|_2 = \|u(0)\|_2.$$

By interpolation, it follows that, for $0 \leq \theta \leq 1$,

$$\|D^{\theta\alpha/2}s(t) * u(0)\|_{2/(1-\theta)} \leq C|t|^{\theta(1+\alpha)/3}\|u(0)\|_{2/(1+\theta)}.$$

STEP 3. The conclusion of the theorem is equivalent, by duality, to

$$\|D^{\theta\alpha/2}s(t) * g(.,t) \, dt\|_2 \leq C\|g; L^{q'}(L^{p'})\|,$$

for all g, with $p = 2/(1 - \theta)$. But, using the fact that the solution operator is a group, we can transform the L^2 norm above:

$$\|D^{\theta\alpha/2}s(t) * g(.,t) \, dt\|_2^2$$
$$= \int (D^{\theta\alpha/2}s(t) * g(x,t) \, dt)(\int \overline{D^{\theta\alpha/2}s(\tau) * g(x,\tau)} \, d\tau) \, dx$$
$$= \int\int g(x,t)(\int D^{\theta\alpha}s(t-\tau) * \overline{g(.,\tau)} \, d\tau) \, dx \, dt.$$

It therefore suffices to show that the inner integral is estimated by $C\|g; L^{q'}(L^{p'})\|$.

STEP 4. We finish the proof by using Minkowski's inequality, Step 2, and the Hardy-Littlewood (or fractional integration) theorem (exercise 11):

$$\| \int D^{\theta\alpha} s(t-\tau) * g(.,\tau) \, d\tau; L^q(L^p)\|$$

$$\leq \ \left\| \int \|D^{\theta\alpha} s(t-\tau) * g(.,\tau)\|_p \, d\tau; L^q\right\|$$

$$\leq \ C\left\| \int \| \ |t-\tau|^{-\theta(1+\alpha)/3} g(.,\tau)\|_{p'} \, d\tau; L^q\right\|$$

$$\leq \ C\|g; L^{q'}(L^{p'})\|,$$

provided that $1/q = 1/q' - (1 - \theta(1+\alpha)/3)$, which gives $q = 6/[\theta(1+\alpha)]$, QED.

A slightly different type of smoothing effect, in which regularity is gained at the expense of spatial decay, is given in the following:

THEOREM 1.20 *Let k be a positive integer, and let V be a real-valued function in $W^{1,\infty}$. Let u be the solution of*

$$iu_t - \Delta u + Vu = 0$$

with $u(0) = f$. Then, if $(1 + |x|^2)^{k/2} f \in L^2$, we have $(1 + |x|^2)^{-k/2} u(t) \in H^k$ for $t > 0$.

A special case is proved in exercise 5. Note the very weak assumptions at infinity on the potential V, which set very few constraints on the spectrum of the associated Schrödinger operator.

Inhomogeneous problems.

Estimates for inhomogeneous problems are generally obtained by the use of Duhamel's formula. For example

THEOREM 1.21 *If $iu_t + \Delta u = g$ and $u(0) = 0$, we have*

$$\|u; L^{2+4/n}(\mathbf{R}^{n+1})\| \leq C\|g; L^{2(n+2)/(n+4)}(\mathbf{R}^n)\|.$$

Proof: Let $p = 2(n+2)/(n+4)$, so that $q = 2+4/n$ is the conjugate exponent. The solution operator is denoted by $e^{it\Delta}$, so that $u(t) = e^{it\Delta} f$.

Since $\|e^{it\Delta} f\|_\infty \leq C t^{-n/2}\|f\|_1$ and $\|e^{it\Delta} f\|_2 = \|f\|_2$, we have, by interpolation,

$$\|e^{it\Delta} f\|_q \leq C t^{-r}\|f\|_p,$$

with $r = n/(n + 2)$. Duhamel's formula gives

$$\left\|\int_0^t e^{i(t-s)\Delta} g \, ds\right\|_q \leq C \int_0^t |t - s|^{-r}\|g(s)\|_p \, ds.$$

The result now follows by the fractional integration theorem (problem 11).

For the wave equation, we have the following result, which includes the analogue of Th. 1.17:

THEOREM 1.22 *If* $\Box u = h(x,t)$, *and* $u(0) = f$, $u_t(0) = g$, *in* n *space dimensions,* $n \geq 2$, *we have*

$$\|u; L^q(\mathbf{R}^{n+1})\| \leq C(|f|_{1/2} + |g|_{-1/2} + \|h; L^p(\mathbf{R}^{n+1})\|),$$

where $p = 2(n+1)/(n+3)$ *and* $q = p/(p-1) = 2(n+1)/(n-1)$.

We list here for the sake of completeness a few theorems of a slightly different nature.

THEOREM 1.23 *If* $u \in C_0^\infty(\mathbf{R}^n)$, *one has*

$$\sum_{|\alpha| \leq m} \|P^{(\alpha)}(D)u\|_2 \leq \|P(D)u\|_2$$

for any differential operator of order m.

Thus, the H^1 norm of u (in space-time) is bounded by the L^2 norm of $\Box u$, *provided that* u *vanishes at infinity.* It is easy to see that the latter assumption is necessary.

THEOREM 1.24 *Let* $P(D)$ *be a second order operator with principal part* $\Box$; *there is a universal constant* C *such that if* $u \in C_0^\infty(\mathbf{R}^n)$, $n \geq 3$,

$$\|u\|_{p'} \leq C\|P(D)u\|_p,$$

where p *and* p' *are conjugate exponents, and* $1/p - 1/p' = 2/n$.

This is a generalization of Th. 1.17. The point is that the constant is independent of the lower-order terms. Applying it to the product of u by an exponential, one can derive Carleman-type estimates, and therefore obtain unique continuation theorems for suitable variable-coefficient perturbations of $P(D)$.

1.5   PROPAGATION OF SINGULARITIES

In accordance with the historical development of the subject, propagation of singularities will be considered from three perspectives. We first consider the propagation of *jump discontinuities*. We then discuss briefly the method of geometrical optics, which describes the propagation of high-frequency oscillations (closely related to singularities). Finally, we discuss the modern notion of (C^∞-) wave front set. For these last two approaches, results are only stated here to be contrasted with their nonlinear counterparts later.

For the sake of simplicity, the results will be detailed for rather special examples, without striving for maximum generality.

Propagation of discontinuities

Let us consider a single equation of second order, with smooth coefficients:

$$(1.22) \qquad \sum_{ij} a^{ij}\partial_{ij}u + a^i\partial_i u + cu = 0,$$

where $a^{ij} = a^{ji}$. We are also given a smooth hypersurface Σ with equation $\phi(x) = 0$, where $\nabla\phi(x_0) \neq 0$. We are interested in functions u which, near x_0, are smooth and satisfy the equation on both sides of Σ. More precisely, we assume that there are smooth functions u_1 and u_2 such that $u = u_1$ for $\phi > 0$, and $u = u_2$ for $\phi < 0$.

All our considerations are local. We choose an orientation on Σ in the vicinity of x_0 and define $[f]$, for any function f, to be the value of the jump of f across Σ from the negative to the positive side.

We need an additional restriction which ensures that u in fact is a solution of the given equation in a suitable sense. Let us assume that u and ∇u are continuous across Σ.

Remark: This is sufficient to ensure that the equation should hold in the weak sense, *viz.*

$$(1.23) \qquad \int [a^{ij}\partial_i u\, \partial_j v - v(a^i\partial_i u + cu)]\, dx = 0$$

for any smooth v with compact support. Indeed, this equation holds if $v \equiv 0$ near Σ. If now v is arbitrary, we may find a smooth function η which vanishes on $\{dist(x,\Sigma) < \varepsilon\}$ and equals one on $\{dist(x,\Sigma) > 2\varepsilon\}$. We may assume also that $|\eta| \leq 1$ and $|\nabla\eta| \leq C/\varepsilon$. Eq. (1.23) is therefore valid with v replaced by ηv. Passing to the limit $\varepsilon \to 0$, we obtain the desired statement. This type of argument proves in fact a "removable singularity theorem" in the sense that the equation was found to hold everywhere, even though it was initially only satisfied off Σ.

The main result of this section is

THEOREM 1.25 *If such a solution exists, we must have*
   (a) $[\partial_{ij}u] = \lambda\phi_i\phi_j$ for some function λ defined on Σ,
   (b) $\sum_{ij} a^{ij}\phi_i\phi_j \equiv 0$ on Σ at any point where $\lambda \neq 0$ (i.e., Σ is characteristic at those points).

Remark: The function λ is further constrained by the "transport equation" derived below. Condition (a) is known as the kinematic compatibility condition, and condition (b) as the dynamic compatibility condition, since only the latter involves the equation that u satisfies (i.e. the "dynamics" governing the evolution).

Proof: (a) The assumptions imply that all the tangential derivatives of u are continuous across Σ. Since $\phi_l \partial_i - \phi_i \partial_l$ is such a derivative, it follows that for any i, j, k and l,

$$\phi_k[\phi_l u_{ij} - \phi_i u_{lj}] = 0$$

and

$$\phi_l[\phi_j u_{ki} - \phi_k u_{ji}] = 0.$$

Adding, we find $\phi_i \phi_k[u_{jl}] = \phi_j \phi_l[u_{ik}]$. Multiplying by $\phi^i \phi^k$ and summing, we get $|\nabla \phi|^4[u_{jl}] = \phi_j \phi_l \{[u_{ik}]\phi^i \phi^k\}$, QED.

(b) Since the equation holds on both sides of Σ, $0 = [a^{ij} u_{ij}] = \lambda(a^{ij}\phi_i \phi_j)$. The claim follows. QED

Let us now derive a further constraint that λ must satisfy. First, we observe that one may assume $a^{ij}\phi_i \phi_j \equiv 0$; it suffices indeed to solve this Hamilton-Jacobi equation with suitable data on a hypersurface transverse to Σ. We are then guaranteed that Σ is still represented by $\phi = 0$. By a suitable choice of coordinates, we assume $\phi = x^n$. This implies $a^{nn} = 0$ and $\lambda = [u_{nn}]$. Applying ∂_n to the equation, we find

$$\sum_{i<n} 2a^{in}\partial_i u_{nn} + a_n u_{nn} = M,$$

where M only involves $\partial_i u$, $\partial_i \nabla u$ and $\partial_i u_{kl}$ with $i < n$. This means $[M] \equiv 0$ and we therefore find the *transport equation*

$$(1.24) \qquad\qquad \frac{\partial \lambda}{\partial s} + a_n \lambda = 0,$$

where $\partial_s = 2\sum_{i<n} a^{in}\partial_i$.

Define the *principal symbol* associated with the equation to be $p(x, \xi) = a^{ij}\xi_i\xi_j$. Then $\partial p/\partial \xi_n = 2a^{in}\xi_i$ vanishes if ξ is proportional to the gradient of ϕ. Define the *bicharacteristic flow* by

$$\dot{x}^i = \partial p/\partial \xi^i; \quad \dot{\xi}_i = -\partial p/\partial x^i.$$

Take as initial conditions $(x_0^i, \phi_i(x_0))$. Since Σ is characteristic, it is easy to see that x remains on Σ while ξ_i remains normal to Σ. Finally,

$$\partial \lambda(x(s))/\partial s$$

is precisely what we called $\partial_s \lambda$ above. We therefore see that the propagation of discontinuities is effected along the spatial projection of the solution curves of the Hamiltonian system associated to the principal symbol.

Propagation of high-frequency oscillations

It is convenient to start first from a single first-order system

$$(1.25) \qquad Lu := u_t + \sum_i A^j(x)\partial_j u + Bu = 0.$$

Let us seek a formal solution of the form

$$(1.26) \qquad u = e^{i\omega\phi(x,t)}\Big(u_0 + \frac{u_1}{\omega} + \cdots + \frac{u_\nu}{\omega^\nu} + \cdots\Big),$$

with $u(0,x) = \psi(x)$ prescribed. Here the parameter ω is assumed to be "large," so that we are interested in "high-frequency" solutions. Let $A = \phi_t I + \sum_j A^j \phi_j$, where subscripts denote derivatives. Substitution of the series into the equation and identification of like powers of ϕ gives

$$\begin{cases} Au_0 = 0; \\ iAu_{\nu+1} + Lu_\nu = 0 \quad (\nu \geq 0). \end{cases}$$

In order for the solution to be non-trivial, we need A to be singular. This means that *the surfaces defined by $\phi = const.$ are characteristic.* Let us make the further assumption that there are right and left eigenvectors l and r which are smooth and uniquely determined upto scaling. We must then have $u_0 = \lambda(x,t)r(x,t)$ and, in order to be able to solve for u_1, we need $lL(u_0) = lL(\lambda r)$, which is a transport equation for λ:

$$(1.27) \qquad \lambda_t + l(Lr)\lambda = 0.$$

Similarly, for $j \geq 1$, $u_j = \lambda_j r + h_j$ where h_j depends on $u_0, \ldots, u_{j-1}$, and λ_j solves a transport equation.

If we apply the same procedure to the second order equation, we will find that the equations we obtain are essentially the same as those which described the propagation of jump discontinuities.

Propagation of C^∞ singularities

A smooth function u can be characterized by the property that for any smooth ϕ with compact support, and any integer N,

$$|\mathcal{F}(\phi u)| \leq C_N(1 + |\xi|)^{-N}.$$

This proves that the Fourier transform can be used to study local regularity. Now one can go one step further and investigate, for functions that are not smooth, the directions in which the above Fourier transform does not decay rapidly. It turns out that these directions are normal to surfaces of discontinuity in the case the singularities of u are not more complicated than jumps. For this reason, the set of these directions will be called the *wave front set*. Remarkably enough, this construction can be localized in space, as we show below. Regularity properties pertaining to the neighborhood of such a pair (x_0, ξ_0) will be referred to as *microlocal*.

We now turn to a more technical definition.

For any compactly supported distribution u, we say that $\xi_0 \notin \Sigma(u)$ if there is a conic neighborhood of ξ_0 in which, for every N,

$$|\hat{v}|(1 + |\xi|)^N \leq C_N.$$

As we pointed out, this notion can be localized consistently, thanks to

THEOREM 1.26 *If ϕ is smooth with compact support, $\Sigma(\phi u) \subset \Sigma(u)$.*

Proof: Since u has compact support, it has finite order, and $|\hat{u}(\xi)| \leq C(1 + |\xi|)^M$. Let $\xi_0 \notin \Sigma(u)$. We must estimate in a small conic neighborhood of ξ_0 the integral

$$\int \hat{u}(\eta) \hat{\phi}(\xi - \eta) \, d\eta.$$

Let Γ be a slightly bigger conic neighborhood of ξ_0 which is still outside $\Sigma(u)$. If $\eta \in \Gamma$, then the integrand is estimated by

$$C(1 + |\xi - \eta|)^{-N}(1 + |\eta|)^{-N} \leq C(1 + |\xi|)^{-N+n}(1 + |\eta|)^{-n},$$

since $(1 + |\xi|) \leq (1 + |\eta|)(1 + |\xi - \eta|)$. If $\eta \notin \Gamma$, we have $|\xi - \eta| \geq c(|\xi| + |\eta|)$ and the integrand is estimated by

$$C(1 + |\xi| + |\eta|)^{-N}(1 + |\eta|)^M \leq (1 + |\xi|)^{-N+M+n}(1 + |\eta|)^{-n}$$

if N is large enough.

The result follows.

We define the wave front set of u to be

$$WF(u) = \{(x, \xi) \in \mathbf{R}^n \times \mathbf{R}^n \setminus \{0\} : \xi \in \bigcap_{\phi(x) \neq 0} \Gamma(\phi u)\}.$$

One can show that for any differential operator with smooth coefficients, $WF(u) \subset WF(Pu) \cup \{P_m(x, \xi) = 0\}$, where P_m is the principal part of P. Furthermore, $WF(u) \setminus WF(Pu)$ is invariant under the Hamiltonian flow (on the cotangent bundle of $\mathbf{R}^n$) defined by the principal part for a large class of operators. Thus, in all three approaches to the propagation of singularities, the final description is given in terms of the bicharacteristics.

Before giving an idea of why this result is true, let us define the H^s version of the wave front set.

We say that $u \in H^s_{ml}(x_0, \xi_0)$, or $(x_0, \xi_0) \notin WF^s(u)$ if there is a smooth compactly supported function ϕ with $\phi(x_0) \neq 0$, and a conic neighborhood Γ of ξ_0 such that

$$(1 + |\xi|)^s \chi_\Gamma(\xi) \mathcal{F}(\phi u) \in L^2(\mathbf{R}^n).$$

Here χ_Γ is the characteristic function of the cone Γ. The invariance of the wave-front set under the bicharacteristic flow is still valid for this notion of micro-local regularity, for strictly hyperbolic operators. We now give an idea of the justification of the latter statement for the wave equation.

Now this definition means that there is an operator a of a very particular form, such that $au \in H^s$. Suppose that we are interested in the regularity of a solution of the wave equation. It is easy to see that if $\square u \in H^s$ and u is compactly supported, then $u \in H^{s+1}$. To get at microlocal regularity, it is however au that we must estimate. Since $\square(au)$ will generally involve derivatives of u, there is a potential loss of one derivative. It is however possible to construct a such that $[\square, a]$ has order 0 in a suitable sense, instead of 1. The statement $au \in H^s$ for this particular a translates into the result that the "H^s wave front set" is invariant under the Hamilton flow associated with $\square$.

The complete proof requires some background on pseudo-differential operators and is therefore omitted. It is nevertheless useful to remember that the general result on propagation of singularities follows from the energy estimate applied to the product of the unknown by a (pseudo-differential) operator, which is chosen so as to make its commutator with the operator under consideration as simple as possible.

We will come back to H^s microlocal regularity in chapter 3.

1.6 WHAT IS A WAVE?

A wave is generally defined as the propagation of a disturbance. We discuss briefly in this section the motivation for, and a few consequences of this definition. This brief incursion into the physical concept of a "wave" explains what one looks for when studying nonlinear waves; furthermore, some acquaintance with these ideas helps in reading much of the literature on the subject.

The most general disturbance is nothing but an initial state; we are therefore immediately led to the issue of well-posedness of the initial-value problem. This point of view is, however, not precise enough. An abrupt disturbance is best viewed as a discontinuity in the initial conditions, and these propagate according to the rules we described in §1.5. But since the initial-value problem is not well-posed in C^k spaces, and one might therefore prefer to use a definition of singularities which refers to spaces which *are* preserved by the given evolution. Such is the modern definition of the wave front set.

On the other hand, a steadily oscillating source will create a wave-train with certain periodicity properties. The simplest example is the exponential (or sinusoidal) plane wave. Fourier analysis enables one to decompose a general solution into a superposition of such plane waves, called a *wave packet:*

$$u = \int e^{i((k,x)-\omega(k)t)} \hat{u}(k) \, dk.$$

Such a representation assumes that we are dealing with a problem invariant under space and time translations. A more general representation would involve for instance a nonlinear phase function $\varphi(x,t,k)$ replacing $((k,x)-\omega(k)t)$. Each component of the wave packet has a different *wave number* k and a definite frequency $\omega(k)$. The relation $\omega = \omega(k)$ is called the *dispersion relation*. For instance, in the case of electromagnetic waves, each exponential represents a monochromatic wave-train. Of course, the solution of the wave equation is the sum of *two* such integrals.

Let us now restrict our attention to wave packets where $\hat{u}$ is supported near $k = k_0$. Let us expand the dispersion relation near $k = k_0$; letting $\kappa = k - k_0$, one finds:

$$\omega(k) = \omega_0 + \omega_0'\kappa + \frac{1}{2}\omega_0''\kappa^2 + O(|\kappa|^3),$$

so that

$$u = e^{i(k_0 x - \omega_0 t)} \int e^{i((\kappa,x)-\omega_0'\kappa t)} \hat{u}(k_0 + \kappa) e^{-i[(\omega_0''\kappa^2/2+O(|\kappa|^3)]t} \, d\kappa.$$

The packet is therefore the product of a plane wave by an amplitude factor, which, if the cubic term in κ were absent, would solve

$$(1.28) \qquad (\partial_t - (\omega_0', \nabla)u = (i/2)\omega_0''(\nabla, \nabla)u.$$

Eq. (1.28) is essentially a Schrödinger equation. A similar argument will account for the universality of the nonlinear Schrödinger equation in chapter 5. If we further neglect ω_0'', we find a simple convection with velocity equal to the *group velocity* $v_g := \nabla_k \omega$ (for $k = k_0$). The quantity $v_p := \omega/|k|$ is called the *phase velocity*.

One speaks of *dispersive waves* if ω has nonlinear dependence on k, and of *dissipative waves* if $\text{Im}(\omega) < 0$. . The phrase *pure dispersion* refers to that case when $\text{Im}(\omega) = 0$ and $\nabla_k v_p \neq 0$ for real k. Finally, *instability* is defined by $\text{Im}(\omega) > 0$. .

It is a rule of thumb that dispersion should cause localized wave packets to spread and eventually decay. As we saw in §1.4, this is essentially correct, but there are several distinct precise versions of this statement, and they can be difficult to prove.

The splitting of wave packets into sums of exponentials is however somewhat unsatisfactory, since plane waves have infinite extension.[1] The above considerations also indicate that a pulse, even initially well-localized, will spread out. One

[1]This can be obviated in some cases by replacing the Fourier transform by the Radon transform.

would like to establish anyhow a correspondence between waves and particle-like objects which keep their individuality in interactions, the way most macroscopic objects do. For linear phenomena, such a correspondence is provided by the approximation of geometrical optics, in which high-frequency wave packets are governed by a Hamilton-Jacobi equation at leading order. The semi-classical approximation proposes a comparable correspondence in quantum mechanics, in the limit when Planck's constant vanishes.

Several recent attempts have repeatedly aimed at identifying particles, or other coherent structures, with localized, pulse-like solutions of nonlinear PDE, which should contribute corrections to "linear" wave packets. They will be met with later.

Thus, to summarize, a wave is a disturbance which propagates, and may interact with others; at the same time, it behaves like a material particle in very particular conditions. It has, however, more characteristic parameters than a particle; it can in particular be associated to several speeds. Its precise features can be captured in particular by suitable expansions which are adapted to different regimes.

1.7 FURTHER RESULTS AND PROBLEMS

1. $\|\hat{u}\|_{p'} \leq (2\pi)^{n/p'} \|u\|_p$ if $1 \leq p \leq 2$ (Hausdorff-Young) [see W. Beckner (1975) Inequalities in Fourier Analysis, *Ann. Math., 102*: 159–182, for further results].

2. Does the Fourier transformation map L^p to any L^q if $p > 2$? [*Hint:* Assume there is an estimate

$$|(\hat{u}, \varphi)| \leq \sum_{k \leq k_0} C \|u\|_p \|D^k \varphi\|_\infty$$

for some $k_0 \geq 0$, and for all φ supported in a given compact set. Define, for $u \in S$, $u(t) = \mathcal{F}^{-1} \exp(it|\xi|^2) \mathcal{F} u$. Note that $u(t) = u * (Ct^{-n/2} \exp(-i|x|^2/4t))$. Assume $\mathcal{F}u$ is supported in K and take $\varphi = \overline{\mathcal{F}u(t)}$. A contradiction results if $k_0 < n(1/2 - 1/p)$. In fact, $\mathcal{F}L^p \subset H^{-s}$ if $s > n(1/2 - 1/p)$ for $p > 2$.] The Fourier transform of an element of L^p has a restriction to smooth surfaces of nonzero Gaussian curvature if $1 \leq p \leq (2n + 2)/(n + 3)$, as a locally square-summable function (see Stein (1986)).

3. Find the eigenfunctions of the Fourier transform on S by seeking a differential equation invariant by $\mathcal{F}$.

4. What is the Fourier transform of $\sum_{m \in \mathbf{Z}^n} c_m \exp(im.x)$?

5. Consider an evolution equation of the form $u_t = Lu$, and assume we are given an operator M such that $[L, [L, M]] = 0$. We write as usual $[A, B] = AB - BA$. We assume L and M are independent of t.

(a) Show that $\Gamma(t) := M + t[L, M]$ commutes with $\partial_t - L$.

(b) If $L = p(D)$, show that one can take $M = x_j$ (which is a multi-plication operator). As an example, show that for the Schrödinger equation $(p(D) = -i\Delta)$, $\|x_j u - 2it\partial_j u\|_2$ is independent of t for every solution u. Thus, the Schrödinger equation has a *smoothing property*. What happens in the case $p(D) = -\partial_x^3$, $x \in \mathbf{R}$?

(c) Work out a generalization to systems of two equations.

6. Let u solve the wave equation in, say two space dimensions. Is the following statement true or false: "If u is of class C^∞ precisely on one side of an analytic hypersurface in $\mathbf{R}^3$, then this surface must be characteristic"? [*Hint:* Consider solutions of the form $\sum_{k=1}^\infty \varepsilon_n n^{-k} J_n(nr) \exp(in(\theta + t))$. (See F. John (1961) Proc. Symp. Pure Math., vol. 4.)]

7. Derive the jump conditions $n \cdot [B] = 0$ and $n \times [E] = 0$ for solutions of the vacuum Maxwell's equations

$$E_t = \operatorname{curl} B, \quad B_t = -\operatorname{curl} E;$$

$$\operatorname{div} B = 0, \quad \operatorname{div} E = 0,$$

for which the electric and magnetic fields E and B have a jump across a smooth surface with normal n. The surface must move in time so as to form a hyper-surface in four-dimensional space which will be "characteristic" in a sense to be made precise.

8. Derive the jump conditions for the equations of linear elasticity:

$$2\mu \partial^k e_{ki} + \lambda \partial_i e_k{}^k - \rho \partial_t^2 u_i = 0,$$

where u_i represents the displacement, $e_{ij} = \frac{1}{2}(\partial_i u_j + \partial_j u_i)$ is the strain tensor, and λ, μ, and ρ are positive constants (Lamé moduli and density respectively). All indices are raised and summed using the Kronecker delta.

9. Prove the commutation relations for the generators of the Lorentz and conformal groups given in §1.3.

Derive the further relation $[\Box, S] = 2\Box$. What is $[\Box, K_a]$?

10. Show that the function λ in theorem 1.24 is equal to $[\partial_\phi \partial_\phi u]$ where $\partial_\phi = \phi^i \partial_i$ provided that $|\nabla \phi| \equiv 1$.

11. Let $I_r f := \int_{\mathbf{R}^n} |x - y|^{-n/r} f(y) \, dy$.

(a) I_r maps L^p to L^q if $1 < p < q < \infty$ and $1/q = 1/p - (1 - 1/r)$ (Hardy-Littlewood-Sobolev, or fractional integration theorem).

(b) The inverse Fourier transform of $(1 + |\xi|^2)^{-s/2}$ is a function, and is bounded by $C_N |x|^{s-n}(1 + |x|)^{-N}$ for every integer N if $s > 0$. Conclude that $L_s^p \subset L^{p^*}$ if $0 < s < n$ and $1/p^* = 1/p - s/n$.

(c) Derive the Sobolev inequality for $W^{1,p}$, $1 < p < n$, from the fractional integration theorem (see Aubin (1982)).

12. Show that for appropriate functions,

$$\sum_{m \in Z^n} u(x + 2\pi m) = (2\pi)^{-n} \sum_{m \in Z^n} \hat{u}(m) e^{i(x,m)}.$$

Suppose that a function $f(t) \in \S$ has Fourier transform with support in $[-M, M]$. How should one choose K to ensure that one can recover f from its values at the points j/K, where $j = 0, \pm 1. \pm 2 \ldots$?

13. Discuss the optimality of the decay rates of Th. 15 and 16. (See Littman (1970)).

14. Find a fundamental solution for $\partial_t^2 - \partial_x^2 - \partial_y^2 - 1$ (see Hadamard (1939)).

15 Let ϕ be a real and smooth function of one variable, with $|\phi^{(k)}| \geq 1$ in $[a, b]$.

(a) Show that $|\int_a^b \exp(it\phi(x))\, dx| \leq c_k t^{-1/k}$ if $k \geq 2$, or if ϕ' is monotone and $k = 1$. [*Hint:* Use induction on k. If $k > 1$, Separate the integral into two parts: $\int_{|x-c|<\delta}$ and $\int_{|x-c|>\delta}$, where $|\phi^{(k-1)}|$ attains its minimum at c. Use the induction hypothesis on the outer part, and take $\delta = t^{-1/k}$.] Show $c_{k+1} \leq 2c_k + 2$.

(b) Show that if ψ is smooth,

$$\left| \int_a^b e^{it\phi(x)} \psi(x)\, dx \right| \leq c_k t^{-1/k} \left(|\psi(b)| + \int_a^b |\psi'(x)|\, dx \right).$$

[*Hint:* View the exponential term as the derivative of $\int_a^x e^{it\phi(y)}\, dy$ and integrate by parts.] This result is the van der Corput lemma.

(c) Use (b) to obtain a decay result for the Airy equation.

(d) If $\phi(x_0) = \cdots = \phi^{(k-1)}(x_0) = 0$, $\phi^{(k)}(x_0) \neq 0$, for some $k \geq 2$, then, if ψ is supported in a small neighborhood of x_0, the integral in (b) has an asymptotic expansion $\sum_{j \geq 0} a_j t^{-(j+1)/k}$.

16. If u solves the wave equation with data $(0, f)$, one has $u(., 1) \in L^p$ for $|1/p - 1/2| = 1/(n-1)$ if $n \geq 2$ and $f \in L_0^p$. (See J. Peral (1980) and M. Beals (1980).)

17. Let $\phi(x)$ be compactly supported in $\mathbf{R}^3$. Find the solution u of the wave equation $\Box u = \phi(x) \exp(i\omega t)$ which vanishes for large negative t. Find $\lim_{t \to \infty} u(x, t)$ for fixed x. (This limit is related to a particular solution of the reduced wave equation, namely that solution which satisfies the "Sommerfeld radiation condition.")

18. For a solution $u(x, t)$ of the wave equation which is even in t, define

$$v(x, r) = \frac{(n-1)\omega_{n-1}}{n\omega_n} \int_{-1}^1 u(x, r\mu)(1 - \mu^2)^{(n-3)/2}\, d\mu.$$

Assume u is even in t.

(a) Show that v is the spherical mean of $g = u_t(0)$.

(b) Show that v solves the Euler-Poisson-Darboux equation

$$v_{rr} + (n-1)v_r/r = \Delta v.$$

(c) Show that the relation $v(r) = \int_{-1}^{1} \phi(r\mu)(1 - \mu^2)^{(n-3)/2} \, dr$ implies, if n is odd, that

$$(\frac{n-3}{2})! \psi(\sqrt{s}) = (d/ds)^{(n-1)/2} w(s),$$

where $r = \sqrt{s}$, $r\mu = \sqrt{\sigma}$, $w(s) = v(\sqrt{s})s^{(n-2)/2}$, and $\psi(\sigma) = \phi(\sqrt{\sigma})/\sqrt{\sigma}$. Conclude that

$$u(x,t) = Ct(\frac{d}{dt^2})^{(n-1)/2}[r^{n-2} I_g(x,t)].$$

(See Courant-Hilbert, Ch. VI). What happens for even n? What can one say about solutions of the wave equation that are not even with respect to t?

19. Show that

$$\frac{1}{4\pi|x|}\delta(t - |x|) - \frac{m}{\sqrt{t^2 - |x|^2}}J_1(m\sqrt{t^2 - |x|^2})H(t - |x|)$$

is a fundamental solution of the Klein-Gordon equation in 3 space dimensions, where J_1 the Bessel function of order 1 and H is the Heaviside function. Show that there are two fundamental solutions with support in the (closure of) the outside of the forward light cone.

20. Assume, with the notation of §1.2, that

$$\| \exp(tP(\xi))\| \le c(t) \exp(-a|\xi|^\rho t)$$

for suitable a and ρ positive. Show that if $\|Q(\xi)\| \le b|\xi|^\sigma$, with $\sigma < \rho$, then $u_t = (P(D) + Q(D))u$ is well-posed.

21. (Friedlander radiation field) Let $\Box u = 0$ in three space dimensions. Show that

$$u(x,t) = k(x/r, r - t) + O(1/t^2),$$

as $t \to \infty$ while $r - t$ remains bounded; here, $r = |x|$, and k can be expressed in terms of the Cauchy data. (See Friedlander (1964) and Lax-Phillips (1989)).

NOTES

The material in §§1.1, 1.2 and 1.3 is mostly classical. We have followed Hörmander for the exposition of the Fourier transform and hyperbolicity, and the reader is referred to this volume, and Stein (1970) for a more thorough treatment. A further discussion of representation formulae for the wave and Klein-Gordon equation can be found in Courant-Hilbert (1962) (vol. 2) and

John (1955). The nonexistence of L^p energy estimates is due to Littman (1963) (see Rauch (1986) for a generalization). Further negative results are proved in Littman-McCarthy-Rivière (1968). A complete discussion of the Sobolev embedding theorem is in Nirenberg (1959), Aubin (1982) and Adams (1975); the best constant is discussed in Aubin (1982). These references also contain more inequalities of Gagliardo-Nirenberg type. Th. 1.6 goes back to Moser (1963). The global Sobolev inequality was introduced in Klainerman (1985), with a simplified argument in Klainerman (1987); a generalization is due to Hörmander (1985). Another simple proof is due to John. There are a few more comparison principles for hyperbolic equations beyond Th. 1.8, see Protter-Weinberger (1967).

For the L^1-L^∞ decay estimates of theorems 1.15 and 1.16, we combined the arguments of von Wahl with those in Klainerman, with a few modifications. Some of these estimates can also be proved by the methods of Marshall-Strauss-Wainger (1980). A refinement in three space dimensions is due to Sideris; see also the references in Strauss (1989).

Th. 1.17 and 1.21 are from Strichartz (1977), which also contains an estimate for inhomogeneous problems (such as Th. 1.23 for $P(D) = \Box$). See Peral (1980), Beals (1982), Oberlin (1989) for related results. Th. 1.18 is taken from Marshall-Strauss-Wainger (1980), which contains many other useful results; this paper was further generalized in Marshall (1986). For related results, see Brenner (1975), Pecher (1976). The interpolation theorems we used go back to Stein (see also the paper of Fefferman-Stein on BMO). A few decay estimates of this type are also available for variable, smooth coefficients; see Littman (1973), Brenner (1977), Beals (1982), Bezard (1992), Sogge (1992), Kapitanskii (1990) and their references. Th. 1.23 is due to Hörmander, and Th. 1.24 to Kenig-Ruiz-Sogge (1987).

Th. 1.19 is from Kenig-Ponce-Vega (1989). Earlier results of this type are due to Kato (1983), Sjölin (1987), Vega (1988), Jensen (1986) and Constantin-Saut (1988); the last paper considers systems as well. Th. 1.20 is from Jensen (1986).

Since the literature on L^p and local smoothing estimates for dispersive equations is rapidly growing, we made no effort at completeness in the references, we have however mentioned recent papers for the reader's convenience. One should mention, that special cases of the decay and space-time estimates for the wave and Klein-Gordon equations were proved or suggested by Segal, Littman, Nelson, Morawetz-Strauss, among others, and that, as is clear from the proofs, advances in harmonic analysis (Stein's restriction and interpolation theorems and the discovery of BMO in particular) played a decisive role.

The three types of propagation of "singularities" are now classical, and were all generalized to nonlinear situations. A modern reference for propagation of singularities in linear problems is Taylor (1982).

Boundary-value problems create special difficulties, especially for the propagation of singularities (see again Taylor (1982) or Beals (1989)) and were omitted to keep the length of this chapter within bounds.

REFERENCES

ADAMS, R. (1975) *Sobolev Spaces,* Academic Press.

AUBIN T. (1982) *Nonlinear Analysis on Manifolds. Monge-Ampère Equations,* Springer.

BEALS, M. (1982) L^p boundedness of Fourier integral operators, *Memoirs of the AMS, 264.*

BEALS, M. (1989) Nonlinear Mocrolocal Analysis, Birkhäuser.

BEZARD, M. (1992) Une version générale de l'inégalité de Strichartz, *C. R. Acad. Sci. Paris Ser. 1, 315:* 1241–1244.

BRENNER, P. (1975) On $L_p - L_{p'}$ estimates for the wave-equation, *Math. Z., 145:* 251–254.

BRENNER, P. (1977) $L_p - L_{p'}$ estimates for Fourier integral operators related to hyperbolic equations, *Math. Z., 152:* 272–286.

CONSTANTIN AND SAUT, J.-C. (1988) Local smoothing property of dispersive equations, *J. of the AMS, 1:* 413–446.

COURANT, R., AND HILBERT, D. (1962) Methods of Mathematical Physics, vol. 2, Wiley-Interscience.

FRIEDLANDER, F. G. (1964) On the radiation field of pulse solutions of the wave equation, II, *Proc. Roy. Soc., 279A,:* 386–394.

GAGLIARDO, E. (1958) Proprieta di alcune classi di funzioni in piu variabili, *Ric. Mat., 7,:* 102–137.

HADAMARD, J. (1939) *Lectures on Cauchy's problem,* Dover.

HÖRMANDER, L. (1983) The Analysis of Linear Partial Differential Operators, vol. 1, Springer.

HÖRMANDER, L. (1985) On Sobolev spaces associated to certain Lie algebras Institute Mittag-Leffler Report # 4.

JENSEN, A. (1986) Commutator methods and a smoothing property of the Schrödinger evolution group, *Math. Z., 191:* 53–59.

JOHN, F. (1955) *Plane Waves and Spherical means applied to Differential Equations,* Interscience, NY,

JOHN, F. (1989) *Partial Differential Equations,* 4th edition, Springer.

KAPITANSKII, L. V. (1990) Some generalizations of the Strichartz-Brenner inequality, *Leningrad Math. J., 1, 3*: 693–726.

KATO, T. (1983) On the Cauchy problem for the (generalized) Korteweg-de Vries equation, *Studies in Appl. Math,* (V. Guillemin, ed.), *8*: 93–128.

KENIG, C., PONCE, G., AND VEGA, J. (1989) On the (generalized) Korteweg-de Vries equation, *Duke Math. J., 59, 3*: 585–610.

KENIG, C., RUIZ., AND SOGGE, C. D. (1987) Uniform Sobolev inequalities and unique continuation for second order constant coefficient differential operators, *Duke Math. J., 55, 5*: 329–347.

KLAINERMAN, S. (1980) Global existence for nonlinear wave equations, *Comm. Pure and Appl. Math., 33*: 43–101.

KLAINERMAN, S. (1985) Uniform decay estimates and the Lorentz invariance of the classical wave equation, *Comm. Pure and Appl. Math., 38*: 321–332.

KLAINERMAN, S. (1987) Global Sobolev inequalities in the Minkowski space $\mathbf{R}^{n+1}$, *Comm. Pure and Appl. Math., 40*: 111–117.

LAX, P. D. (1957) Asymptotic solutions of oscillatory initial-value problems, *Duke Math. J., 24*: 135–169.

LAX, P. D., AND PHILLIPS, R. (1989) *Scattering Theory,* Academic Press.

LITTMAN, W. (1963) The wave equation and L^p norms, *J. Math. and Mech., 12,*: 55–68.

LITTMAN, W. (1963) Fourier transforms of surface-carried measures and differentiability of surface averages, *Bull. AMS, 69*: 766–770.

LITTMAN, W. (1970) Maximal rates of decay of solutions of partial differential equations, *Arch. Rat. Mech. Anal, 37, 1*: 11–20.

LITTMAN, W. (1973) $L^p \rightarrow L^q$ estimates for singular integral operators, *Proc. Symp. Pure Math., 23*: 479–481.

LITTMAN, W., McCARTHY, C., AND RIVIÈRE, N. (1968) The non-existence of L^p estimates for certain translation invariant operators, *Studia Math., 30*: 219–229.

MARSHALL, B., STRAUSS, W., AND WAINGER, S. (1980) $L^p - L^q$ estimates
for the Klein-Gordon equation, *J. Math. pures et appl., 59*: 417–440

MARSHALL, B. (1986) Mixed norm estimates, *Canadian Math Bull., 29*: 11–
19.

MEYER, Y. (1973) Trois problèmes sur les sommes trigonométriques, *Astéri-sque, 1*: 1 (pp. 1–17).

NIRENBERG, L. (1959) On elliptic partial differential equations, *Ann. Sc.
Norm. Sup. Pisa, 13,*: 116–162.

OBERLIN, D. (1989) Convolution estimates for some distributions with singu-larities on the light cone, *Duke Math. J., 59, 3*: 747–757.

PECHER, H. (1976) L^p-Abschätzungen und klassische Lösungen für nichtlin-eare Wellengleichungen, I, *Math. Z., 150*: 159–183.

PERAL, J. (1980) L^p estimates for the wave equation, *J. Funct. Anal., 36*:
114–145.

PROTTER, M., AND WEINBERGER, H. (1967) *Maximum Principles in Differ-ential Equations,* Prentice-Hall, Englewood Cliffs, NJ.

RAUCH, J. (1986) *BV* estimates fail for most quasilinear hyperbolic systems
in dimensions greater than one, *Comm. Math. Phys, 106*: 481–484.

SIDERIS, T. (1989) Decay estimates for the three dimensional inhomogeneous
Klein-Gordon equation and applications, *Comm. PDE, 14 (10)*: 1421–
1455.

SJÖLIN (1987) Regularity of solutions to the Schrödinger equation, *Duke Math.
J., 55*: 699–715.

SOGGE, C. D. (1992) *Fourier Integrals in Classical Analysis,* Cambridge Uni-versity Press.

STEIN, E. (1970) *Singular Integrals and Differentiability Properties of Func-tions,* Princeton University Press, Princeton, NJ; see also *Harmonic Anal-ysis,* Princeton (1993).

STEIN, E., AND WEISS, G. (1971) *Introduction to Fourier Analysis on Eu-clidean Spaces,* Princeton University Press, Princeton, NJ.

STEIN, E. (1986) Oscillatory integrals in Fourier analysis, in *Beijing Lectures in
Harmonic Analysis,* E. M. Stein ed., Princeton University Press, 307–355.

STRAUSS, W. (1989) *Nonlinear Wave Equations,* CBMS lecture notes, *73,* Amer. Math. Soc., Providence, RI.

STRICHARTZ, R., (1970) Convolutions with kernels having singularities on a sphere, em Trans. AMS, 148: 461–471.

STRICHARTZ, R. (1970) A priori estimates for the wave equation and some applications, *J. Funct. Anal., 5*: 218–235.

STRICHARTZ, R. (1977) Restriction of Fourier transform to quadratic surfaces and decay of solutions of wave equations, *Duke Math. J. 44*: 705–714.

TAYLOR, M. (1991) *Pseudo-differential operators and Nonlinear PDE,* Birkhäuser, Boston.

TRÈVES, F. (1975) *Basic Linear Partial Differential Equations,* Academic Press, New York.

VEGA, J. (1988) Schrödinger equations: Pointwise convergence to the initial data, *Proc. AMS, 102*: 874–878.

VON WAHL, W. (1970) Über die klassische Lösbarkeit des Cauchy-Problems für nichtlineare Wellengleichungen bei kleinen Anfangswerten und das asymptotische Verhalten der Lösungen, *Math. Z., 114*: 281–299

VON WAHL, W. (1971) L^p decay rates for homogeneous wave-equations, *Math. Z., 120*: 93–106.

Chapter 2

Local and Global Existence

This chapter presents a few current methods for proving local and global existence for nonlinear wave equations. We focus on perturbations of the wave equation, which are rather representative of the difficulties one may expect. Nonlinear evolution equations are *not* ODE's in Banach spaces, in the sense that spatial differentiations are unbounded operators, even in an analytic set-up. One may say that all the following methods are sophisticated ways of getting around this problem. The various approaches on the existence issue can be summarized as follows .

The analytic Cauchy problem is the first to consider, not only because it was addressed first historically, but because it provides the motivation and basis for many modern methods. The first result in this area is the classical Cauchy-Kowalewska theorem, proved by the method of majorants. This technique can be refined in order to reduce the amount of guesswork involved in the choice of majorants. More recently, a number of proofs using an iteration in a scale of Banach spaces were given, motivated by applications involving ill-posed problems, such as the vortex sheet problem. Such proofs essentially contain the results of the majorant method, and are sometimes sharper. They allow to relax the analyticity hypotheses in the time variable. Both of these approaches are based on the fact that the Taylor expansion of the solution of the Cauchy problem converges whenever it can be computed to all orders. This is very similar to well-known results for solutions of ODE's with regular singular points, and in fact, there is a class of PDE, termed Fuchsian for that reason, which share this property. Any system of Cauchy-Kowalewska type can be transformed into a Fuchsian system by multiplication by the time variable, so that results on Fuchsian equations contain strictly the results mentioned above. The analytic Cauchy problem therefore reduces to the construction and convergence of a formal solution; this procedure is very close to the one followed in the applications, and is applicable in situations which would be intractable in any other way. Also, the process of singularity formation seems to lead naturally to the consideration of complex analytic solutions (see Ch. 3). These help understand the continuing advances made even today on this issue.

In the non-analytic situation, it is not sufficient any more that the initial surface be non-characteristic for the Cauchy problem to be solvable. We already saw this in the linear case: the equation must be hyperbolic with respect to the initial surface. In other words, this surface must be "space-like".

This line of argument actually works for more general systems, called "symmetric hyperbolic." However, the solution will not in general persist for all time. One can find specific norms, the boundedness of which implies persistence of smooth solutions.

There are however many non-hyperbolic initial value problems which are well posed, such as the IVP for the nonlinear heat or Schrödinger equations. It turns out that all of these equations can be treated in a unified framework: the solution operator corresponding to the linear part defines a semi-group of bounded operators in a suitable Banach space. Using a variant of the formula of variation of parameters, the nonlinear problem can be recast as an integral equation. One can then prove the existence of a solution by iteration. Despite its sometimes technical character, this procedure actually gives precise smoothness information in a natural way.

Even without using semi-groups, it is often possible to re-cast a semi-linear evolution equation in the form of an integral equation. The issue is then to set-up a convergent iteration, in a suitable space. When applicable, this is certainly the simplest approach. It sometimes gives global existence (i.e., for all time) as well. More complicated iterations, in the spirit of the Nash-Moser theorem, have often preceded such simpler proofs.

We now describe the contents of this chapter.

Section 2.1 gives a version of the majorant method applied to the Cauchy-Kowalewska theorem, and proves an existence result for generalized Fuchsian equations, which will be used in Ch. 3. The latter is an illustration of iterative methods in an analytic set-up.

Section 2.2 presents a version of the energy method, with simple applications.

Section 2.3 states and proves the Hille-Yosida theorem, and includes a few typical applications.

Section 2.4 presents results which pertain more specifically to perturbations of the wave equation, for the case of small data. It details the application of the conformal method to equations which satisfy the null condition. Other roads to this result are outlined.

Section 2.5 lists other special results and methods. Although we focus on semilinear hyperbolic equations, many of these have analogues for other types. Some brief ideas on the method of "nonlinear scattering" are included.

More existence results are contained in Chapters 4 and 6. In particular, we defer the discussion of hyperbolic problems on manifolds to Ch. 6.

2.1 THE ANALYTIC CAUCHY PROBLEM

We are interested in finding local holomorphic solutions to initial-value problems with holomorphic data. In other words, we are interested in a class of problems the formal solutions of which always converge. The prototype of such equations is an ODE with regular singular points, that is, of Fuchsian type. After a brief discussion, we prove the Cauchy-Kowalewska theorem by a version of the majorant method. We then introduce a class of PDE of "Fuchsian type" for which an existence theorem will be proved by an iterative method. Further examples of equations of this type will come up in the study of blow-up.

An example.

Consider the equation

$$(2.1) \qquad\qquad u_t = f(x, t, u, u_x),$$

where u is an analytic function of $x \in C$, and f is entire in all its arguments. Let us write the expansion of u as

$$\sum_{j \geq 0} u_j(x) t^j.$$

Let us also write $\{X\}_j$ for the coefficient of t^j in the expansion of any given expression X. We then find, for every j,

$$(2.2) \qquad\qquad \{t u_t\}_j = j u_j(x) = \{t f\}_j = \{f\}_{j-1}.$$

Therefore, u_0 is arbitrary, and $u_1, u_2, \ldots$ can be computed recursively. Thus, it is natural to consider the initial-value problem for (2.1) with data on $t = 0$.

We wish to prove that the series for u found in this way converges for all choices of u_0, in an appropriate domain. It will then be the only analytic solution of Eq. (2.1) such that $u(x, 0) = u_0(x)$.

Note that one may assume $u_0 = 0$ by replacing u by $u - u_0$.

Two methods can be used to achieve this:

1. Seek an estimate of the form $|u_j| \leq C M^j$;

2. Rewrite the equation as

$$u = u_0 + \int_0^t f(x, s, u, u_x)\, ds,$$

and try to use a fixed point theorem.

Both methods can be used successfully in a variety of situations.

The majorant method, briefly treated next on a few special cases, gives a way to obtain the estimates required by the first method. We will then turn to the iterative procedure, which will be carried out for a much more general class of problems, in view of later applications.

Method of majorants.

Recall that power series $\sum_I b_I x^I$ in several variables is a majorant for another series $\sum_I a_I x^I$ if $|a_I| \leq b_I$ for all I. We write

$$\sum_I a_I x^I \ll \sum_I b_I x^I.$$

If this only holds if the length of I does not exceed N, we speak of majorants upto order N, and replace $\ll$ by $\ll_N$. Any function analytic near 0 has a majorant of the form $C/(1 - (\sum_i x_i)/R)$.

Examination of the calculation of the formal solution shows that u_j is obtained from the previously found coefficients by differentiations and application of nonlinear functions. If these functions and the coefficients are all replaced by majorants, we find that one can find a majorant for u_j by performing finitely many operations with nonnegative coefficients on the majorants of the previously computed approximations.

Let us now, to fix ideas, consider a linear equation of order m:

$$\partial^m u/\partial t^m = a(\partial_t, \partial_1 \ldots, \partial_n)u + v(x, t).$$

Here, a is a differential operator of order m with "constant coefficients." We assume, as usual, that a does not contain $\partial^m u/\partial x_1^m$. We assume, as we may, that $v \ll C/(1 - s/r)$, where

$$s = \frac{x}{\alpha R} + (x_2 + \cdots + x_n)/R.$$

The positive number α will be chosen later. We seek a majorant U for u as a function of t. This leads to the ODE

$$d^m U/ds^m = (\alpha R)^m [a(p \frac{d}{ds})U(s) + \frac{C}{1 - s/r}],$$

where we have set $p = (1/(\alpha R), 1/R, \ldots, 1/R)$. In view of our assumption on a, we see that one can assume that the total coefficient of the top derivative $d^m U/dt^m$ is positive by taking α sufficiently small. One then solves the ODE, which has an analytic solution with nonnegative coefficients, as we check by inspection of its Taylor expansion. This is the desired majorant. Note that the introduction of s reduced the problem not to a simpler PDE, but to an ODE.

A variant is given in the exercises.

Fuchsian equations.

We consider now two generalizations of (2.1).
The first is

$$(2.3) \qquad\qquad tu_t + A(x)u = tf(x,t,u,u_x),$$

where $u = (u_1,\ldots,u_m)$ is a function of $(x_1,\ldots,x_n) \in \Omega \subset \mathbf{C}^n$, while t is close to zero in $\mathbf{C}$ and f is holomorphic in its arguments in a neighborhood of $\Omega \times \{0\}^{1+m+mn}$. We seek, without loss of generality, solutions close to 0. The matrix A will be constant in most applications. It is crucial that the right-hand side be $O(t)$.

Equations of the above type will be called *Fuchsian PDE*.

A still wider class of equations which we will need later is that of *generalized Fuchsian equations*. These involve several "time" variables and have the general form

$$(2.4) \qquad\qquad (N + A)u = \sum_{p=0}^{l} t_p f_p(x, t_0, \ldots, t_p, u, u_x),$$

where $N = \sum_{i,j} n_{ij} t_i \partial/\partial t_j$.

Example 1. Let u satisfy equation (2.1). Then, after multiplication by t, we see that it also solves a Fuchsian equation. Further, if $u = u_0 + tv$, we find

$$tv_t + v = f(x,t,u_0 + tv, \partial_x(u_0 + tv)) - f(x,t,u_0,u_{0x}).$$

The right-hand side is clearly $O(t)$. More generally, if u solves a linear first-order Cauchy problem, u/t^θ, where θ is a real number, will solve a Fuchsian problem. This explains why Fuchsian equations are ubiquitous.

Example 2. Let us consider an arbitrary system of partial differential equations, of order M. Let us also single out a hypersurface Σ in the space of independent variables. We may assume locally that Σ is represented by $t = 0$, using an appropriate change of coordinates, provided the surface is smooth. It is well-known that by introducing partial derivatives of u as unknowns, one can re-cast the equation in the form of a first-order system

$$Qu_t = f(x,t,u,u_x)$$

where Q is a function of x, t and maybe u. If Q is nonzero, Σ is said to be *non-characteristic*, and the equation has the form considered here. However, the reduction is not unique, and all solutions of the first-order system do not give rise to solutions of the original problem, see the Notes.

Example 3. Assume that $v(x, t, y)$ solves the generalized Fuchsian equation

$$(t\partial_t + (t + y)\partial_y)v = tf(x, t, v, v_x).$$

Then $u(x, t) = v(x, t, t \ln t)$ solves

$$\partial_t u = f(x, t, u, u_x).$$

Similarly, if

$$(t\partial_t + \alpha y\partial_y)v = tf(x, t, v, v_x).$$

Then $u(x, t) = v(x, t, t^\alpha)$ solves

$$\partial_t u = f(x, t, u, u_x).$$

By adding as new dependent variables derivatives of sufficiently high order, one can always convert it to a first-order system

Basic existence theorem.

We now give fairly general existence theorems which imply in particular the usual Cauchy-Kowalewska theorem in its general form.

We consider the problem

$$(2.5) \qquad\qquad Nz + Az = f(t', x, z, Dz),$$

where $N = N_l = \sum_{k=0}^{l}(t_k + kt_{k-1})\partial/\partial t_k$, A is constant, and f is analytic near $(0,0,0,0)$ without constant term in t'. The unknown z has m components, and $t' = (t_0, \dots, t_l)$. All functions are analytic in their arguments unless otherwise specified. One may, by introducing new dependent variables, assume, as we will, that f is linear in Dz.

We prove the following result.

THEOREM 2.1 *If A has no eigenvalue with negative real part, (2.5) has near the origin exactly one analytic solution which vanishes for $t' = 0$.*

Remark: Instead of requiring A to be constant, one may require that there exists a matrix-valued function $P(x)$ such that $P(x)^{-1}AP(x)$ is constant, since the latter case reduces to the former by a redefinition of u. It is likely that N may be replaced by a more general first-order operator, with similar proofs, but the present set-up is sufficient for most applications.

A special case is the following:

$$(2.6) \qquad\qquad tu_t + Au = tf(t, x, u, Du)$$

where u and f are vector-valued, with m components, and are analytic near $(t, x, u, Du) = (0, 0, 0, 0)$, subject to the same conditions as before.

THEOREM 2.2 *This problem has a unique holomorphic solution.*

Proof: We first treat the case of two time variables ($l = 1$), and indicate the modifications in the general case afterwards. We let $(t_0, t_1) = (T, Y)$. The problem has therefore the form:

(2.7)
$$\begin{cases} (N + A)z = f(T, Y, X, z, Dz) \\ z(0, 0, X) = z_0(X) \in Ker(A), \end{cases}$$

where $D = D_X$ and $f \equiv 0$ for $T = Y = 0$. The unknown has m components. Replacing z by $z - z(0, 0, X)$, we see that, since here $Az(0, 0, X) = 0$, we may assume

$$z(0, 0, X) = 0.$$

This means that the solution of (2.7) depends on the choice of one function, which has been incorporated into the right-hand side.

We define

$$F[z] := f(T, Y, X, z, Dz).$$

The argument is in five steps.

Step 1. We first observe that

(2.8)
$$\begin{cases} (N + A)\, z(T, Y) = k(T, Y); \\ \qquad z(0, 0) = 0, \end{cases}$$

where k is analytic, independent of X, and vanishes for $T = Y = 0$, has a unique analytic solution, given by

(2.9)
$$z(T, Y) = H[k] := \int_0^1 \sigma^{A-1} k(\sigma T, \sigma(T \ln \sigma + Y))\, d\sigma.$$

Indeed, let

$$g(\sigma) = z(\sigma T, \sigma(T \ln \sigma + Y))$$

for $0 < \sigma < 1$. We find

$$\frac{d}{d\sigma}(\sigma^A g(\sigma)) = \sigma^{A-1} k(\sigma T, \sigma(T \ln \sigma + Y)),$$

and since $g(\sigma)$ must tend to zero as σ goes to zero, while σ^A remains bounded, we have

$$\sigma^A g(\sigma) = \int_0^\sigma \tau^{A-1} k(\tau T, \tau(T \ln \tau + Y))\, d\tau,$$

Since k vanishes at the origin, the contribution from k to the integral is $O(\tau^{1-\varepsilon})$ for any $\varepsilon > 0$, and the integral converges.

Equation (2.9) follows by letting $\sigma = 1$ in the last equation. One checks directly, using the Cauchy-Riemann equations, that this does provide an analytic solution to the problem.

Eq. (2.7) can therefore be rewritten as the integral equation $z = H[F[z]]$. We let instead $u = F[z]$, and consider

$$(2.10) \qquad\qquad u = G[u] := F[H[u]],$$

which will be solved by a fixed point argument. The desired solution will then be given by $z = H[u]$.

Step 2. We define two norms. Assume f is analytic for $X \in C^n$ and $d(X, \Omega) < 2s_0$ and $u \in C^m$ with $|u| < 2R$, for some positive constants s_0 and R, where Ω is a bounded open neighborhood of 0.

For any function $u = u(X)$ we define the s-norm

$$(2.11) \qquad\qquad \|u\|_s := \sup\{|u(X)| : d(x, \Omega) < s\}.$$

For any function $u = u(T, Y, X)$, and a a sufficiently small positive number, to be chosen later, we define the a-norm

$$(2.12) \quad |u|_a := \sup_{\substack{\delta_0(T,Y)<a(s_0-s) \\ 0 \le s < s_0}} \left\{ \delta_0^{-1}\|u\|_s(T,Y)(s_0-s)\sqrt{1 - \frac{\delta_0}{a(s_0-s)}} \right\},$$

where $\delta_0 = \delta_0(T, Y) := |T| + \theta|Y|$, and $0 < \theta < 1$ is fixed. We wrote $\|u\|_s(T, Y)$ for the s-norm of $u(., T, Y)$.

We also let $\delta(\sigma) = \delta_0 \sigma(1 - \theta \ln \sigma)$. The main properties of $\delta(\sigma)$ are

1. $\delta(\sigma)$ increases strictly from 0 to δ_0,

2. $\delta_0(\sigma T, \sigma(Y + T \ln \sigma)) \le \delta(\sigma)$ if $0 < \sigma < 1$,

3. $d\delta(\sigma)/d\sigma \ge \delta(\sigma)/(C_0\sigma)$. One can take $C_0 = 1 - \theta$.

It follows in particular that if $|u|_a < \infty$,

$$\|u\|_s(\sigma T, \sigma(T \ln \sigma + Y)) \le \frac{\delta(\sigma)|u|_a}{s_0 - s} \left(1 - \frac{\delta(\sigma)}{a(s_0 - s)}\right)^{-1/2}.$$

This is how the a-norm comes into the argument.

Step 3. We prove that one can estimate the s-norm of Hu in terms of the a-norm of u. From the definitions of our various norms, it follows that

$$(2.13) \qquad \|Hu\|_s(T,Y) \le \frac{|u|_a}{s_0 - s} \int_0^1 |\sigma^A| \frac{\delta(\sigma)}{\sigma} \left\{ 1 - \frac{\delta(\sigma)}{a(s_0 - s)} \right\}^{-1/2} d\sigma$$

if $\delta_0 < a(s_0 - s)$. We estimate σ^A by a constant C_1. Let us define

$$\rho = \frac{\delta(\sigma)}{a(s_0 - s)}$$

so that, by property 3. above,

$$\frac{d\sigma}{d\rho} \leq C_0 \frac{\sigma}{\delta(\sigma)} a(s_0 - s).$$

It follows that

$$\|Hu\|_s(T,Y) \leq \frac{|u|_a}{s_0 - s} \int_0^1 C_0 C_1 \frac{a(s_0 - s)\, d\rho}{\sqrt{1 - \rho}} = 2 C_0 C_1 a |u|_a,$$

or

$$(2.14) \qquad \|Hu\|_s(T,Y) \leq C_2 a |u|_a.$$

Step 4. Next, we observe that, since we have taken f to be linear in the spatial derivatives, Cauchy's inequality gives

$$(2.15) \qquad \|F[u] - F[v]\|_{s'}(T,Y) \leq \frac{C_3 \delta_0(T,Y)}{s - s'} \|u - v\|_s$$

for $0 < s' < s < s_0$, if $\|u\|_s \leq R$ and $\|v\|_s \leq R$. We should of course also require $s' < s_0 - \delta_0/a$ if we wish to use the a-norm.

Step 5. Let us now assume $|u|_a, |v|_a < R/(2C_2 a)$. Let $G[u] = F[Hu]$. We prove that

$$(2.16) \qquad |G[u] - G[v]|_a \leq C_4 a |u - v|_a$$

for some constant C_4.

To this end, let $\sigma_j = j/n$, for $0 \leq j \leq n$, and

$$w_j = \int_0^{\sigma_j} \sigma^{A-1} u(\sigma T, \sigma(T \ln \sigma + Y))\, d\sigma - \int_{\sigma_j}^1 \sigma^{A-1} v(\sigma T, \sigma(T \ln \sigma + Y))\, d\sigma,$$

and observe that

$$(2.17) \qquad G[u] - G[v] = \sum_{j=1}^n F[w_j] - F[w_{j-1}].$$

One checks, using the argument of Step 3, that $\|w_j\|_s \leq R$ for $\delta_0(T,Y) < a(s_0 - s)$, so that $F[w_j]$ is indeed defined.

We can now use (2.15): if $s_j \in (s, s_0 - \delta_0(T,Y)/a)$ for every j, we find from Step 4 that

$$\|F[w_j] - F[w_{j-1}]\|_s \leq \frac{C_3 \delta_0}{s_j - s} \|w_j - w_{j-1}\|_{s_j}.$$

On the other hand,

$$\|w_j - w_{j-1}\|_{s_j} \leq \int_{\sigma_{j-1}}^{\sigma_j} |\sigma^{A-1}|\, \|u - v\|_{s_j}(\sigma T, \sigma(T \ln \sigma + Y))\, d\sigma.$$

This suggests the choice:

$$s_j = \min\{s(\sigma) : \sigma_{j-1} \le \sigma \le \sigma_j\},$$

where

$$s(\sigma) = \frac{1}{2}\left[s + s_0 - \frac{\delta(\sigma)}{a}\right].$$

Observe next that $\sum_j s_j \chi_{[\sigma_{j-1},\sigma_j)} \to s(\sigma)$ as j tends to infinity, uniformly and *from below* on $(0,1)$, if $\delta_0(T,Y) < a(s_0 - s)$. Furthermore, if $\sigma_{j-1} \le \sigma \le \sigma_j$,

$$
\begin{aligned}
\|u - v\|_{s_j}(\sigma T, \sigma(T\ln\sigma + Y)) &\le \|u - v\|_{s(\sigma)}(\sigma T, \sigma(T\ln\sigma + Y)) \\
&\le \frac{\delta(\sigma)|u - v|_a}{s_0 - s(\sigma)}\left(1 - \frac{\delta(\sigma)}{a(s_0 - s(\sigma))}\right)^{-1/2}.
\end{aligned}
$$

We therefore find, since $|\sigma^{A-1}| \le C_1/\sigma$, letting $j \to \infty$,

$$
\begin{aligned}
&\|G[u] - G[v]\|_s(T,Y) \\
&\le C_3\delta_0 \int_0^1 \frac{\delta(\sigma)|u - v|_a}{(s(\sigma) - s)(s_0 - s(\sigma))}\left(1 - \frac{\delta(\sigma)}{a(s_0 - s(\sigma))}\right)^{-1/2} C_1\frac{d\sigma}{\sigma}.
\end{aligned}
$$

Since

$$s(\sigma) - s = \frac{s_0 - s}{2}\left(1 - \frac{\delta(\sigma)}{a(s_0 - s)}\right)$$

and

$$s_0 - s(\sigma) = \frac{s_0 - s}{2}\left(1 + \frac{\delta(\sigma)}{a(s_0 - s)}\right)$$

we let again $\rho = \delta(\sigma)/[a(s_0 - s)]$. As σ varies from 0 to 1, ρ varies from 0 to $\delta_0/[a(s_0 - s)]$ (which is always less than 1); note also that

$$1 - \frac{\delta(\sigma)}{a(s_0 - s(\sigma))} = \frac{1 - \rho}{1 + \rho}.$$

Performing this change of variables, we find, using $d\sigma/d\rho \le C_0\sigma/\rho$,

$$
\begin{aligned}
&\|G[u] - G[v]\|_s(T,Y) \\
&\le C_1C_3\delta_0 \int_0^{\delta_0/[a(s_0-s)]} C_0\delta(\sigma)\frac{4|u - v|_a}{(s_0 - s)^2}(1 - \rho^2)^{-1}\sqrt{\frac{1+\rho}{1-\rho}}\,\frac{a(s_0 - s)\,d\rho}{\delta(\sigma)} \\
&= 4a(s_0 - s)^{-1}C_0C_1C_3\delta_0|u - v|_a \int_0^{\delta_0/[a(s_0-s)]} d\rho/(1 - \rho)^{3/2} \\
&= C_4\delta_0 a(s_0 - s)^{-1}|u - v|_a\left(1 - \frac{\delta_0}{a(s_0 - s)}\right)^{-1/2}.
\end{aligned}
$$

Therefore

$$|G[u] - G[v]|_a \le C_4 a|u - v|_a,$$

QED.

End of proof ($l = 1$). Let us define $u_0 = 0$ and $u_1 = G[u_0]$. There is a constant R_0 such that

$$\|u_1\|_{s_0} \le R_0 \delta_0(T, Y)$$

if $|T| + \theta|Y| = \delta_0$. Assume a is chosen so small that

$$C_4 a < 1/2 \quad and \quad R_0 s_0 < R/(4C_2 a).$$

This ensures in particular $|u_1|_a \le R_0 s_0 < R/(4C_2 a)$. The mapping G is now a contraction in the a-norm on the set $\{|u|_a \le R/(2C_2 a)\}$. The existence (and uniqueness) of the desired solution now follow from the contraction mapping principle.

This ends the proof of the theorem in this case.

End of proof (general case).

The proof parallels the one given for (2.7), and we therefore only indicate the differences.

We write

$$N = \sum_j m_{ij} t_j \partial_i$$

for suitable coefficients m_{ij} forming a matrix M. One must then replace $H[k]$ by

$$\int_0^1 \sigma^{A-1} k(\sigma^M t') \, d\sigma,$$

and use $\delta_0(t_0, \ldots, t_l) = \sum_{k=0}^l \theta^k |t_k|$, where $\theta \in (0, 1/l)$ is fixed. We then use for $\delta(\sigma)$ the quantity $\delta_0 \sigma (1 - \theta \ln \sigma)^l$.

We need to check the three properties of δ in Step 2, §3.

The first follows from the assumption $0 < \theta < 1/l$.

The second is checked as follows: Since $t'(\sigma) := \sigma^M t'$ solves $\sigma \, dt'/d\sigma = Mt'$, $t'(1) = t'$, we find

$$(\sigma^M t')_k = \sum_{j=0}^k t_j \binom{k}{j} \sigma (\ln \sigma)^{k-j}.$$

We then compute

$$
\begin{aligned}
\delta_0(\sigma^M t') &= \sum_k \theta^k |t_k(\sigma)| \le \sigma \sum_{0 \le j \le k \le l} t_j \theta^k \binom{k}{j} |\ln \sigma|^{k-j} \\
&= \sigma \sum_{0 \le j \le k \le l} \theta^j |t_j| \binom{k}{j} (-\theta \ln \sigma)^{k-j}.
\end{aligned}
$$

$$= \sigma \sum_{0 \le j \le k \le l} (-\theta \ln \sigma)^j \binom{k}{j} \theta^{k-j} |t_{k-j}|$$

$$\le \sigma \sum_{j=0}^{l} (-\theta \ln \sigma)^j \binom{l}{j} \sum_{k \ge j} \theta^{k-j} |t_{k-j}|$$

$$\le \sigma \delta_0(t') \sum_{j=0}^{l} (-\theta \ln \sigma)^j \binom{l}{j}$$

$$= \sigma \delta_0(t')(1 - \theta \ln \sigma)^l,$$

which is the desired result.

The third property follows from

$$\frac{\delta'(\sigma)}{\delta(\sigma)} \ge \frac{1 - l\theta}{\sigma}.$$

The rest of the proof proceeds verbatim. We obtain existence on domains of the form

$$\{\delta(t') < a(s_0 - s); \quad d(x, \Omega) < s\}.$$

This completes the proof of the existence result.

2.2   THE ENERGY METHOD

Local existence for evolution equations cannot be deduced from existence theorem for ODE's, because differential operators are unbounded in all usual function spaces. Even in the analytic case, the operator d/dz is not bounded on the space of functions analytic in a strip, with the sup-norm. We overcame this difficulty in §2.1 by iterating in a scale of spaces, as in the Nash-Moser theorem (Ch. 5). We here present a different way out. We first solve a regularized problem, which *is* an ODE in a Banach space. We then pass to the limit, using the *energy method*. This method is based on the observation that even if an operator L is unbounded, one *can* bound (Lu, u) provided that $L + L^*$ is bounded. This idea is illustrated on a few examples.

Symmetric hyperbolic systems.

Generalities. We consider systems of the form

$$A_0(x, t, u)\partial_t u = \sum_j A_j(x, t, u)\partial_j u + B(x, t, u).$$

We wish to solve the initial-value problem when one requires $u(x, 0) = f(x)$. Regularity assumptions are spelled out below.

Such a system is called *symmetric hyperbolic* if $A_0 = I$ and the matrices A_j are all (real and) symmetric.

A system is called *symmetrizable* if it can be cast in the above form after multiplication by suitable matrix operator (a "symmetrizer"), where in addition A_0 is symmetric, and there is a positive constant c such that for every constant vector ξ, $(A_0\xi) \cdot \xi \geq c|\xi|^2$.

Examples are given at the end of the section.

Assumptions and main result. We turn to the assumptions on f and the A_k.

We assume $f \in H^s(\mathbf{R}^n)$ for some $s > n/2 + 1$. We take s to be an even integer, for simplicity; see however the Notes. Also, the case of compact (smooth, oriented) manifolds could be handled by similar methods. We require:

1. The (Moser-type) estimate

$$|\varphi(.,t,u)|_s \leq C(\|u\|_\infty)(1 + |u|_s),$$

where φ stands for A_0, A_j or B;

2. The commutator estimate

$$\|[\Lambda^s, \varphi]v\|_{L^2} \leq C(|\varphi|_s\|v\|_\infty + \|\varphi\|_{C^1}|v|_{s-1}),$$

where $\Lambda := (1 - \Delta)^{1/2}$,

We also require the analogous estimates for differences of the form $\varphi(.,t,u) - \varphi(.,t,v)$.

For smooth nonlinearities, both of these assumptions follow from the Gagliardo-Nirenberg inequality (exercise).

We then have

THEOREM 2.3 *Under the above hypotheses, the initial-value problem has a unique solution in $C(0,T; H^s) \cap C^1(0,T; H^{s-1})$, for some $T > 0$. If $\|u(t)\|_{C^1}$ is bounded on $(0,T)$, the solution extends beyond $t = T$.*

Remark: Note that this provides a criterion for singularity formation. It is not possible for, say, the C^4 norm of u to blow-up without the C^1 norm blowing up as well. A similar theorem holds in the symmetrizable case, and the (somewhat lengthy) details are omitted.

Idea of proof: We first introduce a family $\{J_\epsilon\}$ is of standard Friedrichs mollifiers, by $\mathcal{F}(J_\epsilon u)(\xi) = \phi(\epsilon\xi)\hat{u}(\xi)$ with ϕ compactly supported and $\equiv 1$ near 0. We proceed in 4 steps.

1. Regularized equation. We then consider the equation

$$\partial_t u_\epsilon = J_\epsilon L_\epsilon J_\epsilon u_\epsilon + B_\epsilon,$$

where

$$L_\epsilon = L(J_\epsilon u_\epsilon) = \sum_j A_j(x, t, J_\epsilon u_\epsilon)\partial_j$$

and

$$B_\epsilon = J_\epsilon B(x, t, J_\epsilon u_\epsilon),$$

is an ODE in H^s which has a unique solution with $u_\epsilon(0) = f$.

2. H^s estimate. We derive, using the commutator estimate, an estimate on the u_ϵ in H^s. Multiplying the equation by $\Lambda^s u_\epsilon \Lambda^s$, we find

$$\frac{d}{dt}|\Lambda^s u_\epsilon|^2 = 2(\Lambda^s J_\epsilon L_\epsilon J_\epsilon u_\epsilon, \Lambda^s u_\epsilon) + 2(\Lambda^s B_\epsilon, \Lambda^s u_\epsilon).$$

The basic observation of the energy method is that for any v,

$$2(L_\epsilon v, v) = ((L_\epsilon + L_\epsilon^*)v, v) = -((\sum_j v\partial_j A_j, v),$$

and therefore does not involve the derivatives of v. It does involve derivatives of u, but since s will be taken greater than $n/2 + 1$, these derivatives will not create any difficulties. To use this observation, however, with $v = \Lambda^s u_\epsilon$, we need to exchange J_ϵ and L_ϵ. This can be achieved thanks to the commutator estimate.

The net result is the inequality

$$\frac{d}{dt}|u_\epsilon|_s^2 \le C(\|\Lambda^s u_\epsilon\|_{C^1})(1 + |J_\epsilon u_\epsilon|_s^2).$$

An estimate in H^s for the approximations follows from Gronwall's inequality, on some interval $[0, T]$. Note already the role of the C^1 norm.

3. Convergence in a weak norm. To pass to the limit, one can prove the convergence of the approximations in the L^2 norm. The main point is that for any two ϵ, η, the difference $J_\epsilon L(J_\epsilon u_\epsilon)J_\epsilon - J_\eta L(J_\eta u_\eta)J_\eta$ can be decomposed into a sum of terms which can all be estimated in terms of $J_\epsilon - J_\eta$. Now $J_\epsilon - J_\eta$ as *an operator from H^r to L^2 has a norm which is small when both ϵ and η are* small. We therefore have an inequality of the form

$$\frac{d}{dt}|u_\epsilon - u_\eta|^2 \le C(t)|u_\epsilon - u_\eta|^2 + D(t)$$

where $D(t)$ is uniformly small on $[0, T]$. The convergence follows.

4. Regularity of the solution. The solution thus obtained is *a priori* only bounded with values in H^s. To prove its continuity, one can derive a Gronwall-type inequality for $(d/dt)|J_\epsilon pu(t)|_s^2$, from which, by *time-reversal*, an estimate for the absolute value of this derivative follows. Thus we have the continuity of the H^s norm. Since weak continuity is clear, the desired continuity follows from a well-known theorem in functional analysis.

Remark: Symmetric systems enjoy an improved energy estimate: Let us for definiteness consider the first-order symmetric hyperbolic system

$$Lu := Qu_t + Au = f,$$

where $A = \sum_j A^j \partial_j u$. Assume, for simplicity, that the coefficients are independent of t. Let $u = v e^{\lambda t}$, where λ is a positive constant. One then has

$$e^{2\lambda t}[Lv + 2\lambda Qv] = f.$$

Therefore, multiplying by v and integrating over both space and time, we find that if λ is large enough, we have

$$(2.18) \qquad \|v\|_2 \le \frac{C}{\lambda}\|f\|_2.$$

Note that we are now using space-time norms.

Examples.

(a) SINGLE SECOND-ORDER EQUATION. Consider

$$u_{tt} + 2a^i u_{it} - a^{ij}u_{ij} + f(x,t,u,Du) = 0.$$

Assume that the a^{ij} form a symmetric and uniformly positive-definite matrix. Consider now the system

$$\begin{aligned}
\partial_t u &= u_t; \\
\partial_t u_t &= -2a^k \partial_k \partial_k u_t + a^{ij}\partial_j u_i + f(x,t,u,Du); \\
a^{ij}\partial_t u_i &= a^{ij}\partial_i u_t.
\end{aligned}$$

This is now a symmetric system for $v = (u, u_t, u_1, \ldots, u_n)$. Note that only those solutions of this system for which u_i is initially the gradient of u correspond to solutions of the original equation.

It can be shown that any strictly hyperbolic equation is in fact symmetrizable.

(b) MAXWELL'S EQUATIONS. Consider Maxwell's vacuum equations:

$$(2.19) \qquad E_t = \text{curl } B; \quad B_t = -\text{curl } E;$$
$$(2.20) \qquad \text{div } E = 0; \quad \text{div } B = 0.$$

The last two equations are in fact satisfied for all t if they are for $t = 0$. We therefore focus on the other six equations. (A similar situation occurs with the constraint equations in General Relativity, see Ch. 6). If we consider (E, B) as an unknown with six components, the system becomes symmetric-hyperbolic.

There is actually another, more geometric way of recovering the energy estimate from the canonical energy-momentum tensor τ_{ab} defined in the example of the nexst paragraph.

Leray systems.

We consider here a class of systems which are natural extensions of higher-order hyperbolic equations. They do not require a preliminary reduction to first-order form. Consider a system

$$A(x, u, \partial)u = B(x, u).$$

Consider for simplicity the case of a linear system.

We assume that $u = (u_\sigma)$ is a vector, and that $B = (b_\tau)$, $Au = (a_{\tau\sigma}u_\sigma)$, where summation over σ is understood. We assume that we have assigned an index $s(\sigma)$ to each unknown, and an index $t(\tau)$ to each equation, in such a way that

1. The $a_{\tau\tau}$ are strictly hyperbolic of order $m(\tau) = s(\tau) - t(\tau) + 1$.

2. For $\sigma \neq \tau$, $a_{\tau\sigma}$ has order at most $s(\sigma) - t(\tau)$, and is zero if $s(\sigma) \leq t(\tau)$.

3. The forward light cones of the operators $a_{\tau\tau}$ have an intersection with non-empty interior.

4. All indices are greater than or equal to 1.

The latter condition ensures that the $a_{\tau\tau}$ satisfy a "global" hyperbolicity condition; a discussion of this point is found in Leray (1951).

It is then possible to find a solution with $u_\sigma \in H^{s(\sigma)}$ if $b_\tau \in H^{t(\tau)}$. The idea of proof is as follows: The equation can be written

$$a_{\tau\tau}u_\tau = b_\tau - \sum_{\sigma \neq \tau} a_{\tau\sigma}u_\sigma,$$

and therefore we need to consider

$$(u_\sigma) \mapsto (u_\sigma - a_{\tau\tau}^{-1}[\sum_{\sigma \neq \tau} a_{\tau\sigma}u_\sigma]).$$

This operator, due to our condition on the indices, maps $\prod_\sigma H^{s(\sigma)}$ to $\prod_\tau H^{t(\tau)}$. But it turns out that $a_{\tau\tau}$ satisfies an estimate similar to (2.18), which, if λ is very large, also makes this operator close to the identity in the space L^2 with a suitable exponential weight.

EXAMPLE: Leray systems have been prominently used in fluid mechanics, magnetohydrodynamics, and relativity. Here is an application to the Einstein-Maxwell equations. The notation is that of Chapter 6.

Consider Einstein's equations with a charged dust energy-momentum tensor:

$$G_{ab} = \chi(\rho u_a u_b + \tau_{ab}),$$

where

$$\tau_{ab} = \frac{1}{4} g_{ab} F^{cd} F_{cd} - F_a{}^c F_{cb}$$

is the canonical tensor associated with the electromagnetic tensor F_{ab}. The latter obeys Maxwell's equations:

$$\nabla_a F^{ab} = \mu u^b; \quad \nabla_a F^{*ab} = 0,$$

where μ represents the charge density of the fluid. The latter half of Maxwell's equations imply the existence, locally, of a potential A_a such that $F_{ab} = \partial_a A_b - \partial_b A_a$. We take the Lorentz gauge: $\nabla_a A^a = 0$. The continuity equation and the equation for the conservation of charge, which follow from the field equations, read:

$$\nabla_a(\rho u^a) = 0; \quad \nabla_a(\mu u^a) = 0.$$

Finally, the action of the electric field on matter gives:

$$u^a \nabla_a u_b = (\mu/\rho) F_{ba} u^a.$$

The unknowns are: The ten components g_{ab} of the metric, the potential A_a, the vector u^a, and the quantities $k = \mu/\rho$ and ρ. We assume that we are working in harmonic coordinates (see Ch. 6), and that therefore $G_{ab} = G_{ab}^{(h)}$.

The difficulty here is that we are coupling second-order equations with first-order equations, so that there seems to be a problem as to what the principal part is. Indeed, it would be unwise to assume that the structure of the equations is contained entirely in the second-order terms, as we would for the pure Einstein equations. Furthermore, it seems that the directions orthogonal to u^a appear in several equations, making the system apparently "non-strictly hyperbolic" if first-order terms were taken into account.

All these difficulties are circumvented by considering the following system, where one has applied $u^c \nabla_c$ to Einstein's equations and to the conservation of charge:

$$
\begin{aligned}
u^c \nabla_c G_{ab} &= \chi(-u_a u_b \rho \nabla_c u^c + \rho u^c \nabla_c(u_a u_b) + u^c \nabla_c \tau_{ab}); \\
u^c \nabla_c(\Box A)_a &= k(-u_a \rho \nabla_c u^c + \rho u^c \nabla_c u_a); \\
u^c \nabla_c u_b &= k F_{cb} u^c; \\
u^c \nabla_c k &= 0; \\
\nabla_c(\rho u^c) &= 0.
\end{aligned}
$$

Numbering these equations from 1 to 5, we obtain a Leray system using the assigment:

$$s(g_{ab}) = s(A_a) = s(\rho) = 3; \quad s(u^a) = s(k) = 2;$$
$$t(1) = t(2) = 1; \quad t(3) = t(4) = 2; \quad t(5) = 3.$$

Hyperbolicity requires the vector u^a to be timelike for the metric g_{ab}, as one would expect on physical grounds. Note that the principal part of the third equation has order $2 - 2 + 1 = 1$ with respect to u^a; the terms in $u^a \nabla_a$ are therefore in this framework on a par with the third order terms in the first equation (since $3 - 1 + 1 = 3$): both are terms of order $s(\sigma) - t(\tau) + 1$ for the associated values of σ and τ.

2.3 SEMI-GROUP TECHNIQUES

We prove an abstract theorem which enables one to construct the "exponential of an unbounded operator." In other words, although the Taylor series for the solution of $u_t = Bu$ is divergent for unbounded B, it can be "summed" indirectly in certain cases. Semi-group techniques apply to hyperbolic and non-hyperbolic evolution equations, and give very easily the existence of weak solutions in linear and nonlinear problems. Its main strength is the fact it provides a rather transparent proof of the continuous dependence of the solution on initial data; in this sense, it is more than an abstract reformulation of the energy method. A few results on nonlinear scattering, which are conveniently introduced in this set-up, are included.

The solution operator for the Cauchy problem for, say, a hyperbolic equation is, we have seen, a nonlinear bounded operator in Sobolev spaces. It has no compactness properties: indeed, it is *invertible*, if the equation is time-reversible. For this reason, flows defined by hyperbolic equations are not expected, in the absence of dissipative or damping terms, to behave like finite-dimensional dynamical systems. There are however general and nontrivial properties of flows in Hilbert (or Banach) spaces which follow from simple continuity requirements; without any compactness assumptions. Such techniques are part of semi-group theory. Semi-groups of operators with smoothing properties are usually generated by parabolic equations, and will therefore not be studied here. These smoothing properties should not be confused with space-time smoothing theorems such as Th. 1.19, which apply to equations which are well-posed forward and backward in time.

The Hille-Yosida theorem.

A one-parameter family $\{S(t)\}_{t \geq 0}$ is said to be a *continuous contraction semigroup* on a Banach space X if

1. $S(0) = I$, and $S(t + s) = S(t)S(s)$ for $t, s \geq 0$.

2. For all $t > 0$, $|S(t)| \leq 1$.

3. For all $x \in X$, $\lim_{t \downarrow 0} S(t)x = x$.

We define an unbounded operator B associated to this semi-group by saying that $x \in D(B)$ (the domain of B) if

$$Bx := \lim_{t \downarrow 0} \frac{S(t)x - x}{t}$$

exists.

Remarkably enough, we have

THEOREM 2.4 *The operator B is closed and has dense domain.*

B is called the infinitesimal generator of the semi-group. One can show that if two semi-groups have the same generator, they must coincide.

The generator is related to the semi-group by the following formula for the *resolvent* $(\lambda I - B)^{-1}$:

$$R(\lambda, B)x := (\lambda I - B)^{-1} = \int_0^\infty e^{-\lambda t} S(t)x \, dt$$

valid for all $x \in X$ and all $\lambda > 0$. Thus, the entire positive axis lies in the resolvent set of B, and the resolvent $R(\lambda, B)x$ has norm $1/\lambda$ at most.

The Hille-Yosida theorem gives a converse of this:

THEOREM 2.5 *An unbounded operator B on a Banach space X generates a continuous contraction semi-group if and only if it is closed and densely defined, and satisfies $\|(\lambda I - B)^{-1}\| \leq 1/\lambda$.*

Proof of Sufficiency: Let $R(\lambda) = (\lambda I - B)^{-1}$. We will obtain the desired semi-group as a uniform limit of (semi-)groups generated by suitable bounded operators. More precisely, we show that

$$S(t)x := \lim_{\lambda \to \infty} S_\lambda(t)x,$$

where $S_\lambda(t) := \exp(tB_\lambda)$ and

$$B_\lambda := \lambda^2 R(\lambda) - \lambda I.$$

To motivate this, it may be useful to note that this result is immediate if B is a constant multiple of the identity. B_λ is called the Yosida regularization of B.

Step 1. Since $\lambda R(\lambda) B = \lambda R(\lambda)(B - \lambda + \lambda) = B_\lambda$, the assupmtions of the theorem show that for any $x \in D(B)$,

$$|B_\lambda x| \leq |Bx|.$$

Step 2. Since $\lambda R(\lambda) - I = R(\lambda)(\lambda - (\lambda - B)) = R(\lambda)B$, we see that for any $x \in D(B)$,

$$|(\lambda R(\lambda) - I)x| \leq \frac{|Bx|}{\lambda} \to 0.$$

Since $\|\lambda R(\lambda) - I\| \leq 1$ and $D(B)$ is dense, it follows that

$$\lim_{\lambda \to \infty} \lambda R(\lambda)x = x.$$

Therefore, if $x \in D(B)$,

$$B_\lambda x = \lambda R(\lambda) Bx \to Bx.$$

Step 3. We now consider the approximating semi-groups $S_\lambda(t) = e^{tB_\lambda t}$. We have:

$$S_\lambda(t) = e^{-\lambda t} \sum_0^\infty \frac{(\lambda^2 t)^n}{n!} R(\lambda)^n,$$

so that $|S_\lambda(t)| \leq e^{-\lambda t} \sum_0^\infty \frac{(\lambda t)^n}{n!} = 1$. Now since B and $R(\lambda)$ commute with $R(\mu)$ and B_μ for all μ,

$$
\begin{aligned}
S_\lambda(t) - S_\mu(t) &= \int_0^t \frac{d}{d\tau}[S_\lambda(\tau)S_\mu(t-\tau)]\,d\tau \\
&= \int_0^t S_\lambda(\tau)S_\mu(t-\tau)[B_\lambda - B_\mu]\,d\tau.
\end{aligned}
$$

It follows that

$$|S_\lambda x - S_\mu x|(t) \leq t|(B_\lambda - B_\mu)x|$$

for all $x \in X$.

Step 4. If $x \in D(B)$, we therefore see that $\{S_\mu(t)x\}$ satisfies the Cauchy criterion on bounded time intervals as $\mu \to \infty$. We therefore obtain a mapping $S(t)$ defined for $x \in D(B)$ by $\lim_{\mu \to \infty} S_\mu(t)x = S(t)x$. By density, it extends to a contraction in X. These operators also clearly form a continuous semi-group.

Step 5. We must finally identify the generator of S with B. Let C be this generator. It coincides with B on $D(B)$, as follows from the relation established in point 3. But since B is closed and densely defined, $D(C) = D(B)$ and the theorem is proved.

Two typical examples of operators satisfying these conditions are skew (a) self-adjoint operators, and (2) maximal monotone (or maximal accretive) operators.

Application.

Consider the abstract evolution equation

$$du/dt = Bu + f(u),$$

where B generates a contraction semi-group on X. We seek solutions with $u(0) = u_0$. If there is a solution, it must solve the analogue of formula of variation of parameters:

$$u(t) = S(t)u_0 + \int_0^t S(t - \tau)f(u(\tau))\,d\tau.$$

Any solution of this equation which is continuous in time, with values in X, is called a *mild solution* of the original problem.

The following basic results essentially follow by applying the Banach fixed point theorem to this equation.

THEOREM 2.6 *If f is uniformly Lipschitz continuous on X, there is a unique mild solution for any u_0 in X. The mapping from u_0 to u is Lipschitz continuous from X to $C([0,T]; X)$, for any T.*

THEOREM 2.7 *If, further, f is of class C^1 from X to X, and $u_0 \in D(B)$, then the mild solution is a classical (viz., differentiable) solution.*

For more results of this type, see the Notes.

Nonlinear scattering.

Nonlinear scattering is one of the procedures whereby one can establish detailed asymptotic properties of the solution of a nonlinear wave equation. It is convenient to introduce in an abstract set-up.

Let us consider a linear abstract evolution equation

$$(2.21) \qquad\qquad u_t = iH_0 u,$$

where H_0, is a self-adjoint operator, to fix ideas. Clearly, $u(t) = \exp(itH_0)u(0)$, where the exponential is the semi-group given by the Hille-Yosida (or even Stone's) theorem. Quantum scattering deals with the following question. Consider a perturbed operator $H = H_0 + V$, and the associated semi-group. One would like to know whether the semigroups generated by H and H_0 are asymptotically similar for large t. More precisely, one would like to define

(a) The *wave operators* $W_\pm$ such that

$$W_\pm u(0) = \lim_{t \to \pm\infty} e^{-itH_0} e^{itH} u(0).$$

(b) A *scattering operator* $S = W_+ W_-^{-1}$.

Intuitively speaking, the solution $u(t)$ behaves as if it were a solution of the unperturbed equation as $t \to \infty$, with however a different initial condition, namely $W_+ u(0)$. The wave operators therefore describe the influence of the perturbation for large positive or negative times. On the other hand, a solution which is close to $\exp(itH_0)u_-$ for $t \ll 0$ will be close, for $t \gg 0$, not to $\exp(itH_0)u_-$, but to $\exp(itH_0)u_+$, where $u_+ = Su_-$. If this program works, the solution of the perturbed equation is described by the asymptotics of the unperturbed system, which is presumably better understood than the perturbed system.

Nonlinear scattering attempts to accomplish the same for nonlinear perturbations.

If the solutions of the unperturbed equations satisfy a decay estimate as $t \to \infty$, the existence of scattering operators is incompatible with the existence of periodic solutions, or traveling waves. If however such solutions are ruled out for data which are small in some topology, it is reasonable to expect that $W_\pm$ and S might be defined near zero ("small energy scattering.")

Nonlinear scattering has been studied for nonlinear perturbations of the wave and Klein-Gordon equation, as well as the nonlinear Schrödinger equation, mostly in three or more space dimensions. Two difficulties in establishing nonlinear scattering may be summarized as follows: if wave or scattering operators do exist, we know that the solution of the nonlinear problem behaves like the solution of a linear problem *in some topology;* if however this fails, we have no general model of what happens: it is expected that a "bound state" is somewhat created, but no precise representation for such a state exists apart from the case of solitons, and the case of linear equations, where bound states are defined in terms of solutions of the form $\exp(i\omega t)\psi(x)$.

Let us outline the first few steps for the construction of wave operators for the nonlinear Klein-Gordon equation $u_{tt} - \Delta u + u + f(u) = 0$ in three space dimensions. A more thorough discussion is found in Strauss (1989).

Let $U_0(t)$ be the solution operator for the Klein-Gordon equation, acting on pairs $v = (u, u_t) \in H^1 \times L^2(\mathbf{R}^3)$. Let us write $F(u) = (0 \quad -\lambda u^3)^T$. We then have for any T,

$$v(t)U_0(t - T)u(T) + \int_T^t U_0(t - s)F(v)(s)\,ds.$$

If wave operators do exist, we can let $t \to \pm\infty$, and find,

$$v_\pm(t) = U_0(t)W_\pm v(0) = v(t) - \int_{\pm\infty}^t U_0(t - s)F(v)(s)\,ds.$$

It therefore suffices to show that these integrals are well-defined. This formula also provides an expression for $u_+ - u_-$, which is small for small data. An application of the contraction mapping theorem enables one to conclude that S exists locally.

Kenig, Ponce and Vega have recently shown that nonlinear scattering also applies to some generalized KdV equations. This surprising fact does not contradict the existence of solitary waves: in the function spaces they use, there are no solitary waves of arbitrarily small norm, and therefore "small-energy" scattering makes sense.

2.4 GLOBAL EXISTENCE FOR SMALL DATA

We now turn to more specialized results. We prove here the global existence of smooth solutions with small data in three space dimensions, for a class of perturbations of the wave equation. This class is characterized by the behavior of the quadratic part of the nonlinearity, which must satisfy the "null condition" introduced below. The result is also valid in higher dimensions, but we focus on the more natural three-dimensional situation. We use the conformal compactification of Minkowski space to prove the result; this method has the advantage of making the null condition appear very naturally. It also has been successful in nonlinear scattering problems. We also indicate other ways of recovering these results; the null condition also comes up in the study of the blow-up time in Ch. 3.

Basic issues.

We are interested in the Cauchy problem for

$$\Box u = f(u, Du, D^2u) := f_1(u, Du)D^2u + f_2(u, Du) + c(u, Du),$$

with data on $t = 0$, where c is cubic ($c = O((|u| + |Du|)^3)$), and f_1 and f_2 are first- and second-degree polynomials in Du respectively. All nonlinearities are assumed to be smooth. We assume here that the number of space dimensions is odd, and is at least three.

We seek solutions defined and regular for all values of x and t, corresponding to small data in an appropriate topology.

The above hypotheses are not sufficient to ensure global existence in three space dimensions. In order to deal with that case, we introduce the *null condition*, which states that for any null (co-)vector ξ_a (i.e., whenever $\eta^{ab}\xi_a\xi_b = 0$), one has, for all u,

$$f_1(u, \xi)\xi \otimes \xi = f_2(u, \xi) = 0.$$

In other words, the quadratic part has the form

$$k^{abc}\partial_a u \partial_{bc} u + l^{ab}\partial_a u \partial_b u,$$

with

$$k^{abc}\xi_a\xi_b\xi_c = l^{ab}\xi_a\xi_b = 0$$

whenever ξ is null. It is not excluded that k^{abc} and l^{ab} depend on u as well. We assume f is independent of x and t to simplify matters.

One can describe more explicitly quadratic terms which satisfy the null condition. In fact, l^{ab} must be proportional to η^{ab} as is easily checked. As for k^{abc}, any such object with three indices can be written as the sum of a completely symmetric k_1^{abc}, and a sum of terms which are each antisymmetric with respect to some pair of indices; this follows from the fact that any permutation is a product of transpositions. For antisymmetric terms, the null condition is obvious. For completely symmetric terms, one can prove by induction on n (exercise) that one must have

$$k^{abc} = m^a \eta^{bc} + m^b \eta^{ca} + m^c \eta^{ab}$$

for some m^a.

It is this form of the null condition which will be useful to us.

Compactification of Minkowski space.

To prove global existence under the null condition, we will map conformally Minkowski space into a bounded region of a Lorentzian manifold, in which there is a time function such that $T < \pi$ includes the image of all of Minkowski space. We will work in n space dimensions, but will eventually restrict ourselves to $n = 3$. If fact, it would be sufficient to work with $n \geq 3$, odd. A straightforward, but quite technical calculation will show that an equation of the form

$$g^{ab} \partial_{ab} u = f[u]$$

where g_{ab} is the Minkowski metric, will be transformed into

$$\tilde{g}^{ab} \tilde{u}_{ab} = \Omega^{-(n+3)/2} f[\Omega^{(n-1)/2} \tilde{u}]$$

where $\tilde{u} = \Omega^{-(n-1)/2} u$. If therefore there is no negative power of Ω on the right-hand side, it suffices to solve the Cauchy problem upto time T to conclude—this will be possible under a smallness assumption on the data.

For $n = 3$, this procedure works if $f = u^3$.

If f involves derivatives, this requirement will lead naturally to the null condition.

Another advantage of the conformal method is that it gives a detailed description of the solution near infinity—this is valuable even for the linear wave equation.

Finally, we are led by this method to continue the solution beyond null and future infinity.

We now turn to the details of the argument.

We consider Minkowski space $\mathbf{M}^{n+1}$ with metric

$$g = \eta_{ab} \, dx^a \, dx^b = -dt^2 + dr^2 + r^2 \, dS_{n-1}^2,$$

where dS_{n-1}^2 stands for the usual metric on the $(n-1)$-sphere.

We also define the *Einstein Static Universe* (ESU) to be $\mathbf{R} \times S^n$, endowed with the metric

$$\tilde{g} = -dT^2 + dR^2 + \sin^2 R(dS_{n-1})^2.$$

We are therefore parametrizing S^n by R and points of the sphere S^{n-1} (which can be thought of as the equator of S^n). If a point on the n-sphere is viewed as a unit vector in $\mathbf{R}^{n+1}$, the variable R represents the angle of this vector with the last coordinate axis. The metric g is simply the product metric on the ESU with the obvious signature. In particular, it is a smooth metric.

In order to study the smoothness properties of functions on the ESU, we introduce explicitly an atlas on S^n obtained as follows:

$$X^i \mapsto \left(\frac{X^i}{1 + A^2/4}, -a\right)$$

and

$$\bar{X}^i \mapsto \left(\frac{\bar{X}^i}{1 + \bar{A}^2/4}, -\bar{a}\right),$$

where $i = 1, \ldots, n$, $A = \sum_i (X^i)^2$, $\bar{A} = \sum_i (\bar{X}^i)^2$, and

$$(2.22) \qquad a = \frac{1 - A^2}{1 + A^2}, \quad \bar{a} = \frac{1 - \bar{A}^2}{1 + \bar{A}^2},$$

map all of $\mathbf{R}^n$ to S^n minus the North and South pole respectively. $X^i/2$ and $\bar{X}^i/2$ correspond to the usual stereographic projections. They form an atlas of the n-sphere if they are related via

$$(2.23) \qquad A\bar{A} = 4, \quad X^i/A = \bar{X}^i/\bar{A}.$$

One then has $\bar{a} = -a$. The metric on the sphere becomes that of a standard space-form:

$$|dX|^2/(1 + A^2/4).$$

It is crucial for the following to remember that a function is smooth on the ESU if and only if it has smooth expressions in both coordinate systems of the atlas.

We now introduce a mapping of Minkowski space into the ESU by the formulae

$$(2.24) \qquad T = \arctan(t + r) + \arctan(t - r);$$
$$(2.25) \qquad R = \arctan(t + r) - \arctan(t - r).$$

By inspection, this maps all of Minkowski space into the region $R \geq 0$, $R - \pi < T < \pi - R$. In particular, the new "time" is now bounded; if the equation is transformed into a "smooth" one on the ESU, all we will need is a *local* existence theorem to obtain the desired *global* result. This mapping is checked to be conformal: if we identify $\mathbf{M}^{n+1}$ with its image under the mapping, the Minkowski metric g satisfies $g_{ab} = \Omega^2 \tilde{g}_{ab}$, where $\Omega = \cos T + \cos R$.

In terms of our atlas, this mapping amounts to prescribing $\bar{A}$ and T as functions of r and t. Indeed, a little elementary geometry shows that $\bar{A} = 2\tan(R/2)$, and $a = -\cos R$.

Transformation formulae.

We now show how the given equation transforms under the above mapping.

The calculations involved are essentially straightforward, but quite lengthy. For that reason, before stating the results, we give a few identities which may help the reader in checking these formulae:

$$\bar{A} = 2\tan(R/2) = (\Omega - \sqrt{\Omega^2 - 4r^2})/(2r);$$

$$rt = \Omega^{-2}\sin T \sin R;$$

$$1 + \bar{a} = 1 + \cos R = \frac{2}{1 + \bar{A}^2/4} = 1 - a;$$

$$(1 + \bar{a})\bar{X}^a = \frac{x^a}{r}\frac{2\bar{A}}{1 + \bar{A}^2/4}$$

$$= \frac{x^a}{r}\frac{2A}{1 + A^2/4} = (1 + a)X^a;$$

$$(1 + \bar{A}^2/4)^{-1}\frac{\bar{A}}{r} = \frac{\sin R}{r} = \cos T + \cos R.$$

Note also that

$$\partial T/\partial r = \partial R/\partial t = -rt\Omega^2 = -\sin T \sin R;$$

$$\partial T/\partial t = \partial R/\partial r = \Omega \cos T + \sin^2 T = 1 + \cos R \cos T;$$

$$\partial \Omega/\partial t = -\Omega \sin T \cos R;$$

$$\partial \Omega/\partial r = -\Omega \cos T \sin R = -\frac{\bar{A}}{2}(1 + \bar{a})\Omega \cos T.$$

The variables x^i, X^i and $\bar{X}^i$ are from now on related by (2.23)–(2.25).

Step 1. First of all, one has

$$\partial X^a/\partial x^b = \Omega I_b^a + K^a \Lambda_b,$$

and

$$M_b^a := \partial \bar{X}^a/\partial x^b = \Omega \bar{I}_b^a + \bar{K}^a \Lambda_b,$$

where

$$I_0^0 = \cos T; \quad I_i^0 = I_0^i = 0; \qquad I_j^i = (1 + A^2/4)\delta_j^i - \tfrac{1}{2}X^i X_j;$$
$$\bar{I}_0^0 = \cos T; \quad \bar{I}_i^0 = \bar{I}_0^i = 0; \qquad \bar{I}_j^i = (1 + \bar{A}^2/4)\delta_j^i - \tfrac{1}{2}\bar{X}^i \bar{X}_j;$$
$$K^0 = \sin T; \quad K^i = X^i;$$
$$\bar{K} = \sin T; \quad \bar{K}^i = -\bar{X}^i;$$
$$\Lambda_0 = \sin T; \quad \Lambda_i = -\tfrac{1}{2}(1 + a)X^i \ = -\tfrac{1}{2}(1 + \bar{a})\bar{X}^i.$$

Note that we did not give a name to $\partial X^a / \partial x^b$ because we will only write out the calculations in the $\bar{X}$ variables, the result in the other variables being very similar.

This gives the transformation of tensors of all types, in both systems of coordinates.

Step 2. Let us recall that $\tilde{g}$ is the usual metric on the ESU, and that g is the Minkowski metric written in the $\bar{X}$ coordinates. Let $\tilde{u} = u\Omega^{-(n-1)/2}$. We seek the equation satisfied by $\tilde{u}$ on the ESU.

We denote by $\tilde{u}_{,i}, \tilde{u}_{,ij}, \ldots$ the *covariant* derivatives of $\tilde{u}$ with respect to $\tilde{g}$, and by $u_{;i}, u_{;ij}, \ldots$ the *covariant* derivatives of $\tilde{u}$ with respect to g. Ordinary derivatives with respect to x^i or $\bar{X}^i$ are both denoted by ∂_i.

We then have

$$\partial_a u = \Omega^{(n-1)/2}[M_a^b \tilde{u}_{,b} + \tfrac{1}{2}(n-1)Y_a],$$

where $Y_a = \Omega_a / \Omega$ are given by:

$$\begin{aligned}
Y_0 &= -\bar{a}\sin T = a\sin T, \\
Y_i &= -\tfrac{1}{2}\cos T(1 + \bar{a})\bar{X}^a = -\tfrac{1}{2}\cos T(1 + a)X^a.
\end{aligned}$$

As for the second derivatives,

$$\partial_{cd} u = \Omega^{(n-1)/2}\left\{ M_c^a M_d^b \tilde{u}_{,ab} + [\tfrac{1}{2}(n+1)(Y_a \tilde{u}_b + Y_b \tilde{u}_a) - \eta_{cd}\Omega\Omega_a\Omega^a] \right.$$
$$\left. + \tfrac{1}{2}(n-1)[\tfrac{1}{2}(n+1)Y_c Y_d - \eta_{cd}\Omega_a\Omega^a + \Omega^{-1}M_c^a M_d^b \Omega_{ab}]\tilde{u} \right\}.$$

Step 3. This leads us to the computation of the second derivatives of Ω:

$$M_c^a M_d^b \Omega_{ab} = \partial_{cd}\Omega - 2\Omega Y_c Y_d + \Omega^2(\cos T - \cos R)\eta_{cd} = \Omega Z_{cd},$$

where $Z_{cd} = a\Omega\eta_{cd} - V_c V_d$, with

$$\begin{aligned}
V_0 &= -(1 - a\cos T) = -(1 + \bar{a}\cos T), \\
V_i &= \tfrac{1}{2}\sin T(1 + \bar{a})\bar{X}^a = \tfrac{1}{2}\sin T(1 + a)X^a.
\end{aligned}$$

End of proof.

From exercise 14, we know that

$$\tilde{g}^{ab}\tilde{u}_{,ab} - \frac{n-1}{4n}\tilde{R}\tilde{u} = \Omega^{-(n+3)/2}g^{ab}u_{ab}.$$

Note that $\tilde{R} = n(n-1)$. Since $g^{ab}u_{ab} = f(u, Du, D^2u)$, we must substitute u on terms of $\tilde{u}$ in this expression. *If the resulting right-hand side extends smoothly to the ESU,* it will suffice to use a local existence theorem to solve for $\tilde{u}$ upto time $T = \pi$. This will automatically produce a global solution to the original problem. In fact, since $u = \Omega^{(n-1)/2}\tilde{u}$, we obtain a decay estimate for the desired solution, which compares very favorably to those of Ch. 1.

We therefore proceed to discuss the smoothness of the right-hand side of the equation for $\tilde{u}$. We achieve this by computing it in both coordinate systems in our atlas. We will write out the calculation in the $\bar{X}$ system, and leave the other, entirely analogous, calculation to the reader.

Direct calculation reveals that

$$\partial u/\partial x^a = M_a^c\tilde{u}_{,c} = \Omega^{(n-1)/2}\zeta_a,$$

where

$$\begin{aligned}
\zeta_a &= M_a^c\tilde{u}_{,c} + \frac{1}{2}(n-1)\tilde{u}Y_a. \\
&= (K^q\tilde{u}_{,q})\Lambda_a + \tilde{u}Y_a + \Omega I_a^p\tilde{u}_p.
\end{aligned}$$

Similarly,

$$\partial^2 u/\partial x^a\partial x^b = M_a^c M_b^d\tilde{u}_{,cd} = \Omega^{(n-1)/2}\xi_{ab},$$

where

$$\begin{aligned}
\xi_{,ab} &= M_a^c M_b^d\partial^2\tilde{u}/\partial\bar{X}^c\partial\bar{X}^d + \frac{1}{2}(n+1)(Y_a M_b^d\tilde{u}_{,d} + Y_b M_a^c\tilde{u}_{,c}) - \eta_{ab}\Omega\Omega^c\tilde{u}_{,c} \\
&\quad + \frac{1}{2}(n-1)[\frac{1}{2}(n+1)Y_aY_b - \eta_{ab}\Omega^c\Omega_c + Z_{ab}]\tilde{u} \\
&= (K^p K^q\tilde{u}_{,pq})\Lambda_a\Lambda_b + 2(Y_a\Lambda_b + Y_b\Lambda_a) + \tilde{u}(2Y_aY_b - V_aV_b) + \Omega\tilde{\xi}_{ab},
\end{aligned}$$

with

$$\begin{aligned}
\tilde{\xi}_{ab} &= \Omega I_a^p I_b^q\tilde{u}_{,pq} + (I_a^p\Lambda_a + I_b^p\Lambda_a)K^m u_{,pm} \\
&\quad + 2(I_a^pY_a + I_b^pY_a)\tilde{u}_{,p} - \eta_{ab}\Omega^p\tilde{u}_{,p} - \eta_{ab}\tilde{u}\cos T.
\end{aligned}$$

Now, if our nonlinearity is cubic, the expression $f(u, Du, D^2u)$ produces a factor of $[\Omega^{(n-1)/2}]^3$ which cancels the singular term $\Omega^{-(n+3)/2}$ for $n \geq 3$. This is not true any more for general quadratic terms (and indeed, the result is false for them). If however the null condition holds, we find by inspection of the quantities $l^{ab}\zeta_a\zeta_b$ and $k^{abc}\xi_{ab}\zeta_c$ (the two possible quadratic terms), that one additional factor of Ω appears, which again cancels the singularity of the term $\Omega^{-(n+3)/2}$ in three space dimensions. Checking this involves the computation of the dot products of Y, Λ and V with each other, and their antisymmetric products $(\Lambda_a V_b - \Lambda_b V_a \text{ etc} \dots)$.

Thus, the null condition enables us to prove that the equation for $\tilde{u}$ on the ESU involves smooth functions only. If the data are sufficiently small in an appropriate norm, it follows easily that the $\tilde{u}$ will exist at least upto $T = \pi$, which means that the function u exists for all values of t. QED.

Other approaches.

It turns out that quadratic terms satisfying the null condition satisfy slightly better estimates than other quadratic terms. This information can be used to prove global existence. The main point is the following. Consider the quadratic forms $Q(f,g) = \eta^{ab}\partial_a f \partial_b g$ and $Q_{ab}(f,g) = \partial_a f \partial_b g - \partial_b f \partial_a g$. Clearly, $Q(u,u)$ and $Q_{ab}(u,u)$ are typical terms which satisfy the null condition. Now let $M(x,t) = max(|x|,t)$. We use the notation of Ch. 1 for the generators of the Lorentz group. One has

THEOREM 2.8 *If we write $|Mu|$ and $|Du|$ for the sum of all the quantities $|M_{ab}u|$ and $|\partial_a u|$ respectively, we have*

$$|Q_{ab}(f,g)| \leq \frac{C}{M(x,t)}[|Df||Mg| + |Dg||Mf|],$$

and

$$|Q(f,g)| \leq \frac{C}{M(x,t)}[|Df||Mg| + |Dg|(|Mf| + |Sf|)].$$

The proof proceeds by expressing ∂_a in terms of ∂_t, ∂_r and the generators of the Lorentz group (and the scaling operator). This suggests that one may be able to gain one power of decay by using the global Sobolev inequality. This is made possible by the following observation: let, for any vector field,

$$[X,Q](f,g) = XQ(f,g) - Q(Xf,g) - Q(f,Xg)$$

and similarly for Q_{ab}. Then these "commutators" are all expressible in terms of the quadratic forms Q and Q_{ab}.

For the remaining details, we refer to Klainerman (1986).

Remark: There is a version of the null condition which allows for products of second derivatives (see e.g. Klainerman (1986)).

2.5   OTHER ITERATION TECHNIQUES

We give a sample of other useful iterative techniques for semilinear hyperbolic equations. Inverting the linear part produces an integral equation, which can sometimes be iterated in a suitable function space, which incorporates some decay information, suggested by the L^1-L^∞ estimates for the wave and Klein-Gordon equations. It should be noted that such a direct iteration also gives directly an estimate of the domain of dependence in the nonlinear situation.

Nonlinear Klein-Gordon equation.

We consider

$$\Box u + u + f(u) = 0,$$

where f is smooth, and vanishes upto second order (inclusive) near 0.

THEOREM 2.9 *If the data are smooth and small in the norm $[u]$ defined below, the solution is global and decays like $t^{-3/2}$ as $t \to \infty$.*

Remark: One has in fact global existence if $n \geq 3$, and $f = f(x, t, u, Du, D^2u)$ is quadratic in u, Du and D^2u, and the data are small in a suitable norm (Klainerman (1985), Shatah (1985), and Sideris (1989) for the case $n = 3$).

Proof: First rewrite the equation using Duhamel's formula as

$$u(t) = U(t) + \int_0^t S(t-s)(0, f(u(s)))\,ds,$$

where U is the solution of the Klein-Gordon equation with the given data, and $S(t)$ is the solution operator for the Cauchy problem for the Klein-Gordon equation.

One then defines the norm

$$[u](t) := \sup_{[0,t]}\{|u(s)|_2 + |u_t(s)|_1 + (1+s)^{3/2}\|u(s)\|_\infty\}.$$

One easily proves that $|f(u(s))|_1 \leq C(1+s)^{-3}[u](s)^3$ if u is bounded, so that the energy identity for the inhomogeneous equation $\Box u + u = -f$ gives

$$|u(t)|_2 + |u_t(t)|_1 \leq C(1 + [u]^3).$$

On the other hand, $\|f(u(s))\|_{2,1} \leq C(1+s)^{-3/2}[u]^3$, so that using the L^1-L^∞ estimate for the Klein-Gordon equation, we find

$$\|u(t)\|_\infty \leq C(1+t)^{-3/2} + C\int_0^t [u](t)^3(1+t-s)^{-3/2}(1+s)^{-3/2}\,ds.$$

Splitting the integral into two parts, from 0 to $t/2$ and from $t/2$ to t, we find $\|u(t)\|_\infty \leq C(1+t)^{-3/2}$. Therefore, all in all, we have

$$[u](t) \leq C_0 + C[u](t)^3.$$

Here C_0 is small if the data are. It follows that $[u]$ remains bounded if it starts out small initially. Thus we have a bound on the supremum norm of u. We may therefore truncate f and obtain global existence.

Finite propagation of support.

We have seen that if data for the wave or Klein-Gordon equation are supported in some ball of radius R, the solution at time t is supported in a ball of radius $R + |t|$. A similar result holds for nonlinear perturbations of the wave equation.

Since it would be awkward to give a theorem which encompasses all the situations the reader is likely to encounter, let us mention a few simple ideas which enable one to prove such results in many cases.

(1) The easiest proof, which applies to perturbations of the wave equation by lower-order terms, is to exhibit the solution as a uniform limit of approximations with support in the desired double cone. For instance, for $\Box u = f(u)$ with $f(0) = 0$, we defime u_k by a recurrence relation

$$u_{k+1} = R * f(u_k)$$

where R is the Riemann function for the Cauchy problem. This method can be shown to converge if f does not grow too rapidly at infinity (see Jörgens (1963).

(2) A variant of the above consists in defining Banach spaces of functions with the appropriate support properties and showing that the iteration used in the semi-group approach remains in those spaces.

(3) Since linear symmetric hyperbolic systems are easily proved to exhibit finite propagation speed for supports, this line of argument extends to rather general systems.

Note that these methods will only give bounds on the size fo the support: they do not give a complete description of the evolution of the support.

Perturbations of the wave equation.

For perturbations of the wave equation, the situation is more delicate, because the energy estimates give only L^2 estimates of the derivatives of the solution, and not of the solution itself. Let us quote one result in this situation.

THEOREM 2.10 *Consider equation*

$$\Box u = F(u, Du, D_x(Du)).$$

Assume $F(X) = O(|X|^{1+\alpha})$ for small values of its argument X. Then global existence of small solutions (in a suitable norm) holds if

$$\frac{n-1}{2}(1 - \frac{2}{\alpha n})\alpha > 1$$

and $\alpha \geq 1$.

Idea of Proof $(1 \leq \alpha \leq 3)$: The proof consists in recasting the equation as a fixed point problem for a suitable integral operator, which is shown to be a contraction in the norm

$$\sup_{t\geq 0}(1+t)^{(n-1)(1-\frac{2}{\alpha n})/2}[u]_{s_0,\alpha n} + \sup_{t\geq 0}[u(t)]_{s,2} + \sup_{t\geq 0}[Du(t)]_{s+1,2},$$

where $[u]_{m,p}$ stands for the weighted norm involved in the global Sobolev inequality, $s_0 \geq n + 10$, and $s_0 + n + 1 \leq s \leq 2s_0 - 9$. The details are somewhat lengthy.

2.6 FURTHER RESULTS AND PROBLEMS

1. Solve $u_t = v_x$, $v_t = 0$, with initial conditions $u(x,0) = 0$, $v(x,0) = \sqrt{x_+}$. Is the solution in L^2 for $t > 0$? For $t = 0$? Is the system symmetric?

2. Analyze the Cauchy problem for a fully nonlinear equation of the form

$$u_{tt} = f(x,t,D_x u, D_x u_t, D_x^2 u)$$

under appropriate conditions on f.

3. Prove the global existence of smooth solutions for $\Box u + u^5 = 0$ in three space dimensions (Grillakis, Struwe). (There are also recent results for $\Box u + u^3$ in 4 space dimensions).

4. If $\Box u + f(u) = 0$, where (1) f and the data are smooth, rapidly decaying at infinity, and (2) $F(u) \geq -C|u|^2$ and $|f'(u)| \leq C(1 + |u|)^{p-1}$, for some $p \in (1, (n+2)/(n-2))$ (resp. $1 < p < \infty$ if $n \leq 2$), then the solution is (global and) smooth for $n \leq 9$. (Brenner-von Wahl (1981)).

5. Consider the Cauchy problem for $\Box u = f(u)$ on a bounded smooth domain Ω, in n space dimensions, with Dirichlet boundary conditions. We assume $n \leq 3$. The function f is of class C^1 at least.

(a) There is a unique local solution if the data are in $H^2 \cap H_0^1 \times H_0^1$ if $n = 1$. [*Hint:* The nonlinearity is locally Lipschitz from H_0^1 to L^2.] What about $n = 2$ and 3?

(b) If we assume $f(0) = 0$ and $f \in C^2$, with $n \leq 3$, show that no growth condition is required for local existence in $H^2 \cap H_0^1 \times H_0^1$. [*Hint:* The nonlinearity is locally Lipschitz from $H^2 \cap H_0^1$ to H_0^1.]

(c) If $f(u) = -u^3$, show that $\|u_t(t)\|_2^2 + \|u(t)\|_{H_0^1}$ is bounded uniformly in time. Deduce from this a global existence theorem.

(d) If $f(u) = u^3$, Prove that the conclusions of (c) hold if either $\|u(0)\|_{H_0^1}$ is small, or if $E(0) < 0$, where $E(t) := \|u_t(t)\|_2^2 + \|u(t)\|_{H_0^1} - (1/2)\|u(t)\|_4^4$. [*Hint:* Take meas $\Omega = 1$ to fix ideas. Then $G(t) := \|u(t)\|_2^2$ satisfies $G'' \geq G^2 - 4E(0)$.]

If there $\|u(t)\|_q$ remains bounded for some $q > n$, the solution cannot cease to exist.

6. Prove the commutator estimate mentioned in §2.2, using the Gagliardo-Nirenberg inequality. What happens for general real values of s?, for L^p norms? (See Kato and Ponce, or Taylor).

7. Prove global existence for small data for $\Box u = f(u)$ with $f = O(|u|^p)$ near $u = 0$, with $p > 1 + \sqrt{2}$, in three space dimensions. What about $n = 2$? (F. John (1979) *Manus. Math., 28*: 235–268, R. Glassey (1981) *Math. Z., 178*: 233-261.)

8. Discuss the following problems, for an equation of the form

$$u_{xy} = a(x,y)u_x + b(x,y)u_y + c(x,y)u.$$

Find out in particular whether the data determine the solution uniquely under reasonable conditions on the coefficients).

(a) Find u when $u(x,0)$ and $u(0,y)$ are prescribed.

(b) Find u when it has prescribed values on the lines $y = 0$ and $y = x$.

(c) Find u when it is prescribed on two non-characteristic curves intersecting at the origin.

9. Show the existence of a solution to

$$u_t = \varepsilon(\partial_z^{1/2}u)^2 + \varepsilon(1 - \exp(iz + 1 - t))^\beta,$$

with u periodic of period 2π in z, $u(0) = 0$, provided that ε is small enough. The number β is given, in $(0,1)$. The operator $\partial_z^{1/2}$ multiplies the kth Fourier coefficient of u by $\sqrt{|k|}$. The solution is defined on the domain $|\text{Im}\, z| < 1 - t$. (See Caflisch (1990)).

10. Let $\theta(t) = \sum_{j=0}^\infty t^j/(j+1)^2$.

(a) Show that $\theta^2 \ll K\theta$ where $K = 16\pi^2/3$.

(b) If $U(y) \ll b\theta(Ry)$ and $y(t) \ll at\theta(rt)$, with $r \geq aRK$, then

$$U(y(t)) \ll b\theta(rt).$$

(c) Let $dy_i/dt = U_i(y_1,\ldots,y_n)$ for $i = 1,2,\ldots n$. Assume $y_i(0) = 0$ for all i. Assume further that the U_i are analytic and satisfy

$$U_i(y_1,\ldots,y_n) \ll b\prod_{i=1}^n \theta(Ry_i)$$

for all i. Show that if a and r are chosen so that $a \geq bK^{n-1}$ and $r \geq aRK$, then $y_i(t) \ll at\theta(rt)$. (Lax (1953)).

11. (Garabedian) Consider a quasi-linear system of Cauchy-Kowalewska type:

$$(2.26) \qquad u_t = \sum_{j=1}^n A_j(z,u)\frac{\partial u}{\partial z^j} + B(z,u),$$

where $z \in \mathbf{C}^n$. The nonlinearities are analytic. Reduce this system to a symmetric-hyperbolic system for the real and imaginary parts of u. [*Hint:* Let $z^j = x^j + iy^j$, $\partial_{z_j} = (1/2)[\partial_{x^j} - i\partial_{y^j}]$, and $\bar{\partial}_{z_j} = (1/2)[\partial_{x^j} + i\partial_{y^j}]$. Add to (2.26) the quantity

$$\sum_j \bar{a}_j^T \bar{\partial} j u,$$

where ^{T} denotes the transpose. Indeed, this quantity necessarily vanishes for analytic solutions. Separate real and imaginary parts.]

12. Define the *uniformly local* L^p norm by

$$\|u\|_{p,\mathrm{ul}} := \sup_{x \in \mathbf{R}^n} \Big(\int_{|x-y|<1} |u(y)|^p \, dy \Big)^{1/p}.$$

Define similarly for integral s the uniformly local Sobolev space H^s_{ul} by the condition that $|u|_{s,\mathrm{ul}} := \sup_{|\alpha| \le s} \|D^\alpha u\|_{2,\mathrm{ul}} < \infty$. Show that $H^s_{\mathrm{ul}} H^t \subset H^r$ if $r = \min(s, t, s + t - [n/2] - 1)$ is nonnegative.

13. (Kato) Consider the solution u^λ of the equation $u_t + uu_x = 0$ with inital condition

$$u_0^\lambda = (\lambda + x_+^{\alpha+1})\varphi(x),$$

where $|\lambda| \le 1$, φ is smooth, compactly supported, and equals 1 for $|x| \le 2$. Assume that s is an integer greater than 1, and that $s - 3/2 < \alpha < s - 1/2$. Show that for x and λ small, on can write $u^\lambda = \lambda + (x - \lambda t)_+^{\alpha+1} p(t(x - \lambda t)_+^\alpha)$, where p is analytic. Conclude that $|u^\lambda(t) - u_0^\lambda| \ge C|\lambda t|^{\alpha - s + 3/2}$ for small t. What does this mean as to the dependence of solutions on their initial data?

14. If $g_{ab} = \Omega^2 \tilde{g}_{ab}$, show that

$$\tilde{R}^{ab}{}_{cd} = \Omega^2 R^{ab}{}_{cd} + \delta^{[a}{}_{[c}\Omega^{b]}{}_{d]},$$

where

$$\Omega^a{}_b = 4\Omega\Omega_{;be}g^{ae} - 2\Omega_{;c}\Omega_{;d}g^{cd}\delta^a{}_b.$$

Compute $g^{ab}u_{ab} - (n-1)Ru/(4n)$ in terms of $\tilde{u} = u\Omega^{-(n-1)/2}$.

15. Apply the conformal method to the large-time behavior of solutions of the wave equation in three space dimensions.

NOTES

The analytic Cauchy problem was studied by Cauchy and Kowalewska in the late nineteenth century, via the method of majorants, which is still useful today. Two directions spurred further developments: the need to pin down the size of the maximal domain of existence, and the desire to avoid having to guess a suitable majorant. Refinements were given by Goursat, Schauder and Petrowsky. One should note that the analytic characteristic (or Goursat) problem can also be solved by a majorant method. Leray (1957), Wagschal (1979) are good examples of the power of this method. Among further refinements, the use of majorants θ such that $\theta^2 \ll K\theta$ (Gevrey, Lax(1953), Wagschal(1957)) should also be noted. A version can also be found in Hörmander. Leray and Ohya (1968) show how to use the majorant method to obtain local existence in Gevrey classes, with a view towards non-strictly hyperbolic problems. These methods give a procedure to construct counter-examples to the uniqueness in the Cauchy problem. Next, iterative proofs were devised, starting with Ovcjannikov, and others. Nirenberg (1972) recovers these as applications of the Nash-Moser IFT (see Ch. 5). (An improved estimate on the domain of existence was given by Tutschke). This also enables one to relax the assumption of analyticity in time (also considered by Nagumo). This argument was simplified by Walter (1985) and by Caflisch (1990) in the quasi-linear case; the latter suggested that one should follow the path of complex singularities in order to obtain sharp results on the formation of singularities in the real domain. This view further implies that analytic continuation beyond singularities is possible, a conclusion supported by the results of Chapter 3. These theorems were useful in proving existence for very ill-posed systems arising in the applications (see Kano and Nishida (1979) for water waves, Caflisch and Orellana (1988), Sulem, Sulem, Bardos and Frisch (1981) for vortex sheets, Nishida (1978) for the Boltzmann equation in the fluid dynamic limit, to name a few). At about the same time, Fuchsian equations were tackled by analogous methods. An early proof via the majorant method, for the case of one time variable, is due to Rosenbloom. The extension introduces a few technical problems. The paper of Baouendi-Goulaouic gives a rather comprehensive result of this kind. We presented the generalization of this argument in Kichenassamy-Littman, since we will need the full strength of that result in Ch. 3. We refer to these two papers for more references. There is a good theory of (generalized) Fuchsian equations in H^s spaces, see Kichenassamy (1995a) [reference in chapter 3].

A quite detailed treatment of the Cauchy problem, including the construction of distinguished fundamental solutions, a discussion of the theory of symmetric and symmetrizable systems, and Leray(-Volevich) systems can be found in Leray (1951) to which the reader is referred for a discussion of earlier work. Symmetric hyperbolic systems were introduced by Friedrichs, and soon proved to be very commonly encountered in applications. The use of the Gagliardo-Nirenberg inequality was emphasized by Moser. The proof we presented is very close to that of Taylor (1991), which in turn adapts the approach of Klainerman and Majda (see Majda (1984)), where bounds in a strong norm are combined with convergence estimates in a weaker norm. This line of thought is further traceable to Kato and Lax.

The Hille-Yosida theorem is classical. The work of Kato showed how to unify most known existence-uniqueness results by semi-group techniques; further, the correct regularity of the solution is proved in a fairly natural way. The case of variable coefficients in the leading part is the most difficult in this approach. Further applications are detailed in the monographs of Pazy and Haraux. See also Reed (1976), which includes bounds on the domain of dependence. One should also point out that it is not necessary to assume spatial decay of the initial data for the hyperbolic case. This may be seen via a domain of dependence argument, but one can also directly use function spaces adapted to this purpose: the *uniformly local H^s* spaces (see Kato (1975b), Majda (1984) and exercise 12). The commutator estimates also extend to Sobolev spaces based on L^p (Kato and Ponce). Further results on nonlinear scattering, with many references, can be found in Strauss (1989); note that the conformal method can be used in this context as well. The possibility of defining scattering operators for equations with solitons was studied by Kenig-Ponce-Vega, and more recently, for fifth-order equations, by Y. Choi.

There is no nonlinear counterpart to Holmgren's theorem, as was shown by Métivier; this is related failure of uniqueness in the C^∞ (linear) Cauchy problem.

The null condition was introduced by Klainerman, and the proof we presented follows Christodoulou (1986). F. John also encountered the null condition in the course of his work on the blow-up time (see Ch. 3). For a recent variable-coefficient version of the null condition and an application to *local* existence, see Sogge (1993), generalizing results of Klainerman and Machedon.

Proofs via direct iteration are fairly old; we followed the presentation of Strauss (1968, 1989), and Li and Chen (1988); the latter generalize several earlier works, in particular that of Klainerman and Ponce. The "method of normal forms" due to Shatah can sometimes be used to reduce a quadratic nonlinearity to a cubic one by a change of dependent variable, much as in the work Poincaré-Dulac-Siegel on the linearization of ODE's.

REFERENCES

BRENNER, P., AND VON WAHL, W., (1981) Global classical solutions of nonlinear wave equations, *Math. Z., 176*: 87–121.

CAFLISCH, R. E., (1990) A simplified version of the Cauchy-Kowalewski theorem with weak singularities, *Bull. AMS, 23*: 495–500

CAFLISCH, R. E. AND ORELLANA, O. F., (1988) Singularity formation and ill-posedness for vortex sheets, *SIAM J. Math. Anal., 20*: 293–307.

CHRISTODOULOU, D., (1986) Global solutions of nonlinear hyperbolic equations for small data, *Comm. Pure and Appl. Math., 39*: 267–282.

GEORGIEV, V. AND POPIVANOV, P., (1991) Global solutions to the two-dimensional Klein-Gordon equation, *Comm. PDE, 16 (6& 7)*: 941–995.

GRILLAKIS, M. G., (1990) Regularity and asymptotic behavior of the wave equation with a critical nonlinearity, *A.. Math., 132*: 485–505.

JÖRGENS, K., (1961) Das Anfangswertproblem im Grossen für eine Klasse nichtlinearer Wellengleichungen. *Math. Z., 77*: 295–308.

HÖRMANDER, L., (1986) On global existence of solutions of nonlinear hyperbolic equations in $\mathbf{R}^{1+3}$, in *Lecture Notes in Math., 1256.*

JOHN, F., (1976) Delayed singularity formation in a homogeneous isotropic elastic solid, *Comm. Pure Appl. Math., 29*: 649–682.

JOHN, F., (1987) Existence for large times of strict solutions of nonlinear wave equations in three space dimensions for small data, *Comm. Pure Appl. Math., 40*: 79–109.

KANO, T. AND NISHIDA, T., (1979) Sur les ondes de surface de l'eau avec une justification mathématique des équations en eau peu profonde, *J. Math. Kyoto Univ., 19*: 335–370.

KATO, T., (1975a) Quasilinear equations of evolution with applications to partial differential equations, *Lect. Notes in Math., 448*: 25–70.

KATO, T., (1975b) The Cauchy problem for quasilinear symmetric hyperbolic systems, *Arch. Rat. Mech. Anal., 58*: 181–205.

KICHENASSAMY, S. AND LITTMAN, W., (1993a) Blow-up Surfaces for Nonlinear Wave Equations, Part I, *Comm. in P. D. E., 18*: (3&4) 431–452.

KICHENASSAMY, S. AND LITTMAN, W., (1993b) Blow-up Surfaces for Nonlinear Wave Equations, Part II, *Comm. in P. D. E., 18*: (11) 1869–1899.

KLAINERMAN, S., (1985) Global existence of smal amplitude solutions to nonlinear Klein-Gordon equations in four space dimensions, *Comm. Pure and Appl. Math., 38*: 631–641.

KLAINERMAN, S., (1986) The null condition and global existence to nonlinear wave equations, *Lect. in Appl. Math., 23*: 293–326.

LAX, P. D., (1953) Nonlinear hyperbolic equations, *Comm. Pure and Appl. Math., 6*: 231–258.

LERAY, J., (1951-1952) *Hyperbolic Differential Equations,* Princeton, Institute for Advanced Study.

LERAY, J., (1957) Uniformisation de la solution de problème linéaire analytique de Cauchy près de la variété qui porte les données de Cauchy, *Bull. SMF, 85*: 389–429.

LERAY, J., AND OHYA, Y., (1967) Équations et systèmes non linéaires, hyperboliques non stricts, *Math. Ann., 170*: 167–205.

LI, T.-T., AND CHEN, Y.-M., (1988) Initial-value problems for nonlinear wave equations, *Comm. in PDE, 13*: (4) 383–422.

LI, T.-T. AND YU, X., (1991) Life-span of classical solutions to fully nonlinear wave equations, *Comm. PDE, 16 (6& 7)*: 909–940.

LIONS, J.-L., (1969) *Quelques Méthodes de Résolution des Problèmes aux Limites Non Linéaires*, Dunod, Gauthier-Villars, Paris.

MAJDA, A., (1984) *Compressible Fluid Flow and Systems of Conservation Laws,* Applied Math. Sci., *53*, Springer.

NIRENBERG, L., (1972) An abstract form of the nonlinear Cauchy-Kowalewski theorem, *J. Diff. Geo., 6*: 561–576.

NISHIDA, T., (1978) Fluid dynamical limit of the nonlinear Boltzmann equation to the level of the compressible Euler equation, *Comm. Math. Phys., 61*: 119–148.

OVCJANNIKOV, L. V., (1971) A nonlinear Cauchy problem in a scale of Banach spaces, *Sov. Math. Doklady, 12*: 1497–1502.

REED, M., (1976) *Abstract non-linear wave equations,* Lecture Notes in Math., *507*, Springer.

SEGAL I., (1963) The global Cauchy problem for a relativistic scalar field with power interaction, *Bull. SMF, 91*: 129–135.

SEGAL, I., (1968) Dispersion for nonlinear relativistic equations, *Ann. ENS, 1*: 459–497.

SHATAH, J., (1985) Normal forms and quadratic nonlinear Klein-Gordon equations, *Comm. Pure and Appl. Math., 38*: 685–696.

SHATAH, J. AND STRUWE, M., (1993) Regularity results for nonlinear wave equations, *Ann. Math., 138*: 503–518.

SIDERIS, T., (1989) Decay estimates for the three dimensional inhomogeneous Klein-Gordon equation and applications, *Comm. PDE, 14 (10)*: 1421–1455.

SOGGE, C. D., (1993) On local existence for nonlinear wave equations satisfying variable coefficient null conditions, *Comm. in P. D. E., 18 (11)*: 1795–1821.

STRAUSS, W., (1968a) Decay and asymptotics for $\Box u = F(u)$, *J. Funct. Anal., 2*: 409–457.

STRAUSS, W., (1968b) *The energy method in nonlinear differential equations,* Notas de Mat.

STRAUSS, W., (1989) *Nonlinear Wave Equations,* CBMS lecture notes, *73,* Amer. Math. Soc., Providence, RI.

STRUWE, M., (1988) Globally regular solutions to the u^5-Klein Gordon equation, *Ann. Sc. Norm. Sup. Pisa, (ser. 4) 15*: 495–513.

STRUWE, M., (1992) Semilinear wave equations, *Bull. AMS, 26*: 53–85.

SULEM P., SULEM C., BARDOS, C., AND FRISCH, U., (1981) Finite time analyticity for the two and three dimensional Kelvin-Helmholtz instability, *Comm. Math. Phys., 80*: 485–516.

TAYLOR, M., (1991) *Pseudodifferential Operators and Nonlinear P. D. E.,* Progress in Mathematics **100** Birkhäuser, Boston.

WAGSCHAL, C., (1979) Le problème de Goursat non linéaire, *J. de Math. Pures et Appl., 58*: 309–337.

WALTER, W., (1985) An elementary proof of the Cauchy-Kowalewsky theorem, *Amer. Math. Monthly, 92*: 115–125.

Chapter 3

Singularity Formation

The initial-value problem for a nonlinear evolution equation, when well-posed, defines a solution for small times. However, the solutions obtained in this fashion may be inadequate in several respects; first, a solution may develop a singularity in finite time; second, it may contain singularities in its data, of which one would like to follow the location; finally, a solution may develop a singularity in a given function space, but may still remain "smooth" according to a weaker measure of regularity.

Singularities can be studied from four standpoints.

In terms of time evolution, we may start from solutions with smooth data and then try and determine (i) whether breakdown occurs; (ii) where and when it occurs; (iii) the nature of the breakdown (e.g.: does the solution remain bounded while its derivatives become infinite?); (iv) whether the solution can be continued after breakdown as a weak solution.

If on the other hand singularities are present in the data, we may ask for their future evolution.

Finally, we may wish to construct solutions with singularities prescribed at the outset.

The occurrence of singularities implies the breakdown of the estimates which ensure well-posedness. Because of this, blow-up has been compared with ill-posed problems, and indeed, the method of differential inequalities is useful in both contexts.

We saw in Ch. 1 that in linear hyperbolic problems, the spontaneous generation of singularities was usually ascribed to a poor choice of functional setting, because energy estimates generally give a global solution in H^s spaces. On the other hand, the evolution of singularities could be studied (i) by describing the

evolution of the wave-front set of the solution; (ii) by postulating that the singularities are jumps across a hypersurface; (iii) by considering highly oscillatory,
but regular, solutions given by an asymptotic series for large frequencies. These
procedures all generalize to nonlinear problems, under appropriate restrictions,
as described in §3.2 for (i) and (ii), and in Ch. 5 for (iii). In addition to singularities propagated along characteristics of the linear part of the equation, new,
weaker singularities can be created (or annihilated) by "interaction." These
results require a minimum of overall regularity of the solution. Jump discontinuities in top order derivatives are generalized into conormal distributions.

In addition to the above, stronger singularities can appear spontaneously in
nonlinear problems.

Recall that for a first order hyperbolic system, local existence theorems produce a local solution in a Sobolev space, which can be extended as long as it
remains bounded in C^1. This leads to two possible types of breakdown from
smooth data: a solution may remain bounded, while the first derivatives become unbounded, or the pointwise value of the solution grows without bound.
We will say that a *gradient blow-up* occurs in the first case, and that the solution
blows up in the second.[1]

One can still make sense of the equation in the sense of distributions in some
important cases of gradient blow-up; if such an interpretation is possible, we
will say that we are dealing with a *shock wave solution*. A very detailed analysis
of such solutions is available in one space dimension, and is presented in §3.3.
Parallel to the construction of conormal solutions is the solution of the "shock
front problem" where the data have a jump across a prescribed surface, for which
some information in several dimensions is available.

For blow-up, the singularity involves a balance of higher- and lower-order
terms, so that the bicharacteristics of the linear part are not relevant directly.
However, singularities seem to be restricted to characteristic surfaces of an appropriate Fuchsian equation. In fact, there is a general procedure for relating
singular solutions of PDE with polynomial nonlinearities to Fuchsian equations.
They are the analogue of the conormal solutions for this case. Analytic continuation has been successful in providing a continuation of singular solutions.
These results and other information on the mechanism of singularity formation
in perturbations of the wave equation are presented in §3.4.

It seems that strong singularities can be unfolded or uniformized by suitable
changes of variables in the space of dependent and independent variables. This

---

[1]The terminology used in the literature is quite variable. Some authors use exclusively, as
we will, the word 'shock' to denote weak, discontinuous, solutions of hyperbolic conservation
laws, while others equate shock with breakdown. The difference between blow-up and breakdown was stressed by Ball (1977). A finer classification of jump discontinuities (which are
cases of gradient blow-up) is current in fluid mechanics. Of course, for higher-order equations,
higher derivatives may become singular while the solution remains C^1.

seems to be confirmed by the fact that several independent researchers arrived at this conclusion in different equations, both for blow-up and gradient blow-up.

The case of singular data has been addressed, for weak singularities, by the techniques of §3.3. Stronger singularities are not well understood. Such problems arise for instance in the vortex sheet problem, and lead to ill-posed initial-value problems, which are only handled at present in an analytic set-up.

We now turn to the contents of this chapter. We will use nonlinear perturbations of the wave equation as our main model, although it should be understood that the methods described have a somewhat wider scope. For references regarding results given in the text without proof, see the Notes.

Section 3.1 gives criteria which guarantee the breakdown of smooth solutions.

Section 3.2 describes the microlocal theory of weak singularities, and includes a discussion of the evolution of singular data.

Section 3.3 describes briefly the theory of hyperbolic conservation laws.

Section 3.4 describes what is known on the "blow-up mechanism" at this time.

3.1 CRITERIA FOR BREAKDOWN

We discuss below a few procedures to prove the necessary breakdown of smooth solutions of nonlinear evolution equations. These procedures rest usually on the proof that the solutions of a certain ODE become infinite in finite time. Since a geometric picture of this process is reasonably complete for scalar equations of first order only, we discuss this case first. The case of quasilinear equations (and first-order systems) in several space dimensions is then treated by reduction to a one-dimensional problem. In both cases, the relevant ODE is related to the characteristics of the equation. Another procedure is to prove the blow-up of integral expressions involving the solution; it is presented next. Other special methods are also briefly indicated.

Scalar equations.

We consider first a single equation

$$(3.1) \qquad\qquad \phi(x, u, p) = 0,$$

where $x = (x_0, \ldots, x_n) \in \mathbf{R}^{n+1}$, $x_0 = t$, and $p = \nabla u$. Let us briefly recall the solution of the initial-value problem for (3.1) via the method of characteristics:

we solve the *characteristic system*

$$(3.2) \qquad \frac{dx_i}{ds} = \frac{\partial \phi}{\partial p_i}; \quad \frac{dp_i}{ds} = -\frac{\partial \phi}{\partial x_i} - p_i \frac{\partial \phi}{\partial u}; \quad \frac{du}{ds} = \sum_k p_k \frac{\partial \phi}{\partial p_k},$$

with (x, u, p) prescribed for $s = 0$ such that (3.1) holds initially. One then has $\phi = 0$ for all time as a consequence of (3.2).

We now assume that $\partial \phi / \partial p_0 \neq 0$ which enables us to solve locally for the derivative $\partial \phi / \partial p_0$ in terms of u and its remaining derivatives; it means that the surface $t = 0$ is non-characteristic for the linearization of the equation near the desired solution. Characteristics depend on the solution.

If u prescribed on $t = 0$, we take data of the form

$$(0, y_1, \ldots, y_n, u(0, y_1, \ldots, y_n), \nabla u(0, y_1, \ldots, y_n)),$$

for the characteristic system, where $(y_1, \ldots, y_n)$ is a set of coordinates on the initial surface ($\{t = 0\}$ in this case), and the derivative $\nabla_0 u (= u_t)$ is computed using (3.1). The desired solution is obtained by eliminating s and $y_1, \ldots, y_n$ from the expression of u and x to obtain a relation giving u directly as a function of x. It is well-known that if this relation is substituted into $p = p(s; y_1, \ldots, y_n)$, we find $p_i = \partial u / \partial x_i$ and that u solves the initial-value problem.

Now there are two possible sources for singularities in this process: The Jacobian $\partial(x_0, \ldots, x_n) / \partial(s, y_1, \ldots, y_n)$ may have vanishing determinant, so that one expects the solution to be "multi-valued;" or u may be singular as a function of the x variables. These generally correspond to shock formation and blow-up respectively. We review a few typical and classical examples below, which are models for the more complicated situation considered next.

Example 1: $u_t + a(u)u_x = 0$. This is generally considered as the prototype equation for which smooth data evolve singularities. The characteristic flow includes the equations

$$du/ds = 0; \quad dt/ds = 1; \quad dx/ds = a(u),$$

and if u is prescribed on $\{t = 0\}$, we find that we may prescribe initial data for the characteristic flow by $t(0) = 0$; $x(0) = y$; $u(0) = u_0(y)$. It is not necessary to consider the other two equations in the characteristic flow. We find $u(s, y) = u_0(y)$ for all s, which leads to $x(s, y) = y + su_0(y)$, and $t(s, y) = s$, so that the characteristics are straight lines in the space of (x, t, u). We may now eliminate s and y and find u in implicit form:

$$u = u_0(x - ta(u)).$$

The implicit function theorem enables us to invert this relation locally near any point of the form $(x, t) = (x_0, 0)$, and a smooth solution is therefore achieved in a neighborhood of $\{t = 0\}$.

To find the first time where a singularity occurs, one notes that, since $u_{xt} + au_{xx} + a'u_x^2 = 0$, the quantity $q(s, y) = u_x(x(s, y), t(s, y))$ solves

$$q_s + a'(u)q^2 = 0.$$

If $a' > c > 0$, we conclude that q must become infinite in finite time.

Even though the characteristics never intersect in the three-dimensional space where (x, t, u) lives, their projections on the (x, t)-plane do, and their envelope represents the locus of points where the method of characteristics breaks down. However, it is in general possible to view (x, u) or (t, u) as the independent variables, in which case t (resp. x) becomes a regular function near the breakdown point.

Example 2: $u_t + x^2 u_x = 0$. Although this equation is linear, it has solutions which are initially smooth and become infinite in finite time. An example is $x/(1 + xt)$. In fact, the characteristics are described by

$$t = y; \quad x = \frac{y}{1 - ys}; u = u_0(y).$$

Note that the set where blow-up occurs is itself a characteristic curve; this is due to the linearity of the equation.

Example 3: $u_t^2 + u_x^2 = 1$. (The eikonal equation in the plane). As is well-known, one can solve for u locally near a smooth curve on which u is required to vanish. The solution then represents the distance to the curve, and will cease to be smooth on the evolute (envelope of normals) of the initial curve. Thus, the solution is singular not on a characteristic, but on an envelope of characteristics.

Example 4: $u_t = u^2$. The characteristics can be parametrized in the form $t = s$, $x = y$, $u = u_0(y)/(1 - su_0(y))$. The (x, t) part of the characteristic system is perfectly well-behaved, but the function u becomes infinite for a finite value of s.

We therefore find that

1. If we define characteristics by (3.2), singularities form on characteristics, envelopes of characteristics, or correspond to points at infinity on characteristics.

2. If we allow x, t and u to play symmetric roles, some of these singularities may be avoided.

3. The presence of singularities may sometimes be ascertained (both for blow-up *and* gradient blow-up) by finding an ODE satisfied along characteristics by a suitable combination of u and its derivatives.

We now examine the possible generalization of some of the above possibilities to systems, and to higher dimensions.

Quasilinear systems.

We consider here quasilinear systems of the form

$$(3.3) \qquad\qquad u_t + \sum_{j=1}^{n} a_j(u)\partial u/\partial x_j = 0.$$

Such systems include quasilinear perturbations of the wave equation, the equations of nonlinear elasticity, and the nonlinear vibrating string, as well as the hyperbolic conservation laws of §3.3.

We study the formation of singularities, for solutions of the special form $u = u(\theta(s,t))$, where $s = k_1 x_1 + \cdots + k_n x_n$. These solutions are sometimes called "simple waves." Letting $a(u) = \sum_j k_j a_j(u)$, we see that we are led to the system

$$\frac{du}{d\theta} = r(u(\theta));$$
$$\theta_t + \lambda(u(\theta))\theta_s = 0,$$

where r is a right eigenvector for a, corresponding to the eigenvalue λ:

$$a(u)r(u) = \lambda(u)r(u).$$

More precisely, any solution of this system generates a solution of the original equation. In view of the situation described for $u_t + c(u)u_x = 0$, we find that it is possible to construct solutions which become singular in finite time if we can guarantee that $d\lambda(u(\theta))/d\theta \neq 0$. It turns out that one can generalize these ideas to systems to some extent:

Let us consider more generally a solution of the system

$$(3.4) \qquad\qquad u_t + a(u)u_s = 0,$$

where u has m components. Assume that we can find N smooth independent right and left eigenvectors for a:

$$a(u)r_k(u) = \lambda_k(u)r_k(u), \quad l_k(u)a(u) = \lambda_k(u)l_k(u).$$

We also assume these eigenfunctions satisfy $l_j(u)r_k(u) = \delta_{jk}$. We may then decompose u_s:

$$(3.5) \qquad\qquad u_s = \sum_k v_k(s,t)r_k(u(s,t)).$$

THEOREM 3.1 *There are functions $c_{ijk}(u)$ such that*

(3.6) $$(\partial_t + \lambda_i(u(s,t))\partial_s)v_i = \sum_{j,k} c_{ijk}(u)v_j v_k.$$

Furthermore, $c_{ijk}(u) = c_{ikj}(u)$, and $c_{ijj}(u) = 0$ if $j \neq i$.

Remark: In the terminology of §3.3, $c_{iii} \neq 0$ if and only if the ith characteristic field is genuinely nonlinear.

Proofs: Let $v = u_s$. We know that $u_t = -a(u)v$ is linear in v. By the chain rule, so are the derivatives of any smooth function $f(u)$. Differentiating the equation with respect to s, it follows that

$$(\partial_t + a(u)\partial_s)v$$

is quadratic in v. Substituting the expansion (3.5), the existence of c_{ijk} follows.

Since we have $v_1(ds - \lambda_1(u)\,dt) = du$ if $m = 1$, it is natural to compute

$$\begin{aligned} d(v_i(ds - \lambda_i(u)\,dt)) &= (v_{i,t} + \lambda_i v_{i,s} + v_i \lambda_{i,x})dt \wedge dx \\ &= (\sum c_{ijk}v_j v_k + v_i \nabla_u \lambda_i \cdot \sum_k v_k r_k(u))dt \wedge dx \\ &:= \sum \gamma_{ijk}(u)v_j v_k dt \wedge dx, \end{aligned}$$

where $v_{k,t} = \partial v_k/\partial t$, and similarly for other derivatives; this relation defines $\gamma_{ijk}(u)$. In the special case of a simple wave, v_k vanishes for all k except one (say, for $k \neq j$), and we find that $v_j(dx - \lambda_j(u)ds) = du_j$ is exact, which proves $\gamma_{ijj} = 0$ for all i. In terms of c_{ijk}, this means that

(3.7) $$c_{ijj} = -\delta_{ij}\nabla_u \lambda_i \cdot r_i(u).$$

The proof is complete.

We now state a blow-up theorem for this class of systems:

THEOREM 3.2 *Assume the system is genuinely nonlinear for u close to a constant value u_0 of u. Consider initial data v_0 which are smooth and differ from u_0 on an interval of length l. Then the solution breaks down in finite time if $l^2\|v_0'' - u_0''\|_\infty$ is small enough.*

More detailed results are available for 2×2 systems. For this and other extensions, see the Notes.

Blow-up via differential inequalities.

We first consider the wave equation

$$\Box u + f(u) = 0$$

in n space dimensions. We let $F(u) = \int_0^u f(s)\,ds$, and

$$E(u, u_t) = \int_{\mathbf{R}^n} \{[u_t^2 + |\nabla u|^2]/2 + F(u)\}\,dx.$$

THEOREM 3.3 *Assume that f is smooth, $f(0) = 0$, $uf(u) \leq (2 + \varepsilon)F(u)$ for some $\varepsilon > 0$ and that u is a smooth solution with $E(u(0), u_t(0)) < 0$. Assume, to fix ideas, that the data have compact support. Then u must develop a singularity in finite time.*

Proof: We deal with classical solutions and E is therefore constant. For the existence of such solutions, see Ch. 2. Let

$$I = \frac{1}{2}\beta(t + \tau)^2 + \frac{1}{2}\int u^2 \, dx,$$

where β and τ are positive constants; β is chosen so that $2E + \beta < 0$, and τ so that $I'(0) > 0$. The objective is to prove that I cannot remain bounded.

Multiplying the equation by u and integrating, we find

$$\int (uu_{tt} + |\nabla u|^2 + uf(u)) \, dx = 0.$$

Adding $-(2 + \varepsilon)E$ to both sides, we find, with $\alpha = \varepsilon/4$, after using the assumption on f,

$$(2 + 4\alpha)E \geq \int [(1 + 2\alpha)u_t^2 - uu_{tt}] \, dx.$$

Therefore,

$$I'' \geq 2(1 + \alpha)\left(\int u_t^2 \, dx + \beta\right) - (1 + 2\alpha)(2E + \beta).$$

This implies

$$II'' - (1+\alpha)I'^2 > (1+\alpha)\left[\left(\int u^2 \, dx + \beta(t+\tau)^2\right)\left(\int u_t^2 \, dx + \beta\right) - \left(\int uu_t \, dx + \beta(t+\tau)\right)^2\right].$$

The left-hand side is nonnegative (use the Cauchy-Schwarz inequality on the product of $(u, \sqrt{\beta}(t + \tau))$ and $(u_t, \sqrt{\beta})$ in $L^2 \times \mathbf{R}$). Let now $J = I^{-\alpha}$. The above inequality becomes $J'' < 0$. On the other hand, the parameters have been chosen so that $J(0) > 0$, and $J'(0) < 0$. It follows that J must vanish in finite time, QED.

The method of differential inequalities is useful to show that singularities do arise for large classes of data. The actual reason for breakdown must of course be investigated separately.

Other methods.

We turn to blow-up theorems based on a skillful use of representation formulae for the wave equation. When applicable, these results can be very powerful.

The following results include in particular a case where *all* compactly supported data lead to breakdown.

THEOREM 3.4 *Let $f(u) = -|u|^p$ with $1 < p \le \gamma(n-1)$, where*

$$\gamma(n) = 1/2 + 1/n + \sqrt{(1/2 + 1/n)^2 + 2/n}$$

is the positive root of $n(\gamma - 1)/2 = \gamma + 1/\gamma$. Then there are solutions with small energy which fail to exist for all time.

In this case, energy refers to the quantity $\int (u_t^2/2 + |\nabla u|^2/2 - u|u|^p/(p+1))\,dx$. Note that $\gamma(3) = 2$, $\gamma(2) = 1 + \sqrt{2}$, $\gamma(1) = (3 + \sqrt{17})/2$; one takes $\gamma(0) = \infty$.

THEOREM 3.5 *For equation $\Box u = u^2$ in three space dimensions, all smooth data with compact support produce singularities in finite time.*

Remark: The result is false if the data are not small at infinity: $\Box u = u^2$ has the solution $6/(t+1)^2$ which is free of singularities in the future. Thus, a global bound on the data is needed to ensure blow-up. Precisely because of finite speed of propagation, the behavior of the data at infinity may become relevant at any point if one waits long enough.

Proof of Th. 3.4 for $n = 3$. Assume $\int u(x,0)\,dx$ and $\int u_t(x,0)\,dx$ are both positive, and that the data are supported by the ball of radius R. Let

$$F(t) = \int u(x,t)\,dx.$$

We prove that F must develop a singularity in finite time.

By finite speed of propagation (see Ch. 2), the solution at time t is supported by the ball of radius $R + t$. Integrating the equation in space, and using Hölder's inequality, we get

$$F''(t) = \int u_{tt}\,dx = \int |u|^p\,dx \ge CF^p(1+t)^{-3(p-1)}.$$

On the other hand, if $u_0(x,t)$ is the solution of the wave equation with the same data as u, we have $\partial_{tt} \int u_0\,dx = 0$, and therefore

$$\int u_0(x,t)\,dx = at + b.$$

Our assumptions ensure that a and b are both positive. Since the strong Huygens principle holds in three dimensions, we also know that u_0 is supported by $\{t - R \le |x| \le t + R\}$.

Kirchhoff's formula implies that $u(x,t) \ge u_0(x,t)$. We therefore have

$$
\begin{aligned}
at + b &\le \int_{\text{supp}\,u_0} u(x,t)\,dx \\
&\le \int_{t-R \le |x| \le t+R} u(x,t)\,dx \\
&\le \left(\int |u|^p\,dx\right)^{1/p} \text{vol}\,(t - R \le |x| \le t + R)^{\frac{p-1}{p}} \\
&\le Ct^{2(p-1)/p}\|u\|_p.
\end{aligned}
$$

Since $F'' = \|u\|_p^p$, we find

$$F'' \geq C(1+t)^{p+2-2p},$$

and, since $F(0)$ and $F'(0)$ are both positive, we find $F \geq C(1+t)^{4-p}$, $F'(t) > 0$, and we still have $F'' \geq CF^p(1+t)^{3-3p}$. It follows that

$$F'' \geq CF^{1+\varepsilon}(1+t)^{-[(p-1)^2+\varepsilon(4-p)]}.$$

The exponent $\delta = [(p-1)^2 + \varepsilon(4-p)]$ can be assumed to be between 0 and 2 if ε is small enough, because we have $1 < p < 1 + \sqrt{2}$.

Multiplying by F and integrating, we find $F' \geq CF^{1+\varepsilon}(1+t)^{-\delta/2}$, hence $(-1/(\varepsilon F^\varepsilon))' \geq C/(1+t)^{\delta/2}$. Integrating once more, we find that $F^{-\varepsilon}$ must vanish in finite time, which is the desired result.

We now give the proof of Th. 3.5.

Let us assume that the data are supported by the ball of radius R, and are not both identically zero. We let $v = u^2$ and denote by $I(r,t)$ and $J(r,t)$ the spherical means of u and v about the origin respectively. We therefore have

$$(3.8) \qquad\qquad (rI)_{tt} - (rI)_{rr} = rJ,$$

for positive and negative values of r.

The objective is to prove that if we assume that a smooth solution exists upto time $6R$, we can derive arbitrarily large lower bounds on I for $0 < r < t - 6R$, and thus derive a contradiction.

STEP 1: PRELIMINARIES.

The Cauchy-Schwarz inequality immediately gives

$$J \geq I^2.$$

Eq. (3.8) has initial data which vanish for $|r| > R$. It follows that if $0 < r < t - R$ (or more generally, if $|t \pm r| > R$), we have

$$I(r,t) = \frac{1}{2r} \int_{R(r,t)} \rho J(\tau,\rho)\, d\rho,$$

where $R(r,t) = \{(\rho,\tau) : \tau \pm \rho < t \pm r, \tau > 0\}$ is the backward solid light cone from (r,t), cut by $\{t > 0\}$. But one say more: since ρJ is odd with respect to ρ, the part of the integral corresponding to the region $\{\tau \pm \rho < t - r\}$, which is symmetric with respect to the τ axis, is zero. One can therefore restrict the integration to the region

$$Z(r,t) = \{(\rho,\tau) : \tau \pm \rho < t \pm r, \tau + \rho > t - r, \tau > 0\}.$$

Letting r go to zero, we find

$$I(0,t) = \int_0^t \rho J(t-\rho,\rho)\,d\rho$$

for $t > R$.

STEP 2: $I(t,0) \not\equiv 0$ FOR $R < t < 3R$.

Indeed, otherwise we would have $J(\tau,\rho) \equiv 0$ for $\tau+\rho \in (R,3R)$, and therefore $v = u = 0$, since $v = u^2 \geq 0$. Letting $t \to R$, we find in particular that $u(r,R) = 0$ for $0 < r < 2R$, which, by a domain of dependence argument, forces $u(x,0) = 0$ for $|x| < R$. The solution would then vanish identically.

There is therefore a point $(0,t_0)$ in the r-t plane, near which I doesn't vanish. Form the above representation for $I(r,t)$, we conclude that $I(r,t) \neq 0$ if $t-r = t_0$ and $r > 0$, since $(0,t_0)$ is then one of the vertices of $Z(r,t)$.

STEP 3: LOWER BOUNDS.

We prove that if $0 < r < t - 6R$, one has

$$I(r,t) \geq C_j \frac{(t-r-6R)^{k_j}}{(t+r)(t-r)^{q_j}}$$

for a sequence (C_j, k_j, q_j) such that

$$C_{j+1} = \frac{C_j^2}{16(k_j+1)^2}, \quad k_{j+1} = 2(k_j+1); \quad q_{j+1} = 2q_j + 1.$$

For $j = 0$, we argue as follows. Consider the set

$$T = \{(\rho,\tau) : 4R < \tau + \rho < 6R, R < \tau - \rho < 3R\}.$$

Since $J(\rho,\tau) > 0$ for $\tau - \rho = t_0$ (with $t_0 \in (R,3R)$) and $\rho > 0$, we have

$$\int_T \rho J(\rho,\tau)\,d\rho d\tau = 2c > 0.$$

Now we certainly have $T \subset Z(r,t)$ if

$$(r,t) \in S := \{(\rho,\tau) : 6R < \tau + \rho, 3R < \tau - \rho < 4R\}.$$

Therefore, if $(r,t) \in S$,

$$I(r,t) \geq c/r.$$

But since $J \geq I^2$, we can further estimate I by

$$I(r,t) \geq \frac{c^2}{2r} \int_{Z(r,t)} d\rho d\tau/\rho \geq \frac{c^2}{2r}\text{Area}\,(S \cap Z(r,t))/(t+r-3R),$$

since $2\rho \leq t + r - 3R$ on S. All in all, this yields

$$I(r,t) \geq c^2/(t+r).$$

Let us assume the result for j. We now use the top part of $Z(r,t)$: define

$$V = \{(\rho,\tau): t - r < \tau + \rho < t + r, 6R < \tau - \rho < t - r\}.$$

Dropping the subscript j on all constants for the time being, we find

$$I(r,t) \geq \int_V \frac{\rho(\tau - \rho - 6R)^{2k}}{(\tau + \rho)^2(\tau - \rho)^{2q}} \, d\rho d\tau.$$

Letting $a = \tau + \rho$, $b = \tau - \rho$, the domain of integration is described by $a \in (t - r, t+r)$, $b \in (6R, t-r)$. Since $\tau - \rho \geq t - r$, and $\rho = (a-b)/2 \geq (t-r-(\tau-\rho))/2$,

$$
\begin{aligned}
I(r,t) &\geq \frac{C^2}{8r(t-r)^{2q}} \int_{t-r}^{t+r} \frac{da}{a^2} \int_{6R}^{t-r} (t - r - b)(b - 6R)^{2k} \, db \\
&= \frac{C^2(t - r - 6R)^{2k+2}}{4r(2k + 1)(2k + 2)(t^2 - r^2)(t - r)^{2q}}.
\end{aligned}
$$

A fortiori, we have the desired estimates with the announced values of C_{j+1}, k_{j+1}, q_{j+1}.

STEP 4: END OF PROOF.

Form the previous step, it is easy to see that there is a positive constant A such that for every j, we have

$$I(r,t) \geq \frac{(t - r)\exp(A2^i)}{(t + r)(t - r - 6R)^2}.$$

The result follows.

In contrast to the above theorem, if one is only interested in proving that some special solutions blow-up, one may use a simpler comparison argument; it has also the advantage of using only local information on the data to give its conclusion.

Consider the solution of

$$\square u = f(u)$$

where f is nondecreasing, with $u(0,x) = u_0(x)$, and $u_t(0,x) = u_1(x)$. We assume that $x \in \mathbf{R}^n$ with $n \leq 3$, and that $u_1 \geq b$ for $|x| \leq T$, where b is a constant. We also require that there is a constant a such that $u_0 \geq a$ (resp. $u_0 = a$) if $n = 1$ (resp. $n = 2$ or 3), again for $|x| \leq T$. One can then conclude that the solution becomes unbounded before time T in each of the following two cases:

(1) $b \geq 0$ and $\int_a^\infty [b^2 + \int_a^z f'(s) \ ds]^{-1/2} \ dz < T$, where the square root is always defined;

(2) $b < 0$ and

$$\int_v^\infty [b^2 + \int_a^z f'(s) \ ds]^{-1/2} \ dz - \int_a^v [b^2 + \int_a^z f'(s) \ ds]^{-1/2} \ dz < T$$

where v is the largest root below a of the equation

$$b^2 + \int_a^z f'(s) \ ds = 0.$$

These simply mean that the solution of the ODE $w_{tt} = f(w)$ with data (a, b) becomes infinite before time T. One can show that the assumptions guarantee that $u \geq w$ in a region in which w blows up, so that u must blow-up before w does.

3.2 PROPAGATION OF WEAK SINGULARITIES

This section is devoted to the propagation of the H^r wave-front set of solutions of a nonlinear single equation of order m:

$$p_m(x, D)u = f(x, u, Du, \ldots, D^{m-1}u).$$

Starting with some overall regularity: $u \in H^s_{\text{loc}}(\mathbf{R}^N)$, one considers the propagation of H^r singularities for some $r > s$. It turns out that they propagate along the null bicharacteristics of the linear part if s is not too small, *and $r \leq 2s - s_0$* for a suitable s_0. After giving the appropriate constraints on s and r, we proceed to show that the constraints are necessary. If r is too large, new singularities can appear when two bicharacteristic pass over the same point of space. We then turn to a class of solutions generalizing the case of functions with jumps in some derivative, across a smooth hypersurface, and for which more precise results on the "interaction of singularities" are available.

We are interested in the possible extension of the results on propagation of singularities via Fourier analysis to nonlinear equations. Significant results are available for weak singularities of solutions which already have a minimum of overall regularity. In particular, blow-up and gradient blow-up are usually excluded at the outset, and, in terms of the Cauchy problem, the singularities are present in the data. We are interested in a precise description of their future evolution.

Fourier analysis has been useful in two respects: in helping understand the action of nonlinear function on Sobolev spaces, and in providing an extension of theorems on the evolution of the wave-front set to nonlinear equations.

Since these results have been described in detail in several recent surveys and monographs, we will merely provide some orientation for the reader to find his way through the growing literature.

Dyadic decomposition and paraproduct.

We decompose any $u \in \mathcal{S}'$ into a sum of functions with Fourier transform supported in increasingly large annuli. More precisely, let $\varphi(x)$ be a smooth function equal to 1 for $|\xi| \leq 1/2$, and to zero for $|\xi| \geq 1$. Let

$$S_j(u) = \mathcal{F}^{-1}[\varphi(2^{-j}\xi)\hat{u}]; \quad \Delta_j(u) = S_{j+1}(u) - S_j(u).$$

The *dyadic* or *Littlewood-Paley* decomposition of u is

$$(3.9) \qquad u = S_0(u) + \Delta_0(u) + \Delta_1(u) + \cdots := \sum_{k=0}^{\infty} u_k.$$

Note that $\Delta_j(u) = \mathcal{F}^{-1}[\psi(2^{-j}\xi)\hat{u}]$, where $\psi(\xi) = \varphi(\xi/2) - \varphi(\xi)$. It follows that $\mathcal{F}\Delta_j(u)$ has support in the annular region $\Gamma_j = \{2^{j-1} \leq |\xi| \leq 2^j\}$. This decomposition depends on the choice of φ.

The usefulness of the Littlewood-Paley decomposition is due to the fact that it gives convenient characterizations of Sobolev and Hölder norms:

THEOREM 3.6 *The norm* $|u|_{s,p}$ *is equivalent, if* $1 < p < \infty$, *to*

$$\|(\sum_{0}^{\infty} |2^{ks}u_k|^2)^{1/2}\|_p.$$

The Hölder spaces C^r, *when* r *is positive and is not an integer, have norm equivalent to*

$$\sup_k 2^{kr}\|\Delta_k(u)\|_\infty.$$

If r is a positive integer, this expression can be taken as the definition of the norm in the Zygmund space C_*^r. It contains C^r. We let $C_*^r = C^r$ when r is not an integer. C_*^1 is the space of bounded continuous functions such that

$$|u(x+y) + u(x-y) - 2u(x)| \leq C|y|$$

for some constant C.

Action of nonlinear functions.

If one allows oneself the use of the properties of operators with symbols in the limiting cases of Hörmander's classes, one can derive a fairly simple proof of the fact that if f is a smooth function with bounded derivative, which vanishes at zero, and if $u \in L^{s,p}$, $1 < p < \infty$, then $f(u) \in L^{s,p}$. Such results are therefore a convenient generalization of Moser-type estimates (which, we recall, were derived for integer s, by application of the Gagliardo-Nirenberg and Sobolev inequalities).

One can however derive a finer result, which expresses the multiplication operator by a merely bounded measurable function as the sum of a "paramultiplication" operator, which has good mapping properties, and a regularizing remainder.

The paraproduct of two elements a and u of $\mathcal{S}'$ is

$$(3.10) \qquad T_a(u) = \sum_{j \geq 2} S_{j-2}(u) \Delta_j(u).$$

The series does converge in the sense of $\mathcal{S}'$. Some authors use $j - 3$ instead of $j - 2$ for convenience. The series $\sum S_j \Delta_j$ would however diverge in general.

One then has

THEOREM 3.7 *If* $f \in C^{\infty}(\mathbf{R})$ *and* $u \in L^{s,p}(\mathbf{R}^n)$ *with* $1 < p < \infty$ *and* $r = s - n/p > 0$, *with* $f(0) = 0$, *one has*

$$f(u) = T_{f'(u)}u + R$$

where $R \in L^{s+r,p}$.

The paraproduct has however the additional virtue of having a "symbolic calculus," which means that there are natural formulae for the composition, commutator, etc ... of operators obtained from differential operators by replacing products by paraproducts. This makes it possible to mimic the proof of linear theorems of propagation of singularities.

Application to nonlinear equations.

All nonlinearities are C^{∞} unless otherwise specified. N is the total number of variables and n the number of space variables.

Recall that a function u is microlocally H^s near $y_0 = (x_0, \xi_0)$ if there is a classical pseudo-differential operator b of order zero which is elliptic at y_0, such that bu is locally H^s. The H^s wave-front set is the set of points y such that $u \notin H^s(y)$; it is denoted by $WF_s(u)$. For any curve Γ in the cotangent bundle, we define $H^s(\Gamma)$ analogously. While $H^s(\mathbf{R}^N)$ is closed under products for $s > N/2$, simple examples show that the wave-front set of a product is larger than the union of the wave-front sets of its factors.

One has more precisely the easy result:

THEOREM 3.8 *Let* $u_i \in H^{s_i}$ *for* $i = 1, 2$, *withe* $s_i > 0$. *Then, for all* $\delta > 0$, $u_1 u_2 \in H^{\min(s_1, 2_2, s_1 + s_2 - N/2 - \delta)}$.

On the other hand,

THEOREM 3.9 *If u_1 and u_2 are in H^s and the second (frequency) component of $WF_s(u_1)$ and $WF(u_2)$ lie in closed cones K_1 and K_2, and if K_3 is a closed cone which does not intersect $K_1 \cup K_2$, then $WF_{2s-N/2-\delta}(u_1u_2)$ does not intersect $\mathbf{R}^N \times K_3$ for every $\delta > 0$.*

(As usual, a cone is said to be closed if its intersection with the unit sphere is.)

In other words, the product u_1u_2 is $H^{2s-N/2-\delta}$ except possibly in $K_1 \cup K_2$.

This result suggests

THEOREM 3.10 *Let Γ be a closed conic subset of $T^*\mathbf{R}^N \setminus 0$. Assume $N/2 < s \leq r \leq 2s - N/2$. Then $H^s(\mathbf{R}^N) \cap H^r(\Gamma)$ is an algebra.*

We are now ready to give extensions of Hörmander's theorem to semilinear equations.

THEOREM 3.11 *Let u be a locally H^s solution of a strictly hyperbolic second-order equation $p_2(x, D)u = f(x, u, Du)$ with n space dimensions (hence, $n + 1$ variables). Assume Γ is the null bicharacteristic through a given point y_0. Then $u \in H^r(y_0)$ implies $u \in H^r(\Gamma)$ if $(n + 1)/2 + 2 < s \leq r < 2s - (n + 1)/2 - 1$.*

If we do not allow the nonlinearity to depend on Du, we have the more precise result:

THEOREM 3.12 *Let u be a locally H^s solution of a strictly hyperbolic second-order equation $p_2(x, D)u = f(x, u)$ with n space dimensions. Assume Γ is the null bicharacteristic through a given point y_0. Then $u \in H^r(y_0)$ implies $u \in H^r(\Gamma)$ if $(n + 1)/2 < s \leq r < 2s - (n + 1)/2 - 1$.*

Idea of Proof: Since the solution is already H^s, we know by Hörmander's theorem that the solution is $H^{s+1}(\Gamma)$. Assume therefore, by induction, that $u \in H^t(\Gamma)$ for $s \leq t < 2s - (n + 1)/2m + 2$ for some m. This implies that the right-hand side of the equation is in

$$H_{\text{loc}}^{s-m+2} \cap H^{t-m+2}(\Gamma),$$

since $(t - m + 2) < 2(s - m + 2) - \frac{1}{2}(n + 1)$. It follows that if $t + 1 \leq r$, $u \in H^{t+1}(y_0)$ and therefore $u \in H^{t+1}(\Gamma)$ by Hörmander's theorem again. QED.

In the second-order case, one can control regularity "upto $3s$:"

THEOREM 3.13 *For an equation of the form $\Box u = f(u)$, Th. 3.12 holds for $(n + 1)/2 < s \leq r < 3s - n + 1$.*

The upper bound on r in these results is not a technical restriction. One example is the following.

THEOREM 3.14 *For any $n \geq 2$ and $s > (n+1)/2$, there is a function $\beta(x,t)$ in $C^\infty(\mathbf{R}^{n+1})$ which vanishes for $t > 0$, and a function $u \in H^s(\mathbf{R}^{n+1})$ such that*

$$\Box u = \beta u^3,$$

but $u \notin H^{3s-n+2+\epsilon}_{loc}$ in any open subset of $\{|x| \leq t\}$, for any $\epsilon > 0$.

If one does not restrict one's attention to second order problems, one can find simpler examples. Thus, we may consider

$$u_t - u_x = 0; \quad v_t + v_x = 0; \quad w_t = uv,$$

which is a degenerate case of the three wave interaction. One can easily construct solutions with $u = H(x+t)(x+t)^k$, $v = H(x-t)(x-t)^k$, where H is Heaviside's function, such that singularities only propagate at speeds ± 1 for $t < 0$, while a new singularity with speed 0 appears for $t > 0$.

To go further, we restrict our class of singularities so that they will be localized on a finite number of hypersurfaces. This is the class of *conormal singularities*, which generalize jump discontinuities in higher derivatives. Given a smooth hypersurface Σ in Euclidean space, we say that a locally H^s function u is conormal of degree k with respect to Σ if $X_1 \ldots X_j u \in H^s_{loc}$ for all $j \leq k$, where the X_j stand for arbitrary smooth fields tangent to Σ. We write $u \in N^{s,k}(\Sigma)$.[2] If $k = \infty$, this implies that $WF(u)$ is a subset of the normal bundle of Σ. This space is invariant under the application of nonlinear functions.

THEOREM 3.15 *If $P_2(x,D)u = f(x,u,Du)$, and u is locally in $H^s(\mathbf{R}^{n+1})$ and in $N^{s,k}(\Sigma \cap \{t < 0\})$, where Σ is a characteristic hypersurface for the strictly hyperbolic operator P_2, then $u \in N^{s,k}(\Sigma)$ if $s > (n+3)/2$. If f is independent of Du, $s > (n+1)/2$ is sufficient.*

3.3 SHOCK WAVES

We discuss here the special results pertaining to hyperbolic conservation laws in one space variable. After a few definitions, we describe the various forms of the entropy condition and list the special properties of the single conservation law. See the notes for similar results for the single conservation law in several space dimensions. A few related results on shock structure are included.

---

[2]Some authors write $u \in H^{s,k}(\Sigma)$.

Definitions.

A *conservation law* is a system of the form

$$(3.11) \qquad \partial_t F_0(u) + \sum_{i=1}^{N} \partial_i F_i(u) = 0,$$

where N is the number of space dimensions. We will be concerned here with one-dimensional systems of conservation laws of the form

$$(3.12) \qquad u_t + f(u)_x = 0$$

where u has n components, and f is a smooth function of u (possibly only in some open region of $\mathbf{R}^n$). Values of u are often referred to as *states*.

The system is hyperbolic if the eigenvalues $\lambda_1(u), \dots, \lambda_n(u)$ of the $n \times n$ matrix $f'(u)$ are real. We always assume here that the eigenvalues are distinct, with

$$\lambda_1(u) < \cdots < \lambda_n(u),$$

and introduce left and right eigenvectors satisfying $l_k(u)[f'(u) - \lambda_k(u)] = 0$ and $[f'(u) - \lambda_k(u)]r_k(u) = 0$ for $k = 1, \dots, n$. They can be chosen to vary smoothly with u by an application of the implicit function theorem.

We say that the kth characteristic field l_k is *genuinely nonlinear* if

$$r_k \cdot \nabla_u \lambda_k \neq 0$$

for all u. For a single equation, this simply means that $f''(u) > 0$. If this property holds for all k, the system itself is said to be genuinely nonlinear. This property will guarantee that all plane wave solutions with bump-like profiles must become singular in finite time. The opposite extreme case is that in which

$$r_k \cdot \nabla_u \lambda_k \equiv 0$$

for all u. One then says that the kth characteristic field is *linearly degenerate*.

A *Riemann invariant* is a smooth function $v(u)$ such that

$$(3.13) \qquad r_k \cdot \nabla_u v(u) \equiv 0.$$

One can prove, using the implicit function theorem, that there exist, locally, $(n-1)$ independent Riemann invariants.

We finally define special types of solutions, which are used as building blocks for more complicated ones.

A plane wave solution is a function of $kx - \omega t$ for some constants k and ω.

A rarefaction wave is a solution of the form $u = v((x - x_0)/(t - t_0))$. It is said to be *centered* at (x_0, t_0).

The basic example of a hyperbolic conservation law is the system of gas dynamics, where the unknown has three components (u, ρ, p) (velocity, density, pressure), and the eigenvalues are 0 and $\pm c$ where c is the speed of sound.

Jump discontinuities.

If a solution u is piecewise C^1, with a jump across a smooth curve with slope $s = dx/dt$ (also called the shock speed), and with limiting values u_l and u_r from the left and right of the curve respectively, we can derive a constraint on these values by expressing that u solves the conservation law in the sense of distributions. The result is the *Rankine-Hugoniot* condition:

$$(3.14) \qquad s[u] = [f(u)],$$

where brackets denote jumps: $[u] = u_l - u_r$. These are n constraints on s and on the $2n$ components of u_l and u_r. For given u_l, there are therefore in general several ways of choosing u_r in accordance with this relation:

THEOREM 3.16 *For a genuinely nonlinear system, there are, for a given constant state u_l, n families of shocks compatible with the Rankine-Hugoniot relation; they can be represented by $u_r = u_r(\varepsilon)$, $u_r(0) = u_l$. The kth family satisfies $du_r/d\varepsilon = r_k(u_l)$ for $\varepsilon = 0$.*

Unfortunately, the Rankine-Hugoniot conditions allow for types of shocks that should be rejected on physical grounds. Thus, there are two weak solutions for

$$u_t + u u_x = 0$$

with the initial value $u(x,0) = H(-x)$, where H is Heaviside's function, namely $u_1 = H(t/2 - x)$, and $u_2 = x/t$ for $0 < x < t$, and zero otherwise. The discontinuous solution is, however, never observed in situations modeled by this equation. The reason is that the shock does not satisfy the *entropy condition*, which, for jump discontinuities requires that, for some index k,

$$(3.15) \qquad \lambda_{k-1}(u_l) < s < \lambda_k(u_l), \quad \text{and} \quad \lambda_k(u_r) < s < \lambda_{k+1}(u_r).$$

These are also called the *shock inequalities*. They express that the kth characteristic impinges on the shock on both sides, instead of diverging from it.

An important example is the system of one-dimensional gas dynamics:

$$
\begin{aligned}
v_t - u_x &= 0 \\
u_t + p_x &= 0 \\
(e + u^2/2)_t + (pu)_x &= 0,
\end{aligned}
$$

where the unknowns are u (velocity), v (specific volume; $v = 1\rho$), and e. The pressure is related to these via the equation of state $p = p(E, v)$, where $E =$

$e+u^2/2$. One can eliminate the shock speed for the Rankine-Hugoniot conditions to obtain the Hugoniot curve (or "shock adiabatic")

$$(3.16) \qquad\qquad e_r - e_l + \frac{1}{2}(p_r + p_l)(v_r - v_l) = 0.$$

The eigenvalues are 0 and $\pm c$, where $c = \sqrt{-p_v}$ is the sound speed.

For this system, one can show that condition (3.2) expresses that the entropy increases across a weak shock, hence the name. It has several versions discussed in the next paragraph. But first, let us include one more result on solutions with jump discontinuities. The *Riemann problem* consists in solving the initial-value problem with $u(x,0) = u_l$ for $x < 0$, and u_r for $x > 0$, where u_l and u_r are given constants. An example is the case of the equations of gas dynamics in a tube, where two states of a gas are separated by a thin membrane which is removed at time $t = 0$. Solutions of Riemann problems are building blocks for more general solutions.

For gas dynamics with a polytropic law, pv^γ =const., we have

THEOREM 3.17 *The Riemann problem has a unique solution which satisfies the entropy condition and which has nonzero density everywhere ("no vacuum") if $u_r - u_l < 2(c(u_l) + c(u_r))/(\gamma - 1)$.*

THEOREM 3.18 *For a gas with an equation of state $p = p(v, S)$ (where v stands for specific volume), with $p_v < 0$, $p_{vv} > 0$, $p > 0$, then the shock inequalities are equivalent to the fact that the entropy increases after the passage of a shock if the shock strength $|u_l - u_r|$ is small. The increase is in fact of the order of the cube of the shock strength.*

Finally, one must point out that a discontinuity in the solution is thought to be an indication that the conditions under which the conservation law was derived break dows; the correct equation ought to be such that the shock is replaced by a smooth profile. To construct this equation is the problem of "shock structure." There are three ways to accomplish this today.

One is to consider a parabolic regularization corresponding, for the case of fluid dynamics, to the inclusion of viscosity terms; it usually leads to a parabolic system

$$u_t + f(u)_x = \nu A u_{xx}$$

for some positive definite matrix A. One then seeks traveling wave solutions $u = u((x - st)/\nu)$ which tend to u_r and u_l respectively as $x - st \to \pm\infty$. This is a "connecting-orbit" problem for a system of ODEs.

The second is to discretize the equation, which is qualitatively similar to adding a dissipation or a dispersive term, depending on the choice of the difference scheme. A good choice leads to a numerical method which reproduces faithfully the sharp profile of the shock wave, without overshoots of other modifications.

The third is to add a dispersive term to the equation. The appearance of oscillations near the shock in some difference schemes motivates this procedure, which has been carried out in detail so far only using inverse scattering.

General form of the entropy condition.

A pair (U, F) is said to be an *entropy pair* if any *classical* solution of the system also solves

$$U(u)_t + F(u)_x = 0.$$

U and F are called and entropy and an entropy flux respectively.

Let us now consider a "viscosity method" for our system: let u_ε solve

$$u_{\varepsilon t} + f(u_\varepsilon)_x = \varepsilon A u_{\varepsilon xx}$$

where $\varepsilon > 0$ and A is a positive definite matrix.

THEOREM 3.19 *Assume that $|u_\varepsilon|_\infty \le C$, $(D^2 U)A$ is positive definite, and that u_ε tends to a function u , as $\varepsilon \downarrow 0$, in the sense of distributions. Then u solves (3.12) and satisfies*

(3.17) $$U_t + F_x \le 0$$

in the sense of distributions.

Note that if $A = I$, we are requiring that the entropy U be *convex*.

THEOREM 3.20 *In the situation of the previous theorem, if u has a jump discontinuity across a line, we need to have (in addition to the Rankine-Hugoniot relations)*

(3.18) $$s[U(u)] - [F(u)] \le 0.$$

THEOREM 3.21 *For a genuinely nonlinear hyperbolic system of conservation laws with a convex entropy, the conditions of the previous theorem are equivalent to the shock inequalities.*

For gas dynamics, the entropy satisfies $S_t = 0$ for smooth solutions, but this equation does not hold in the weak sense for discontinuous solutions.

Single conservation law.

We consider

(3.19) $$u_t + f(u)_x = 0$$

where $f'' > 0$. We consider also the solutions of the associated parabolic equation:

$$(3.20) \qquad\qquad u_{\varepsilon t} + f(u_\varepsilon)_x = \varepsilon u_{\varepsilon xx}$$

for $\varepsilon > 0$, with $u(0) = u_0 \in L^\infty$.

THEOREM 3.22 *The solutions u_ε converge in the sense of distributions to a weak solution of the conservation law. Furthermore, denoting by $S(t)$ the solution operator, mapping u_0 to $u(t)$ for $t > 0$, we have:*

1) $S(t)$ extends uniquely to a contraction in L^1.

2) $S(t)$ is a compact operator in L^1, and its range is in $BV(\mathbf{R})$, the space of functions with bounded variation.

3) For u_0 in L^∞, we have $S(t)u_0$ is Lipschitz continuous; in fact, there is a constant C such that for any positive t and y, and for any $x \in \mathbf{R}$,

$$|u(x + y) - u(x)| < Cy/t.$$

Note that uniqueness follows immediately. The third property is a form of the entropy condition. The compactness of the solution operator is interpreted as a statement about the irreversibility of the process modeled by the conservation law.

The decay properties of solutions are rather peculiar, in that the rate of decay is independent of the solution:

THEOREM 3.23 *There is a constant C such that for all smooth u_0 with compact support, one has $|u(t)| \leq C/\sqrt{t}$, and the size of the support grows like $\sqrt{t}$. If the data are periodic with period p, there is a constant C which depends again only on f, such that $|u(t) - \bar{u}| \leq Cp/t$, where $\bar{u}$ denotes the average of u over one period.*

We thus see that the size of the domain of dependence does not grow linearly. In fact, as a rule, shocks constantly impinge on the boundary of the support, correcting its shape.

Further results on the single conservation law are given in the appendix.

3.4   MODELS FOR SINGULARITY FORMATION

Once the appearance of singularities from smooth data has been ascertained, the next question is to explain where they occur, and how to describe them more precisely. We therefore need *models* of singularity formation. One hopes that there exists a *blow-up set*

$\{t = \psi(x)\}$ such that the solution is regular precisely for $t < \psi(x)$.[3] The first time a singularity occurs is the minimum of ψ.

One strategy is to consider solutions with small data, which should be well-approximated by solutions of the linear part, which are hopefully better understood. Seeking a perturbation series with linear solutions as first term, one can sometimes compute asymptotics of the breakdown time in the limit of small data using, remarkably enough, explicit functionals of the data.

Another strategy is to look for a simple singular solution (e.g., space-independent) and to use it instead of linear solutions as the starting-point of a perturbation series. One then tries to prove that this series does describe accurately the blow-up mechanism for solutions which are not close to zero. This procedure in many cases gives a way to analytically continue solutions after blow-up.

Two such arguments have been successful so far: one, for semilinear hyperbolic equations, gives a complete solution of the blow-up problem for near-constant, smooth data for $\Box u = \exp(u)$; the method seems to be quite robust and general. A second argument, for quasi-linear equations, has been successful in low-dimensional cases, and appears very promising. When it works, it also providees a complete description of the onset of singularities.

The future evolution of singularities present in the past has been considered in §3.2.

A description of singularity formation requires the determination of the maximal domain of existence, or at least the maximal domain of influence of the data, and the proof of an asymptotic expansion of the solution near the boundary of this domain. This boundary will be called the *blow-up surface*, for short, even though we may be dealing with shock wave singularities. A first step is the computation of the breakdown time, which represents the point of the blow-up surface which is closest to the initial surface. Although these results do not give precise information on the blow-up set, they do involve the same expression that we used in the proof of global existence via the null condition. Much more precise results in the semilinear case suggest a pattern of general applicability. This line of argument is confirmed by the success of analogous ideas for quasi-linear equations.

The breakdown time.

In this section, the issue is to find and justify an asymptotic evaluation of the solutions near their breakdown set, which is itself unknown; in this sense,

---

[3]The singularity may well be a gradient blow-up, but this terminology is becoming standard.

we are dealing with a free boundary problem. We concentrate on perturbations of the wave equation:

Let u solve

$$\Box u = G(u, Du, D^2 u),$$

in three space dimensions, with $u(x,0) = \varepsilon f(x)$, $u_t(x,0) = \varepsilon g(x)$, where G is quadratic near the origin, linear in the second order derivatives, and ε is a small parameter. It turns out that the breakdown time $T(\varepsilon)$ can be estimated in the limit $\varepsilon \to$ zero. Before giving an idea of why this should be true at all, let us give two rather comprehensive theorems which summarizes the situation.

THEOREM 3.24 *Denote by $T(\varepsilon)$ the breakdown time (which may be infinite).*

1. *If G is independent of u and has the form $\sum_{a,b=0}^{3} r_{ab}(Du)D_{ab}u$, then*

$$\ln T(\varepsilon) \geq K/\varepsilon$$

   where K can be described explicitly in terms of the data;

2. *For radial solutions of $u_{tt} = c(u_t^2)\Delta u$, $\ln T(\varepsilon) = K/\varepsilon$.*

THEOREM 3.25 *Assume that G is linear in the second derivatives.*

1. *If $G_{uu}''(0,0,0) = 0$, then $\ln T(\varepsilon) \geq C/\varepsilon$;*

2. *If $G_{uu}''(0,0,0) \neq 0$, we have $T(\varepsilon) \geq C/\varepsilon^2$;*

3. *Statement (4) is optimal if $G = u^2 + Q(Du)$, where Q is a positive definite quadratic form;*

4. *If $G = u^2$, $\lim_{\varepsilon \to 0} \varepsilon^2 T(\varepsilon) = T$ exists and is positive; furthermore,*

$$\lim_{\varepsilon \to 0} \varepsilon^4 u(t/\varepsilon^2, x/\varepsilon^2)$$

   exists in the sense of distributions, and solves

$$\Box v = v^2 + \mu$$

   for $t < T$, where μ is a positive homogeneous measure supported by the forward light cone $|x| = t$.

Note that in Th. 3.24, it is expected that the second derivatives of u become infinite; the system satisfied by u and its first derivatives is similar to a quasi-linear conservation law. The solution itself is expected to blow up inthe case of Th. 3.25.

Idea of Proofs and Remarks: (For more details, see John (1990))

(Proof of Theorem 3.24, part 1.) From problem 21 (in Ch. 1), we know that for $\varepsilon = 0$, and large times, $u = t^{-1}k(x/|x|, |x| - t) + O(1/t^2)$. If ε is very small, one may hope that u will, for moderate values of t, satisfy a similar relation. On the other hand, if $|x| - t$ is at the same time bounded, one can show (much as in the proof of the global Sobolev inequalities in Minkowski space), that spatial derivatives of u are controlled by the time derivative $D_0 u$, by formulae involving t and the unit vectors $\xi_i = x_i/r$. If we write

$$G \approx \sum_{abc} z_{abc} D_a u D_{bc} u,$$

and let t grow while $|x| - t = p$ stays bounded, we may hope to find an ODE with a quadratic nonlinearity which $D_0 u$ might satisfy. One finds eventually that if one moves along the curve

$$dr/dt = 1 + \frac{1}{2} \sum_{ab} r_{ab}(Du(t, r\xi)) X_a X_b,$$

where $r = |x|$ and $X_0 = -1$, $X_i = \xi_i$, then $w = tu_{tt}^2$ satisfies

$$w_t \approx Z(\xi) w^2/t,$$

where $Z(\xi) = -(1/2) \sum_{a,b,c} z_{abc} X^a X^b X^c$. If we admit that the blow-up process is very slow, we may integrate this equation using the approximation

$$w \approx \varepsilon k_{pp}(\xi, p)$$

given by Friedlander's theorem, and finally obtain

$$w \approx k_{pp}(\xi, p)/(1 - \varepsilon k_{pp}(\xi, p) Z(\xi) \ln t).$$

This suggests that the breakdown time $T(\varepsilon)$ satisfies

$$\varepsilon \ln T(\varepsilon) \geq 1/H$$

where

$$H := \sup_{\xi, p} Z(\xi) k_{pp}(\xi, p).$$

It is noteworthy that H is zero when G satisfies the null condition; one can check that $H > 0$ in all other cases unless $u \equiv 0$.

(For Th. 3.24, part 2.) The very drastic approximations made in the proof of (1) are justified not so much by the smallness of ε but even more by the expectation that the second derivatives blow up like $\ln t$ and thus vary slowly when compared with t. Even so, it is remarkable that the previous estimate is actually accurate for the radial case. This suggests that the blow-up is due to the behavior of the solutions for large x, and that the pointwise value of u may be very small at the points where the solution breaks down. References for the complete proof is found in the Notes.

(For Th. 3.25.) Here again, the result should be understood intuitively from Friedlander's formula: if we take a solution of the *linear* wave equation and perform on it the present scaling, a solution of and equation of the form $\Box v = \mu$ results. Note that the limit v of rescaled solutions can be extended to a maximal domain which is the rescaled limit of the domains between the initial surface and the blow-up set for each ε, and which will contain (strictly) the support of μ.

The blow-up surface: semi-linear case.

Because nonlinear hyperbolic equations can be solved locally, it is reasonable to assume that if the solution blows up at time T for $x = x_0$, it may remain smooth elsewhere for some time. One is therefore led to define a function $\psi(x)$ as the supremum of those values of t such that u is a smooth solution in a neighborhood of (x, s) for $0 \leq s \leq t$. Let us call the set $t = \psi(x)$ the *blow-up surface*. The minimum of ψ gives the breakdown time. One would like to represent the solution asymptotically as $t \to \psi$.

There are at present only two general methods to study the blow-up surface. The first proceeds in four steps:

1. Given a singular surface $\Sigma : \{t = \psi(x)\}$, construct a formal series representing a putative singular solution blowing up on this surface. This series usually involves one arbitrary function, which we will call $w_0(x)$.

2. Show that for any (ψ, w_0) in high enough Sobolev spaces, there actually exists a singular solution represented by the first few terms of this series. At this point, we have produced solutions with prescribed singularities. The pair (ψ, w_0) will be referred to as the *singularity data*.

3. Considering a nearby initial surface S, define a map sending the singularity data to the Cauchy data on S.

4. Invert this map.

At the end of this process, we have a detailed description of blow-up for all data in the range of the map from singularity to Cauchy data, because the first few terms of the series of 1. give a valid approximation of the solution near blow-up.

It turns out that these solutions can be thought of as conormal solutions (in the sense of §3.2), but their overall regularity is not sufficient for the analysis of §3.2 to apply directly.

The first and second steps have been carried out for quite general hyperbolic semilinear systems (Kichenassamy and Littman (1993) for the analytic case,

Kichenassamy (1995a) for the Sobolev case). The analytic case is actually, like the Cauchy-Kowalewska theorem, not restricted to the hyperbolic case. The analytic case includes the proof of the WTC expansions (§4.6). The third and fourth steps were carried out in Kichenassamy (1995b) for

$$\Box u = e^u,$$

which was singled out because fewer terms of the correspondong series need to be computed to complete the argument. The ideas involved are however readily transposable to other equations.

In all cases, the solutions have natural continuations across their blow-up surface.

The above results are valid in any number of space dimensions.

The second idea, consists in viewing the problem as a free boundary problem: this suggests that an appropriate use of comparison principles might, as in other problems, yield a proof of regularity of the blow-up surface. This method is therefore expected to work in low space dimensions only. When successful, it proves that $\psi \in C^1$, and that the first term of the appropriate expansion is valid. The best results are available in one space dimension. Higher regularity does not follow from this approach at this time, but is nevertheless accessible to the first approach if the data are smooth and close to a known solution.

We begin with the second argument, which was found first.

Blow-up and free boundaries. To illustrate the first idea, we discuss:

THEOREM 3.26 *The solution of*

$$u_{tt} - u_{xx} = u^2$$

with data in $C^4(\mathbf{R}) \times C^3(\mathbf{R})$ *is either finite for all* $t > 0$, *or else will blow-up on a space-like curve* $t = \psi(x)$:

$$\lim_{t \to \psi(x)-} u(x,t) = +\infty$$

A similar result is valid for power-like nonlinearities (see the Notes), and the blow-up rate is the same as that for the ODE $u_{tt} = u^2$.

Proof: We will argue on a fixed cone $\{|x| < R - t\}$. Most of the following does not use the equation itself, but only the fact that $\Box u \geq 0$.

Define the backward and forward cones $K_\pm(x_0, t_0)$ from any point (x_0, t_0). Introduce the characteristic variables $\xi = t + x$ and $\eta = t - x$. It is then easy to see that u, u_ξ and u_η are all bounded below by constants depending only on R and the data. It follows that u_t is bounded below, and that

$$|u_x| \leq u_t + C.$$

Now define a characteristic rectangle (N, S, E, W) (for the cardinal points) within the domain of u. Integrating $\Box u \geq 0$, we obtain the classical result

$$u(N) + u(S) \geq u(E) + u(W).$$

Taking S on $\{t = 0\}$, and using the lower bound on u, it follows that

$$u(N) \geq u(W) - C.$$

Therefore, the value of u at any point also controls it on the whole left (and right) characteristics, and hence in the entire backward cone. This means that if u remains finite for all time at one point in space, it must remain finite everywhere. This proves the first part of the theorem. For the other, one notes that if there were points with $\psi(x_1) > \psi(x_0) + |x_1 - x_0|$, (x_0, t_0) would belong to the backward cone from a point $(x_1 - \varepsilon, t_1)$ at which the solution is still finite. The proof that $|\psi'|$ is actually less that 1 is omitted.

In the case when blow-up does occur, one may view the solution as the limit of solutions of equations with truncated nonlinearities, all of which will have $|u_x| \leq Cu_t$ and satisfy the other properties proved earlier.

We now turn to the second approach.

Step 1: Formal expansion. In order to make the calculations more explicit, we restrict ourselves to the equation

$$\Box u = e^u.$$

We seek a solution of the form

$$u = \ln(2/T^2) + v(x, T, \ln T).$$

Performing the change of variables $X_i = x_i$, $T = t - \psi(x)$, and substituting the formal expansion $u = \ln(2/T^2) + v_0(X) + v_1(X)T + \ldots$ into the equation, we find, after identification of like powers of T, that we must take

$$v_0 = \ln(1 - |D\psi|^2)$$

and

$$v_1 = -\frac{\Delta\psi}{(1 - |D\psi|^2)}.$$

However, there is no equation to determine v_2 and we are left instead with a constraint on ψ, which, after some algebra, reduces to

$$R = 0$$

where R is the scalar curvature of Σ for the metric induced by Minkowski space.

Let us introduce a new unknown

$$(3.21) \qquad v(x,t) := [u - \ln(2/T^2) - v_0(x) - v_1(x)T]/T^2.$$

One finds that v satisfies a Fuchsian equation of second order. In order to use the results of Ch. 2, we reduce it to a first-order system by letting $w = (v, T\partial_T v, T\nabla v)$. It has the form

$$(3.22) \qquad Tw_T + Aw = \alpha(X) + Tf(X, T, w, \nabla w).$$

where A is a constant matrix with eigenvalues 0 and 3, and α is proportional to R. Now this system has, by what we just found, analytic solutions only if $R \equiv 0$. On the other hand, if this constraint does hold, there are infinitely many solutions, as follows from Th. 2.2.

An examination of (3.22) reveals however that one should try solutions with a logarithmic leading part. This suggests the new change of variables

$$(3.23) \qquad v(T, X) = \lambda(X)\ln T + f(T, Y, X),$$

where

$$(3.24) \qquad Y = T\ln T,$$

and

$$\lambda(X) = -\frac{2}{3}R(x)(1 - |D\psi|^2)^{-1}.$$

Our new unknown, f now depends on $n + 2$ independent variables: X, T, and Y. Remarkably, the equation satisfied by f involves only these variables, and not $\ln T = Y/T$.

Let now $N = T\partial_T + (T + Y)\partial_Y$, and define

$$z_1 = f;$$

$$z_2 = Nf;$$

$$z_{2+i} = T\partial_{i'}f \quad (1 \le i' \le n),$$

where $\partial_{i'} = \partial_{X^i}$. Thus, $z_2 = z_3 = \cdots = 0$ for $T = Y = 0$.

We find

$$
\begin{aligned}
(1 - |D\psi|^2)(Nz_2 + 3z_2) &= -T\Big\{\Delta\psi(\lambda + z_2) + 2\psi^i\delta_i^{i'}\partial_{i'}z_2 \\
&\qquad - \delta^{ii'}\partial_{i'}(z_{2+i})\Big\} + O(T) + O(Y); \\
Nz_{2+i} &= (T\partial_T + (T + Y)\partial_Y)T\partial_{i'}f \\
&= T\partial_{i'}(Nf) + T\partial_{i'}f \\
&= T\partial_{i'}(z_2 + z_1).
\end{aligned}
$$

Step 2: Prescribing the blow-up surface. We now construct solutions with prescribed blow-up set. This may be viewed as a semi-linear analogue of the "shock front problem" for (quasilinear) conservation laws. We restrict ourselves here to the analytic situation, for simplicity, although most of the following has been extended to H^s solutions. These solutions are conormal in a weak sense, but not regular enough for the analysis of §3.4 to apply as such. Note also that the blow-up surface is *not* characteristic in these examples, even though it is characteristic for an appropriate Fuchsian equation.

We focus on the case of exponential nonlinearities, for which the algebra is simplest.

Consider the equation

$$\Box u = e^u$$

in n space dimensions. We are also given a hypersurface Σ, with equation $t = \psi(x)$.

THEOREM 3.27 *If Σ is space-like, there are infinitely many solutions defined near p_0 such that u blows up precisely on Σ. More precisely, they have the form $u = \ln(2/T^2) + v(x, T, T\ln T)$, where $T = t - \psi(x)$ and v is analytic in its arguments.*

Observe that the solutions given by this theorem have *ipso facto* an analytic continuation beyond their blow-up surface. It is uniquely determined provided a branch of the logarithm has been chosen. Even this arbitrariness can be avoided in some cases:

THEOREM 3.28 *The function v in the preceding theorem is analytic in x and t if and only if the scalar curvature of Σ for the (Riemannian) metric induced by its embedding into Minkowski space vanishes identically.*

This condition is vacuous if $n = 1$, and reduces itself to the Monge-Ampère equation ("$rt - s^2 = 0$") if $n = 2$.

Proof of Th. (3.28-29): The system (3.25) has the general form

$$(3.25) \qquad Nz + Az = Th_1(T, Y, X, z, D_X z) + Yh_2(T, Y, X, z, D_X z),$$

where A is a constant matrix; in our case, A has eigenvalues 3 and 0, with respective multiplicities 1 and $n + 1$. Using the full strength of Th. 2.1, we obtain the conclusion of the theorem in the analytic case.

Remark: The singularities live on the surface $\Sigma = \{T = 0\}$ which *is* characteristic for the generalized Fuchsian system (3.25).

Steps 3 & 4: From Cauchy to singularity data. We now consider solutions close to the solution $\ln(2/t^2)$. We show that the model of blow-up given by the above calculations is actually the correct one for near-constant data. The reader will be able judge the generality fonthe arguments developed below.

THEOREM 3.29 *There are number $s_0(n)$ and $s_1(n)$, depending on the number of space dimensions, such that Cauchy data on $\{t = -1\}$ which are close to $(\ln 2, 2)$ in $H^s \times H^{s-1}$, for some $s \geq s_0(n)$, lead to blow-up on a hypersurface $\{t = \psi(x)\}$ where $\psi \in H^{s-s_1(n)}$. In fact, $u - \ln[2/(t - \psi(x))]$ is bounded for $|t - \psi(x)|$ small.*

Idea of proof: (The details are in Kichenassamy (1995).) The proof proceeds by application of the Nash-Moser IFT (§5.2). After proving that the map from singularity to Cauchy data is C^2 in appropriate space, the main difficulty is to prove the invertibility of the linearization of this map near any point at which it is defined. This can be accomplished by observing that this linearization involves the solution of the linearization of $\Box u = \exp(u)$ near the singular solution under consideration, rather than the linearization of the Fuchsian equation, which is unwieldy. This linearization itself can be reduced to Fuchsian form. One then shows quite generally that solutions of such linear Fuchsian equations *must* have logarithmic expansions near $T = 0$. There is however another way to derive such an expansion: by taking the variation of the series for u itself. Indeed, the coefficients of this series are themselves differential expressions in ψ and w_0. By identification of these two series, the desired inverse is computed. It however exhibits a loss of derivatives, hence the use of the Nash-Moser theorem.

The blow-up surface: quasi-linear case.

There is in the case of quasi-linear systems a construction which parallels Steps 1 and 2 of the above approach.

Let us consider an $N \times N$ system

$$F[u] := \sum_{j=1}^{n} A_j(x, u)\partial_j u + B(x, u) = 0.$$

We therefore do not distinguish time and space variables any more.

As in the semilinear case, the idea is to show that there is a smooth function from which the singular solution is deduced by a singular transformation. For $\Box u = \exp(u)$ for instance, we saw that there was a smooth function $v(t_0, t_1, x)$, of $n + 2$ variables if n is the number of space dimensions, such that the transformation

$$(t, x, u) \mapsto (T, T \ln T, x, e^{-u}),$$

where $T = t - \psi(x)$, maps the graph of u to the graph of a perfectly regular function. In this sense, we have "uniformized the solution." For the semilinear case, however, we will choose a transformation acting on u alone, and which is non-invertible on the locus of singularities. It will not increase the number of variables.

Let us describe the procedure in more detail: we make a change of variables: $y = \Phi(x)$, and we define a transformed unknown v by

$$u(\Phi(y)) = v(y).$$

However, Φ will not be known at the outset. The singular set will correspond now to the set of points where Φ is *non-invertible*. It is expected that v itself is very regular, and that the singularity is entirely due to the singular behavior of Φ^{-1}. Note that if $v = v(y^1)$, u is a simple wave. For a single equation, the method amounts to straightening out characteristics.

We now define an "unfolded system"[4] or "blowup system" for v and Φ, and give a few illustrative examples. We describe a slightly simplified situation; more details and applications can be found in Alinhac (1995).

Let us limit ourselves to the vicinity of $x = x^0$ and $u = u^0$. Assume that there is a direction $\xi^0 (\neq 0)$ and a real simple eigenvalue $\lambda(x, u, \xi)$ of the matrix symbol:

$$A(x, u, \xi) := \sum_j A_j(x, u)\xi_j,$$

defined near (x^0, u^0, ξ^0). We further assume that

$$\lambda(x^0, u^0, \xi^0) = 0, \quad \nabla_\xi \lambda(x^0, u^0, \xi^0) = \neq 0,$$

and

$$r \cdot \nabla_u \lambda(x^0, u^0, \xi^0) \neq 0,$$

where r and l are smoothly varying right and left eigenvectors for $A(x, u, \xi)$. The latter condition is recognized as the genuine nonlinearity condition for the characteristic field under consideration. Let us write $x = \Phi(y)$, where $\Phi(y_1, \ldots, y_n) = (\varphi(y), y_2, \ldots y_n)$. Define v by the relation $u(\Phi(y)) = v(y)$. The unfolded system is obtained by substitution into the equation for u. Since there are $N + 1$ unknowns, namely φ and the components of v, we need an additional equation: we take

(3.26) $$\lambda(\Phi(y), v(y), \eta(y)) = 0,$$

where $\eta = \nabla_x y_1 = (1, -\varphi_2, \ldots, -\varphi_n)$.

The derivatives of φ are eventually further constrained at the origin, in order to control the geometry of the map, Φ. In particular, one chooses the derivatives of φ at the origin to ensure $\eta(0)$ is proprotional to ξ^0. We also impose the conditions $v(0) = u^0$ and $\partial_1 \varphi(0) = 0$, but $d(\partial_1 \varphi) \neq 0$.

We now write down a form of the unfolded system.

The chain rule yields

$$\partial_j = \frac{\partial}{\partial y_j} - \frac{\varphi_j}{\varphi_1}\frac{\partial}{\partial y_1} \text{ for } j \neq 1; \quad \frac{\partial}{\partial x_1} = \frac{1}{\varphi_1}\frac{\partial}{\partial y_1}.$$

---

[4] *système éclaté* in French

We therefore find, writing v_j for $\partial v / \partial y_j$,

$$(3.27) \qquad A(\Phi, v, \eta)v_1 + \varphi_1 [\sum_{j \neq 1} A_j v_j + B] = 0.$$

Equations (3.26-3.27) form the unfolded system. It is helpful to replace (3.27) by its projection on l and on some complementary subspace.

The type of singularity is controlled by the classification of Φ in the sense of catastrophe theory; thus, a *fold* occurs if $\varphi_1(0) = 0$ but $\varphi_{11}(0) \neq 0$. The locus of singularity is defined by $\det \Phi' = 0$.

Example: The simplest example is Burgers' equation $u_t + uu_x = 0$. The variables "x" are now denoted by (t, x), and the new ("y") variables are called (T, X). We take $\xi^0 = (-1, u_0(x_0))$, $\eta = (-\varphi_T, 1)$, so that $A(x, v, \eta) = \lambda = -\varphi_T + v$. The unfolded system becomes

$$(3.28) \qquad v - \varphi_T = 0$$

$$(3.29) \qquad (v - \varphi_T)v_X + \varphi_X v_T = 0,$$

or more simply:

$$(3.30) \qquad \varphi_T = v$$

$$(3.31) \qquad v_T = 0.$$

Observe the advantages of the unfolded system over the original one, even in this simple example.

3.5 APPENDIX

As an illustration of geometric methods in the study of singularity formation, we prove here a breakdown theorem for the single conservation law in several space variables. The proof relies on the application of the theory of integral invariants applied to the characteristic flow. It clarifies earlier treatments, and gives as a bonus a simple proof of the conservation of total variation in characteristic tubes.

Note: This section uses systematically Einstein's summation condition on repeated indices.

Integral invariants for first order ODEs.

Let us consider an ordinary differential equation

$$(3.32) \qquad dx^i / dt = \xi^i(x^1, \ldots, x^n; t),$$

where ξ is assumed to be C^∞. The value at time t of the solution of (3.32) having the value

$$x_0 = (x_0^1, \ldots, x_0^n)$$

for $t = t_0$ is denoted by $\varphi(t, t_0)(x_0)$. We let

$$\Theta(t)(x_0, t_0) = (\varphi(t, t_0)(x_0), t_0 + t).$$

$\{\Theta(t)\}_{t \in \mathbf{R}}$ forms a group of diffeomorphisms of $\mathbf{R}^n \times \mathbf{R}$.

Let now ω be a smooth k-form on $\mathbf{R}^n \times \mathbf{R}$. We say that ω is an (absolute) integral invariant for (3), if for every $\Omega \subset \mathbf{R}^n \times \mathbf{R}$, smooth, bounded and of dimension k, one has

$$(3.33) \qquad\qquad \int_\Omega \Theta(t)^* \omega = \int_\Omega \omega.$$

Remarks:

1. One speaks of a *relative* invariant if (3.33) only holds for *closed* Ω. The exterior derivative of a relative invariant is an absolute invariant.

2. Poincaré introduced a related notion by requiring that ω be a form in the variables $x^1, \dots, x^n$ alone, and that

$$(3.34) \qquad\qquad \int_\Omega \varphi(t, 0)^* \omega = \int_\Omega \omega$$

for every bounded $\Omega \subset \mathbf{R}^n$ of dimension k. Elie Cartan, seeking to remove the assumption that the points of Ω in the above formula should be *simultaneous* (*i.e.* that one has $\Omega \subset \mathbf{R}^n$ instead of $\mathbf{R}^n \times \mathbf{R}$), introduced what we now call integral invariants. He showed that Poincaré's invariants are actually truncated forms corresponding to "complete" invariants.

Integral invariants are characterized by the following theorem:

THEOREM 3.30 *The form ω is an integral invariant for*

$$\dot{x}^i = \xi^i,$$

where $\dot{x}^i = dx^i/dt$, if and only if the Lie derivative $\mathcal{L}_\eta$ of ω with to $\eta := \xi^i \partial_i + \partial_t$ vanishes.

The proof of the theorem is by direct calculation, since the condition for invariance is $(d/ds)\Theta(s)^* \omega = 0$.

Remarks:

1. We recall that $\mathcal{L}_\eta = i_\eta d + d i_\eta$, where d denotes the exterior derivative and i_η the interior product by η. In local coordinates, if

$$\omega = \frac{1}{k!} \omega_I dx^I$$

where $\omega_I = \omega_{i_1 \ldots i_k}$ is antisymmetric in its indices, we have

$$i_\eta \omega = \frac{1}{(k-1)!} \eta^\lambda \omega_{\lambda i_2 \ldots i_k} dx^{i_2} \wedge \ldots \wedge dx^{i_k},$$

and

$$\mathcal{L}_\eta = \frac{1}{k!} \left[\eta^\lambda \partial_\lambda \omega_{i_1 \ldots i_k} + (\partial_{i_1} \eta^\lambda) \omega_{\lambda i_2 \ldots i_k} + (\partial_{i_2} \eta^\lambda) \omega_{i_1 \lambda i_3 \ldots i_k} \right.$$
$$\left. + \cdots \right] dx^{i_1} \wedge \ldots \wedge dx^{i_k}.$$

2. The conditions
$$i_\eta \omega = 0 \text{ and } i_\eta d\omega = 0$$

imply that ω is invariant. The latter two conditions define what is called an *invariant attached to the trajectories of (3.32)*. The reason for the terminology is that one then has

$$\mathcal{L}_{\alpha\eta} \omega = 0$$

for any function $\alpha(x, t)$; in other words, ω is then invariant under all flows having the *same trajectories* as (3.32).

3. From the properties of the Lie derivative and the interior product, one obtains the following by a direct computation:

   • If ω is invariant, so are $d\omega$ and $i_\eta \omega$.

   • If ω and ϖ are invariant, so is $\omega \wedge \varpi$.

   • If $\omega = i_\eta \varpi$ and $d\omega = 0$, then ω is invariant.

The single first order equation.

We consider here the equation

(3.35) $$u_t + \partial_i (f^i(u)) = 0.$$

$f^1, \ldots, f^n$ are assumed to be C^∞. We first consider smooth solutions, and investigate integral invariants for the characteristic flow

(3.36) $$\dot{x}^i = a^i(u(x, t)); \quad a^i(u) := df^i(u)/du.$$

We then turn to solutions with shocks and show that their total variation is nonincreasing.

We assume $u \in C^\infty$ for all times to be considered. We thus first consider the solution before the first shock occurs.

THEOREM 3.31 *a) The form*

$$\omega = \phi(x,t) \bigwedge_{i=1}^{n} (dx^i - a^i dt)$$

is an integral invariant for (3.36) if and only if ϕ solves the adjoint equation

$$(3.37) \qquad\qquad \phi_t + \partial_i(a^i\phi) = 0.$$

b) Examples of such functions ϕ are $\partial_i a^i$ and $|Du|$.

Remark 1: Let $\Omega \subset \mathbf{R}^n$, and let Ω_t be the image of Ω under the map obtained by solving (3.36) up to time t. One then has

$$\frac{d}{dt}\int_{\Omega_t} dx = \int_{\Omega_t} \partial_i a^i \, dx = \int_{\Omega} \partial_i a^i \, dx,$$

since

$$(\partial_i a^i) \bigwedge_{j=1}^{n} (dx^j - a^j dt)$$

is invariant. It follows that

$$|\Omega_t| = |\Omega| + t\int_{\Omega} \partial_i a^i(x,0)\, dx.$$

Remark 2: If we let $v = \partial_i a^i$, the invariance condition reduces to $v_s + v^2 = 0$, where $\partial_s = \partial_t + a^i\partial_i$, from which it immediately follows that the solution blows up as soon as $v(x,0) \leq 0$.

Proof of Theorem 1:

a) By Example 3, §2.3, it suffices to express that ω is closed. We find that

$$
\begin{aligned}
d\omega &= (\phi_t + a^i\phi_i)dt \wedge dx^1 \wedge \ldots \wedge dx^n \\
&\quad + \phi\sum_i(-1)^i(d(a^i)\wedge dt)\wedge(dx^1 - a^1 dt)\wedge\ldots\wedge\widehat{(dx^i - a^i dt)}\wedge
\end{aligned}
$$

$$(3.38) \qquad \wedge\ldots\wedge(dx^n - a^n dt).$$

Since $d(a^i)\wedge dt = (\partial_j a^i)(dx^j - a^j dt)\wedge dt$, we are left with

$$(3.39) \qquad d\omega = (\phi_t + a^i\phi_i + (\partial_i a^i)\phi)dt\wedge\bigwedge_i dx^i.$$

Q.E.D.

b) i)Let $\varrho = \partial_i a^i$. Differentiate (3.35) with respect to x^j and multiply the result by $a^{j\prime}$ (where the prime denotes the derivative with respect to u). We find:

$$a^{j\prime}[u_{jt} + a^i u_{ij} + a^{i\prime} u_i u_j] = 0,$$

that is,

$$(\varrho_t - (\partial_t(a^{j\prime}))u_j) + a^i(\partial_i\varrho - (\partial_i(a^{j\prime}))u_j) + \varrho\partial_i a^i = 0,$$

or

$$\varrho_t + (a^i\varrho)_i = u_j(\partial_t(a^{j\prime}) + a^i\partial_i(a^{j\prime})).$$

Now, it is clear that any function $v(u)$ satisfies

$$v_t + a^i v_i = 0.$$

In particular, such is the case with $a^{j\prime}(u)$. It follows that

$$\varrho_t + (a^i\varrho)_i = 0,$$

as announced.

ii) Let us now establish the conservation of total variation. Let ψ be a function of the derivatives $u_1, \ldots, u_n$ of u.

Let us again differentiate (3.35) with respect to x_j, and then multiply it by $\psi^j := \partial\psi/\partial u_j$. We find

$$\psi^j[u_{jt} + a^i u_{ij} + a^{i\prime}u_i u_j] = 0,$$

or

$$\psi_t + a^i\partial_i\psi + (\partial_i a^i)\psi^j u_j = 0,$$

that is,

$$(3.40) \qquad \psi_t + \partial_i(a^i\psi) = (\partial_i a^i)[\psi - \psi^j u_j].$$

Take now $\psi = \psi_\alpha := (\alpha^2 + |Du|^2)^{1/2}$ in this equation, where α is some positive number. As

$$(3.41) \qquad \psi_\alpha - |Du| = \alpha^2/(|Du| + \sqrt{\alpha^2 + |Du|^2})$$

and

$$(3.42) \qquad \psi_\alpha - \psi_\alpha^j u_j = \frac{\alpha^2}{\sqrt{\alpha^2 + |Du|^2}},$$

both these quantities lie between zero and α, and it follows that one may let α tend to zero in (3.40) to find that $|Du|$ satisfies (3.37) in the weak sense. As

$$L_\eta(\psi_\alpha \bigwedge_i (dx^i - a^i dt)) \to 0$$

uniformly and as, for every bounded Ω,

$$\int_\Omega \psi_\alpha dx \to \int_\Omega |Du|\, dx$$

as α tends to zero we obtain the conservation of total variation between characteristics as announced.

3.6   FURTHER RESULTS AND PROBLEMS

1. Consider the nonlinear heat equation $u_t - \Delta u = f(u)$ on a smooth bounded domain Ω in $\mathbf{R}^n$, with Dirichlet conditions. Let $E(t) = \int [|\nabla u|^2/2 + F(u)]\, dx$, where $F' = f$ and $F(0) = 0$.

(a) If f is of class C^3, $f(0) = 0$ and $n \le 3$, prove that local solutions exist in $H^2 \cap H_0^1(\Omega)$, and are local unless they fail to remain bounded in H^2 and in L^∞. (It thus suffices to prove breakdown in one of these norms).

(b) If solutions do breakdown, the L^q norm of $u(t)$ cannot remain bounded for any $q > n$ if $n \le 3$.

(c) For the case $f(u) = u^3$, show that solutions breakdown if $E(0) \le 0$.

2. Consider solutions of

$$u_{tt} - u_{xx} + u_t - u^3 = 0$$

for $0 < x < 1$ and $t \ge 0$. Prove that smooth solutions breakdown if $E(0) < -1/8$. [*Hint:* Let $\phi(t) = \int u^2\, dx$ and show that $\phi''(t) \ge \phi^2/2 + C$ for some $C > 0$.]

3. Assume that the solution of the Cauchy problem for $\Box u = f(u)$ becomes infinite along the blow-up surface $t = \sqrt{1 + |x|^2}$, so that the first singularity occurs for $t = 1$, at $x = 0$. Perform a Lorentz transformation with velocity $v \in (-1, 1)$ and compute the blow-up time in the new coordinates. Can this time become zero? Negative? Interpret the results. Same questions with the blow-up surface $t = 1$. How do the results change?

4. Equation $\Box u = u_{tt}^2$ in three space dimensions has no global smooth solutions with compactly supported Cauchy data on $t = 0$.

5. Consider a hyperbolic equation of the form $\sum_{a,b} a_{ab}(Du)\partial_{ab}u = 0$, where indices run form 0 to n. Convert it to a first order system for the components of the gradient of u, assuming $a_{ab} = a_{ba}$. Find the system satisfied by a plane wave solution $u = u(t, (k \cdot x))$, and compute its characteristic speeds. Show that all the eigenvalues are degenerate for all choices of the direction of the plane wave precisely when the original equation satisfies the null condition.

6. For operators of order m, the analogue of Th. 3.18 requires:

$$\frac{n+1}{2} + m - 2 < r \le 2s - \frac{n+1}{2} - m + 3$$

if $f = f(x, u, \ldots, D^{m-2}u)$. How could one apply the results of this chapter to the case of fully nonlinear equations?

7. Assume $\Box u = f(u)$ in n space dimensions, with u locally H^s with $s > (n+1)/2$. Let S be the scaling operator (p. 11). Show that if $M^j u \in H^s_{loc}(\{t < 0\})$, then $u \in C^\infty(\{|x| < |t|\})$.

8. Show that any analytic solution of the Liouville equation $u_{xt} = e^u$ has locally the form $\ln[f'(x)g'(t)/(f(x)+g(t))^2]$ where f and g are analytic functions. Construct a solution which blows up precisely for $xt = 1$. Is it true that any pole of f or g corresponds to a singularity in u? Construct an "entire" solution (*i.e.*, a solution free of singularities for all real x and t.)

9. Consider the "focusing" nonlinear Schrödinger equation:

$$iu_t = \Delta u + u|u|^2.$$

Write $u = v + iw$.

(a) Show that $\int[v_t w - v w_t]\,dx + \int |\nabla u|^2\,dx = \int |u|^4\,dx$.

(b) Show that $(1/2)\int |x|^2|u|^2\,dx + 2\int \sum_i x_i[vw_i - v_iw]\,dx = 0$.

(c) Show that $(1/2)(d/dt)\int \sum_i x_i[vw_i - v_iw]\,dx + (n/2)\int[vw_t - v_tw]\,dx + (1-n/2)\int |\nabla u|^2\,dx = -(n/4)\int |u|^4\,dx$.

(d) Conclude that $(d^2/dt^2)\int |x|^2|u|^2\,dx = \int[8|\nabla u|^2 - 2n|u|^4]\,dx$. Using the fact that $\int[|\nabla u|^2 - |u|^2/2]\,dx$ is independent of time, state and prove a blow-up theorem for the nonlinear Schrödinger equation for $n = 2$.

(e) Generalize to $iu_t = \Delta u - f(u)$, where $f = F'(u)$, $F(0) = 0$ and $uf(u) \leq (2 + 4/n)F(u)$. (see Glassey (1977); the two-dimensional case is relevant for laser propagation and was studied by Zakaraov and Synakh and many others, cf. Papanicolaou et al. (1982), and forthcoming work of Papanicolaou and Malkin for references.)

10. Show that, if $P(u)$ is a polynomial of degree m, and given any noncharacteristic hypersurface $\Sigma : t = \psi(x)$ in Minkowski space and a point on it, one can construct near this point infinitely many solutions of

$$\Box u = P(u)$$

which blow up precisely on Σ. They have the form $u = T^{-2/(m-1)}v(x, \tau, \tau \ln \tau)$ where T is $t - \psi(x)$, $\tau = T^{1/(m-1)}$, and m is the degree of P; v is analytic in its arguments. State and prove an analogous result in case ψ is in a sufficiently high Sobolev space. [*Remark:* Inspection of the series expansion shows that $T^{2/(m-1)}u$ is conormal (as a function of τ and x) with respect ot $\tau = 0$, with $s \leq 2m + 2$. However, u itself is not even locally integrable.

11. Give a sufficient condition on a surface Σ and on the nonlinearity in a system

$$u_t = \sum_j a_j(u)\partial_j u + b(u)$$

for the existence of singular solutions such as one of the components of u becomes infinite precisely on Σ. (See Kichenassamy-Littman (1993b).)

12. Convert the scalar second-order equation

$$u_{xt} + u_x u_{xx} + a u_{yy} = 0,$$

where a is a constant, to a first-order system for u_x and u_y. Write down an unfolded system for this situation.

NOTES

The method of characteristics is classical. The precise description of the singularities of first-order equations (i.e., caustics) is still an active area. We have not stressed the relation to contact geometry since one might object to putting dependent and independent variables on an equal footing; the geometry of caustics has been studied extensively from a geometric viewpoint; see V. Arnold's recent work on the subject. We have nevertheless given in the appendix an idea of the application of old ideas of E. Cartan, and give some improvements of folklore results on the single conservation law.

Even though solutions of first-order equations can be continued in many cases as multi-valued solutions, we now know that (in applications to gas dynamics, aerodynamics and magneto-hydrodynamics in particular) physical processes select a particular, single-valued and discontinuous, solution which can be described mathematically as the limit of a sequence of regularizations. One such regularization is the method of artificial viscosity and the other is discretization. Both seemed to lead to the same result, the "entropy solution." This evolution has been surveyed by many authors (see Oleinik (1963), Lax (1973a,b), and also Gel'fand (1963)). The monograph by Smoller (1983) gives a good idea of the present state of hyperbolic conservation laws in one space dimension. Most of the specific results on the single conservation law mentioned in the text are due to Oleinik, Conway, and Hoff.

More recently, the viscosity method has been applied successfully to fully nonlinear Hamilton-Jacobi equations, starting with work of Crandall and P.-L. Lions. It has been adapted to large classes of fully nonlinear parabolic equations, with applications in computer vision, phases transitions, and control theory among others. The modern presentation bypasses the "vanishing viscosity" limiting process to a large extent. See Kichenassamy, Kumar, Olver, Tannenbaum and Yezzi (1995) and the survey Crandall, P.-L. Lions and Ishii (1989) for details and many references. The structure of the blow-up set for Hamilton-Jacobi equations has been examined in particular in Friedman-Souganidis (1986).

A more recent line of research must be noted: first, by studying old numerical observations by von Neumann, it was realized that dispersive difference schemes lead to a new kind of regularization of conservation laws, based on the zero dispersion limit of the Korteweg-de Vries equation rather than the zero viscosity limit of Burgers' equation. The multi-valuedness of the classical approach resurfaces here, and interesting connections with Whitham's theory are being explored (see Ch. 5). This is the "Lax-Levermore theory," which includes significant contributions by Venakides. Such multi-valued solutions may perhaps be unfolded by the methods of §3.4.

A general reference for Th. 3.1 and 3.2 is the survey by Hörmander (1986), see also Strauss's notes; The results go back to John (1974), with improvements by Klainerman and Majda (1980), T.-P. Liu (1979) (see also Majda (1987)). The relevance of the null condition in this context has been pointed out in John (1990) (who mentions an observation by Shatah (see problem 5)); Hanouzet and Joly (1985a,b) have related it to the type of compatibility occurring in the theory of compensated compactness, thus suggesting analogues of the null condition for systems which are not connected with Minkowski space. Th. 3.3 is from Levine (1974); Glassey (1977) gives the corresponding theorem for the nonlinear Schrödinger equation. The various cases of Th 3.4 are due to Glassey (1981), Schaeffer (1985), Kato (1980), John (1979). Th. 3.5 is a special case of John (1979). The last criterion of §3.1 is from Keller (1957). The special features of low-dimensional equations have proved surprisingly useful in recent years and have accordingly been stressed here.

General references on nonlinear microlocal analysis are Beals (1989), Taylor (1991), and Bony (1991). The first results on propagation of weak singularities "upto $2s$" (Th. 3.9-10) are due to Bony (1978, 1980), Rauch (1979), Lascar (1978). For Th. 3.8-10, see also M. Beals (1985); the limiting case of Th. 3.10 is due to Meyer (1981). For regularity "upto $3s$," see Beals (1985) and further results in Chemin (1988). The example of "self-spreading singularities" in Th. 3.14 is due to Beals (1983). Conormal singularities were studied by Bony, Rauch and Reed (1982b, 1988), Melrose and Ritter (1985), and many others. Many one-dimensional examples were treated in detail by Rauch and Reed (see e.g. Rauch and Reed (1982a)). The paraproduct which Bony introduced in Bony (1980) gave rise to a large literature. Note that this paper also contains results in Hölder spaces in addition to the more familiar Sobolev regularity statements; for the use of C^k regularity assumptions, see also Taylor. A notion of "paracomposition" was defined and studied by Alinhac (1986). Hörmander recently suggested that the loss of regularity overcome by the use of the Nash-Moser implicit function theorem might be circumvented by the use of the paraproduct, and proved a version of the Nash embedding theorem along these lines.

A more detailed presentation of hyperbolic conservation laws can be found in Lax (1971) and Smoller (1983), where the reader will also see the important role of difference schemes in this subject. An important and unconventional existence theorem due to Glimm will also be found there. To avoid duplication with the many existing treatments of this subject, the details have been kept to a minimum. Th. 3.20-21 can be found in Lax (1973a); for the case of several dimensions, see Kruzhkov (1970), Vol'pert (1967), Conway (1977), Hoff (1981, 1983).

Th. 3.22 is from John (1987) and Hörmander (1986), where other references can be found. The fact that there should be a limiting equation in the small amplitude limit is a precise form of the universality of the equation $u_t + u u_x = 0$ in the sense of Ch. 4. Th. 3.23 is due to John (1979) and Lindblad (1990a,b) Th. 3.24 is from Caffarelli and Friedman (1985,1985), who prove further properties of the blow-up surface. This is the first result on semi-linear hyperbolic equations which goes beyond small-amplitude analysis. The other results on the blow-up surface are due to Kichenassamy and Littman (1993a,b), Kichenassamy (1995a) and can be viewed as a semi-linear counterpart to the construction of shock fronts by Majda (1985a,b), Alinhac (1988). The construction of the unfolded system follows Alinhac (see his forthcoming volume for details and many applications). The solution of the blow-up problem for exponential nonlinearities is in Kichenassamy (1995b). A few results on the classification of possible singularities in hyperbolic conservation laws are due to Caflisch *at al.* (1993). As was stressed by Caflisch, the development of singularities from smooth data can, in the analytic case, be viewed as a case of *propagation* of the complex singularities in the data, which travel toward the real domain of the independent variables. Models based on this idea seem to fit numerical calculations, and further support the heuristic value of the study of analytic solutions in singularity formation. Our results on semilinear blow-up are entirely compatible with this view-point.

As for the role of self-similar solutions, we mention only the paper of Le Mesurier, Papanicolaou, Sulem and Sulem (1985) on blow-up for non linear Schrödinger equations, and the recent notes by Cazenave on this equation.

Other types of singularity formation which are very actively studied, but are not directly related to the subject of this book are: blow-up for semilinear parabolic equations (a recent survey is Bebernes and Eberly (1993); see also Levine (1990)), vortex sheet roll-up and the search for singular solutions of the Euler equations (for which a brief introduction is Majda (1991)), and, more broadly speaking the understanding of turbulence and instability in fluid mechanics.

REFERENCES

ALINHAC, S., (1986) Paracomposition et opérateurs para-différentiels, *Comm. PDE, 11(4)*: 87–121.

ALINHAC, S., (1988) Interaction d'ondes simples pour des équations complètement non-linéaires, *Ann Sci. ENS, 21*: 91–132.

ALINHAC, S., (1995) *Blowup for Nonlinear Hyperbolic equations,* Birkhäuser, Boston.

BALABANE, M., (1986) Ondes progressives et résultats d'explosion pour des systèmes non linéaires du premier ordre, *C. R. Acad. Sci. Paris, 302*: 211–214.

BALL, J., (1977) Remarks on blow-up and nonexistence theorems for nonlinear evolution equations, *Quart. J. Math. Oxford, 28*: 473–486.

BEALS, M., (1983) Self-spreading and strength of singularities for solutions of semilinear wave equations, *Ann. Math., 118*: 187–214.

BEALS, M., (1985) Propagation of smoothness for nonlinear second-order strictly hyperbolic equations, *Proc. Symp. Pure Math., 43*: 21–44.

BEALS, M., (1989) *Propagation and Interaction of Singularities in Nonlinear Hyperbolic Problems,* Birkhäuser, Boston.

BONY, J.-M., (1978) Localisation et propagation des singularités pour les équations non linéaires, Journées EDP, Saint-Jean-de-Monts, exposé numero 17.

BONY, J.-M., (1980) Calcul symbolique et propagation des singularités pour les équations aux dérivées partielles non linéaires, *Ann. Sci. ENS Paris, 4ème série, 14*: 209–246.

BONY, J.-M., (1990) Analyse microlocale des équations aux dérivées partielles non linéaires, in "Microlocal Analysis and Applications," L. Cattabriga and L. Rodino eds., *Springer Lect. Notes in Math., 1495*: 1–45.

CAFFARELLI, L. A. AND FRIEDMAN, A., (1985) Differentiability of the blow-up curve for one-dimensional wave equations, *Arch. Rat. Mech. Anal., 91*: 83–98.

CAFFARELLI, L. A. AND FRIEDMAN, A., (1986) The blow-up boundary for nonlinear wave equations, *Trans. AMS, 297*: 223–241.

CAFLISCH, R. E., ERCOLANI, N., HOU, T. Y. AND LANDIS, Y., (1993) Multi-valued solutions and branch point singularities for nonlinear hyperbolic or elliptic systems, *Comm. Pure Appl. Math., 46*: 453–499.

CHEMIN, J.-Y., (1988) Interaction contrôlée dans les équations aux dérivés partielles non linéaires, *Bull. SMF 116*: 341–383.

CONWAY, E. D. (1977) The formation and decay of shocks for a conservation law in several dimensions, *Arch. Rat. Mech. Anal., 64, no. 1*: 47–57.

GEL'FAND, I. M., (1963) Some problems in the theory of quasilinear equations, *AMS Transl., 29*: 295–381.

GLASSEY, R., (1973) Blow-up theorems for nonlinear wave equations, *Math. Z., 132*: 183–203.

GLASSEY, R., (1977) On the blowing-up of solutions to the Cauchy problem for nonlinear Schrödinger equations, *J. Math. Phys., 18*: 1794–1797.

GLASSEY, R., (1981) Finite-time blow-up for solutions of nonlinear wave equations, *Math. Z., 177*: 323–340.

HANOUZET, B., AND JOLY, J.-L., (1985a) *C. R. Acad. Sci. Paris, 301 (10)*: 491–494.

HANOUZET, B., AND JOLY, J.-L., (1985b) Explosion pour des problèmes hyperboliques semilinéaires avec second membre non compatible, *C. R. Acad. Sci. Paris, 301*: 581–584.

HOFF, D. (1983) The sharp form of Oleinik's entropy condition in several space variables, *Trans. AMS, 276, no. 2*: 707–714.

HOFF, D. (1981) Locally Lipschitz solutions of a single conservation law in several space variables, *J. Diff. Eq., 42, no. 2*: 215–233.

HÖRMANDER, L., (1986) The lifespan of classical solutions of non-linear hyperbolic equations, *Springer Lect. Notes in Math., 1256*: 214–280.

JOHN, F. (1979) Blow-up of solutions of nonlinear wave equations in three space dimensions, *Manus. Math. 18*: 235–268.

JOHN, F., (1987) Existence for large times of strict solutions of nonlinear wave equations in three space dimensions for small data, *Comm. Pure Appl. Math., 40*: 79–109.

JOHN, F., (1990) *Nonlinear wave equations, formation of singularities*, University Lecture Series, vol. 2, AMS.

KATO, T., (1980) Blow-up of solutions of some nonlinear hyperbolic equations, *Comm. Pure Appl. Math., 33*: 501–505.

KELLER, J., (1957) On solutions of nonlinear wave equations, *Comm. Pure Appl. Math., 10*: 523–530.

KICHENASSAMY, S., (1995a) Fuchsian equations in Sobolev spaces and blow-up, *J. Diff. Eq.,* to appear.

KICHENASSAMY, S., (1995b) The blow-up problem for exponential nonlinearities, submitted.

KICHENASSAMY, S. AND LITTMAN, W., (1993a) Blow-up Surfaces for Non-linear Wave Equations, Part I, *Comm. in P. D. E., 18*: (3&4) 431–452.

KICHENASSAMY, S. AND LITTMAN, W., (1993b) Blow-up Surfaces for Non-linear Wave Equations, Part II, *Comm. in P. D. E., 18*: (11) 1869–1899.

KICHENASSAMY, S., KUMAR, A., OLVER, P., A. TANNENBAUM AND A. YEZZI, (1995) Conformal curvature flows: from phase transitions to computer vision, *Arch. Rat. Mech. Anal.,* to appear.

KLAINERMAN, S., AND MAJDA, A., (1980) Formation of singularities for wave equations including the nonlinear vibrating string, *Comm. Pure Appl. Math., 33*: 241–263.

KRUZHKOV, S. N. (1970) First order quasilinear equations in several independent variables, *Math. of the USSR — Sbornik, 10*: 217–242.

LASCAR B., (1978) Singularités des solutions d'équations aux dérivées partielles non linéaires, *C. R. Acad. Sci. Paris, Série A, 287*: 527–529.

LAX, P. D. (1973a) The formation and decay of shock waves, *Amer. Math. Monthly, 79, no. 3*: 227–241.

LAX, P. D. (1973b) *Hyperbolic systems of conservation laws and the mathematical theory of shock waves,* Regional Conference Series in Applied Math., SIAM.

LE MESURIER, B., PAPANICOLAOU, G., SULEM, C., AND SULEM, P.-L., (1987) The focusing singularity of the nonlinear Schrödinger equation, in *Directions in PDE,* Academic Press, 159–201 (see also their articles in *Physica D* (1988).

LEVINE, H, (1974) Instability and nonexistence of global solutions to nonlinear wave equations of the form $Pu_{tt} = -Au + F(u)$, *Trans. AMS* **192** 1–21.

LEVINE, H., (1990) The role of critical exponents in blowup theorems, *SIAM Review, 32*: 262–288.

LINDBLAD, H., (1990a) Blow up for solutions of $\Box u = |u|^p$ with small data, *Comm. PDE, 15 (6)*: 757–821.

LINDBLAD, H., (1990b) On the lifespan of solutions of nonlinear wave equations with small initial data, *Comm. Pure Appl. Math, 43*: 445–472.

LIU, T.-P., (1979) Development of singularities in the nonlinear waves for quasi-linear hyperbolic partial differential equations, *J. Diff. Eq., 33*: 92–111.

MAJDA, A., (1983a) The stability of multi-dimensional shock fronts—a new problem for linear hyperbolic equations, *Memoirs of the AMS, 275.*

MAJDA, A., (1983b) The existence of multi-dimensional shock fronts, *Memoirs of the AMS, 281.*

MAJDA, A., (1984) *Compressible Fluid Flow and Systems of Conservation Laws in Several Space Variables,* Applied Math. Sci., 53, Springer.

MELROSE, R., AND RITTER, N., (1985) Interaction of nonlinear progressive waves, Ann. Math., (2) 121: 187–213.

OLEINIK, O. A. (1963) Discontinuous solutions of nonlinear differential equations, *AMS Transl. (2) 26*: 95–172.

MAJDA, A., (1991) Vorticity, turbulence, and acoustics in fluid flow, *SIAM Review, 33*: 349–388.

RAUCH, J., (1979) Singularities of solutions to semilinear wave equations, *J. Math. Pures et Appl., 58*: 299–308.

RAUCH, J., AND REED, M., (1980) Propagation of singularities for semilinear hyperbolic equations, *Ann. Math., 111*: 531–552.

RAUCH, J, AND REED, M., (1982a) Nonlinear microlocal analysis of semilinear problems in one space dimension, *Duke Math. J. 49,*: 397–475.

RAUCH, J, AND REED, M., (1982b) Singularities produced by the interaction of three progressive waves; examples, *Comm. PDE 7*: 1117-1133.

RAUCH, J, AND REED, M., (1988) Classical conormal solutions of semilinear systems, *Comm PDE, 13,*: 1297–1335.

SCHAEFFER, J., (1985) The equation $u_{tt} - \Delta u = |u|^p$ for the critical value of p, *Proc. Roy. Soc. Edinburgh, 101A*: 31–44.

SCHAEFFER, J., (1986) Finite-time blow-up for $u_{tt} - \Delta u = H(u_r, u_t)$ in two space dimensions, *Comm. PDE, 11 (5)*: 513–543.

SIDERIS, T., (1984) Nonexistence of global solutions to semilinear wave equations in high dimensions, *J. Diff. Eq., 52*: 378–406.

SMOLLER, J., (1983) *Shock Waves and Reaction-Diffusion Equations,* Grundlehren der Math. Wiss., Springer.

STRAUSS, W., (1989) *Nonlinear Wave Equations,* CBMS lecture notes, *73,* Amer. Math. Soc., Providence, RI.

TAYLOR, M., (1991) *Pseudodifferential Operators and Nonlinear P. D. E.,* Progress in Mathematics **100** Birkhäuser, Boston.

VOLPERT, A. I. (1967) The spaces BV and quasilinear equations, *Math. of the USSR — Sbornik, 2*: 225–276, [original Russian in *Mat. Sbornik, 73 (115)*].

Chapter 4

Solitons and Inverse Scattering

Solitons are solutions of nonlinear wave equations which possess two permanence properties: (i) they evolve without change of form over large distances; (ii) they "interact" with other solutions in a robust fashion, as if they each represented a localized particle-like entity. The first permanence property is found in traveling waves. It is this permanence that was observed in 1834 by John Scott Russell in a famous observation of a wave of elevation in the Edinburgh canal. The existence of traveling waves can be a taxing problem by itself; they do, however, exist in a wide variety of settings. Solitons on the other hand exist, as far as we know, only in some very particular equations such as

$$u_t + 6uu_x + u_{xxx} = 0 \quad \text{Korteweg-de Vries (KdV)},$$
$$iu_t + u_{xx} \pm u|u|^2 = 0 \quad \text{Nonlinear Schrödinger (NLS)},$$
$$u_{tt} - u_{xx} + \sin u = 0 \quad \text{Sine-Gordon (SG)}.$$

There are several families of equations having soliton solutions, but none of these families is stable under, say, polynomial perturbations.

The importance of solitons in today's literature may therefore cause some surprise at first. It is explained by the *universality* of soliton equations as models for weakly nonlinear wavetrains. In this sense, very general nonlinear wave equations have particular regimes in which their long time behavior is modeled by an equation which has solitons. This modeling process, sometimes called the *reductive perturbation method*, may be compared to a linearization, solitons playing the role of exponential solutions.

135

Once the usefulness of a few model equations has been ascertained, the next step is to give a precise mathematical meaning to the permanence properties of the solutions. We describe the procedure in some detail on the Korteweg-de Vries equation on the real line, not only because it is the first and most famous equation having solitons, but also because it is the one on which the greatest amount of rigorous results are available. For each value of t, one associates to the solution $u(x, t)$ of the KdV equation an operator $L(t)$. To describe a solution of KdV, we need (a) the list of point eigenvalues $\lambda_1, \ldots, \lambda_N$ of $L(0)$; (b) the values of the L^2 norms of special eigenfunctions normalized by their decay properties; (c) the "reflection coefficient" defined in §4.3. One then finds that for large times, the solution behaves like a superposition of a solution of the linear equation $u_t + u_{xxx} = 0$, and N traveling waves which are the solutions with *one eigenvalue* and *vanishing reflection coefficient*. The latter are the one-solitons.

It is widely expected that such a picture is, *mutatis mutandis*, correct for all equations having solitons, but it is not well-established except in isolated instances. Its importance stems from the fact that it is essentially the only sophisticated model for the balance between dispersion and nonlinearity for PDE in an infinite-dimensional context; it plays the role of exact solutions.[1] This is why KdV and its kin are labeled "exactly solvable" or "integrable" models.[2]

While the spectral approach, subsumed under the term "inverse scattering transform," dominates our understanding of solitons, there are other properties of equations possessing them, which make the subject still more intriguing. Existence of infinitely many conserved quantities of a simple form, an algebraic formula for the "addition of one eigenvalue" (the Bäcklund transformation), existence of meromorphic solutions, or solutions expressible in terms of Painlevé transcendents, or even families of rational solutions, complete integrability in the sense of (infinite-dimensional) Hamiltonian systems, existence of more than one Hamiltonian structure, of higher symmetries: several or all of the above have been variously taken as a definition of a "soliton equation." This state of affairs makes it desirable to have a systematic procedure to identify equation liable to be dubbed "integrable." This problem has several candidates for its answer. It is perhaps noteworthy that it forces us to forego for a while a natural tendency to dismiss algebraic properties of PDE as irrelevant to the description of "real" phenomena, just as the study of singularities leads us to investigate further the properties of analytic solutions before we can tackle less regular solutions.

After discussing in §4.1 the universality of soliton equations, we turn to the general framework of isospectral deformations (§4.2) which relates some linear eigenvalue problems to nonlinear evolution equations. §4.3 then works out the

---

[1] Finite-dimensional reductions give considerable insight. Remarkable success has been achieved for water waves and other problems from hydrodynamics. It seems however that such methods are not (yet?) applicable to hyperbolic equations without dissipation.

[2] The general solution of KdV is *not* given in closed form in the elementary sense of the term.

numerous details for the KdV case. The parallel details for the NLS and SG case are outlined in §4.4, while a more recent approach to the reconstruction procedure is given in §4.5. Finally some of the major criteria for complete integrability are discussed in §4.6.

4.1 UNIVERSAL EQUATIONS

After introducing some terminology on instabilities, we discuss the universality of five important model equations.

The NLS equation describes the modulation of nearly monochromatic wavetrains.

The KdV equation describes wavetrains which are weakly dispersive in the limit of long waves, and is replaced, for weakly dispersive wavetrains, by Burgers' equation.

The Ginzburg-Landau, arises in modeling dissipative instabilities, while the sine-Gordon is a special case of a model for wavetrains close to dispersive instability.

The three-wave interaction models the resonant interaction of three waves, and has several generalizations to the case of more than three resonant wavenumbers.

A linear wave-train is a superposition of exponential solutions $\exp[i(kx - \omega t)]$ of a PDE, where k and ω are related by the dispersion relation, which can be computed formally even if the initial-value problem is not well-posed.

We limit ourselves, to fix ideas, to the case of one space dimension, and introduce some classification which is standard in many applications.

Let us consider one branch

$$\omega = \omega(k) = \omega_R(k) + i\omega_I(k)$$

of the dispersion relation. Hyperbolicity would require $\omega_I = 0$ for all branches. It is however possible to solve the forward initial-value problem in cases when $\omega_I < 0$, since these exponentials would then decay as t increases from zero, the heat equation being an example. Superpositions of exponentials are called *dissipative waves* if $\omega_I > 0$, and *dispersive waves* if $\omega_I = 0$; the solutions of the wave equation are in the latter case. If $\omega_I > 0$ for some k, one speaks of an *instability* on account of the exponential growth of the associated exponential solution. Wave packets are defined as arising from clusters of wave numbers close to a fixed value k_0. Finally, the long wave and short wave limits correspond to k tending to zero or infinity respectively.

For nonlinear problems, the same classification can be made relative to the linear part, but if one scales dependent and independent variables suitably, it becomes possible to define regimes where linear and nonlinear terms are equally important. These are the situations of interest here. Despite the variety of model equations, the procedure used to derive them is essentially the same: it consists in introducing new variables involving one or more scale parameters, and expanding in powers of these parameters ("multiple-scales algorithm.") We therefore work out the procedure in more detail for the first example, the others being similar.

Nearly monochromatic wavetrains.

The first and best-known universal equation is probably the nonlinear Schrödinger equation, which will be derived as governing the amplitude of a nearly monochromatic wave packet. We have already seen that the linear Schrödinger equation is obtained in the linear analogue of this situation, and this motivates the line of attack.

Let $U = \begin{pmatrix} u_1 \\ \vdots \\ u_m \end{pmatrix}$ solve an $m \times m$ system

$$(4.1) \qquad\qquad U_t + A(U)U_x + B(U) = 0$$

where A and B are smooth, real functions of U.

We are interested in a formal series solution representing a nearly monochromatic wavetrain superimposed on a constant solution. Let therefore U_0 satisfy $B(U_0) = 0$. The values of A and its higher differentials at U_0 are written A_0, A_0', ..., and similarly for B. We have seen in Ch. 1 that the modulation of a linear wave packet obeys the linear Schrödinger equation, and it is therefore appropriate to introduce the new variables

$$\xi = \varepsilon(x - \lambda t); \quad \tau = \varepsilon^2 t,$$

where ε will be the expansion parameter and λ will be found later to be a group velocity. The plane wave we wish to perturb will be assumed to be real, and to have the form

$$U_0 + V e^{i(kx - \omega t)} + \text{c.c.}$$

with $k = k_0$, c.c. denoting the complex conjugate. We require that this expression solve the linearization of our equation at U_0, so that

$$[i(kA_0 - \omega) + B_0']V = 0.$$

To get a non-trivial solution, we require

$$(4.2) \qquad\qquad \det[i(kA_0 - \omega) + B_0'] = 0,$$

which is the *dispersion relation* for our problem. We choose one solution branch $\omega = \omega(k)$, and will consider values of k close to k_0. We let $W_l = [il(kA_0 - \omega) + B_0]$.

To simplify, we assume that we may solve uniquely for $\omega = \omega(k)$ and that the algebraic multiplicity of zero, as an eigenvalue of W_1, is 1. Also, to prevent resonances, we further assume that

$$(4.3) \qquad \det W_l[il(kA_0 - \omega) + B_0'] \neq 0$$

when the integer l is not ± 1. The necessary properties of the matrix W_1 are summarized in the following theorem, the proof of which is elementary.

THEOREM 4.1 *There are, for k close to k_0, two smooth families of (row, resp. column) vectors, $R_1(k)$ and $L_1(k)$ such that $L_1 W_1 = W_1 R_1 = 0$ and $L_1 R_1 \neq 0$. They are uniquely determined upto scaling.*

1. *$W_1 X = Y$ has a solution if and only if $L_1 Y = 0$.*

2. *$L_1(A_0 - d\omega/dk)R_1 = 0$ and $W_1(dR_1/dk) = -i(A_0 - \omega'(k))R_1$.*

3. *$L_1 R_1 \omega''(k) = L_1(A_0 - d\omega/dk)dR_1/dk$.*

We now turn to our basic problem: consider a formal solution of the form

$$(4.4) \qquad U = \sum_{j=0}^{\infty} U_j(\xi, \tau, \theta)\varepsilon^j,$$

where the U_j are assumed to be real and *periodic* in $\theta := k_0 x - \omega(k_0)t$ with period 2π (like the plane waves we want to perturb). Before substituting into the equation, we first *view U as a function of three independent variables ξ, τ and θ*. This may seem paradoxical, since these are definitely not independent. However, they do correspond to different scales.[3] The chain rule immediately gives

$$(4.5) \qquad (k_0 A - \omega(k_0))U_\theta + B + \varepsilon(A - \lambda)U_\xi + \varepsilon^2 U_\tau = 0.$$

We now substitute the formal expansion and identify like powers of ε.

At order 0, we simply obtain $B(U_0) = 0$, which holds thanks to the choice of U_0.

At order 1, we find that since we have chosen $\omega = \omega(k_0)$ to satisfy the dispersion relation, there are nonzero solutions for the equation in U_1:

$$(4.6) \qquad U_1 = \phi_1(\xi, \tau)R_1 + \text{c.c.},$$

where ϕ_1 is, at this stage, arbitrary.

The universality of NLS is expressed by

---

[3]This is a convenient and systematic way to describe "multiple-scales algorithms."

THEOREM 4.2 *The equations at order 2 and 3 imply that*

$$i\phi_{1\tau} + \frac{1}{2}\omega''\phi_{1\xi\xi} + Q\phi_1|\phi_1|^2 = 0$$

for some complex constant Q depending on A and B.

Idea of Proof: The proof is a straightforward, but somewhat lengthy, calculation. We provide a few intermediate steps for the convenience of the reader. Th. 4.1 is used repeatedly.

Note first that if we write, as we may, $U_j = \sum_l U_{jl}e^{il\theta}$, the recurrence relation for the U_j has the form

$$W_l U_{jl} = F_{jl}(U_0, \ldots, U_{j-1}).$$

Therefore, we have two pieces of information at order j: (1) U_{jl} can be determined from the preceding terms if $l \neq \pm 1$; (2) we need $L_1 F_{j1} = 0$ (and the conjugate relation for $l = -1$) as a solvability condition for U_{l1}; U_{l1} is then determined upto an additive term $\phi_l R_1$. This condition translates into a constraint on ϕ_{l-2}.

In particular, for $l = 2$, we find that all the U_{2l} are determined but for $U_{2\pm1}$:

$$U_2 = (-i\phi_1\xi\frac{dR_1}{dk} + \phi_2(\xi,\tau)R_1)e^{i\theta} + \text{c.c.} + \text{other harmonics},$$

where ϕ_2 is, at this point, left arbitrary. The solvability condition for U_{21} takes the form $L_1(A_0 - \lambda)R_1 = 0$, which forces $\lambda = \omega'(k)$.

For $l = 3$, the solvability condition for U_{31} reduces, after a careful calculation, to the NLS equation for ϕ_1, as announced; it is remarkable that the nonlinearities of the equation *all* generate terms proportional to $|\phi_1|^2\phi_1$.

Thus, in a coordinate system moving at the group velocity, the wave packet appears, in this approximation, as a plane wave with an envelope satisfying a nonlinear Schrödinger equation. Since the solution can contain 'humps' which propagate without attenuation, one will have the impression that a monochromatic wavetrain develops an instability whereby it breaks up into a series of pulses.

Long waves.

We are now interested in the limit of long waves (small wave numbers). We will again seek a 'far-field pattern' which does not vary much if the variation of x and t is of order unity. The scaling will, however, be quite different. In more precise terms, we need

$$U = \sum_{j\geq0} U_j(\xi,\tau)\varepsilon^j$$

where

(4.7)
$$\xi = \varepsilon^a(x - \lambda t), \quad \tau = \varepsilon^b t.$$

This change of variables is called a *Gardner-Morikawa transformation* if $b = a + 1$. We have so far only considered the case $a = b - 1 = 1$. It turns out that the case $a = b - 1 = 1/(p-1)$ is associated with the KdV-type equation

$$\phi_\tau + \alpha\phi\phi_\xi + \beta\partial_\xi^p\phi = 0,$$

while $a = b/3 = r/2$ leads to

$$\phi_\tau + \alpha\phi^r\phi_\xi + \beta\partial_\xi^3\phi = 0.$$

We recognize $a = 1/2$, $b = 3/2$ as leading to KdV, and $a = b/3 = 1$ as leading to the modified KdV (mKdV) equation.

The calculation, being lengthy and quite similar to the derivation of the nonlinear Schrödinger equation, is omitted.

Three-wave interaction.

We are now interested in the perturbation of the superposition of three plane waves. It turns out that an interesting result appears in the case when they satisfy a resonance condition.

It is natural to consider now a *multi-phase expansion*:

$$U = \sum_{j=0}^{\infty} U_j(\xi, \tau, \theta_1, \theta_2, \theta_3)\varepsilon^j,$$

where $\theta_q = k_q x - \omega_q t$, $\xi = \varepsilon x$, and $\tau = \varepsilon t$, $q = 1, 2, 3$. We require periodicity in the θ_j, so that there is a Fourier expansion

(4.8)
$$U = \sum_{k \geq 0} U_k \varepsilon^k = \sum_{k \geq 0} \varepsilon^k \sum_{q_1, q_2, q_3 = -\infty}^{\infty} U_{k, q_1 q_2 q_3} e^{i(q_1\theta_1 + q_2\theta_2 + q_3\theta_3)}.$$

The equation becomes

(4.9)
$$\sum_{q=1}^{3} i(k_q A - \omega_q)\frac{\partial U}{\partial \theta_q} + B(U) + \varepsilon(\partial_\tau + A(U)\partial_\xi)U = 0.$$

For $\varepsilon = 0$, we take $U = U_0$ to be constant, as before.

At first order, we take the superposition of three plane waves: Let $W^{(q)} = i(k_q A_0 - \omega_q) + B_0'$. We assume that there are three wave numbers such that $k_1 + k_2 + k_3 = 0$, *and* $\omega(k_1) + \omega(k_2) + \omega(k_3) = 0$ (resonance condition).

At first order, we must have $\det W^{(q)} = 0$, and we again assume that the associated eigenvectors are uniquely determined upto scaling: $W^{(q)} R_q = L_q W^{(q)} = 0$. We then take

$$U_1 = \sum_{q=1}^{3} \phi_q(\xi, \tau) e^{i\theta_q} + \text{c.c.},$$

expressing the fact that we are interested in the modulation of a system of three pure plane waves. We seek an equation for the ϕ_q.

At second order, we see that because of the resonance condition, terms in $e^{i\theta_1}$ arise from second order terms $e^{-i\theta_2} e^{-i\theta_3}$. Therefore, if we multiply the q equation by L_q, we find:

$$(\partial_\tau + \frac{(L_q A_0 R_q)}{L_q R_q} \partial_\xi)\phi_q = \sum c_{qlm} \phi_l \phi_m,$$

where the summation extends over those l and m such that $k_q + k_l + k_m = 0$ and $\omega_q + \omega_l + \omega_m = 0$. Since $L_q(A_0 - \omega'(k_q))R_q = 0$, we finally obtain

$$(4.10) \qquad\qquad (\partial_\tau + \omega'(k_q)\partial_\xi)\phi_q \;=\; c_{qlm}\phi_l\phi_m$$

$$(4.11) \qquad\qquad (\partial_\tau + \omega'(k_l)\partial_\xi)\phi_l \;=\; c'_{qlm}\phi_m\phi_q$$

$$(4.12) \qquad\qquad (\partial_\tau + \omega'(k_m)\partial_\xi)\phi_m \;=\; c_{qlm}{}''\phi_q\phi_l.$$

This is the general form of the three wave interaction system. Note that this system is non-dispersive in the sense that the wave speeds are constants.

Instabilities.

Instabilities were defined at the beginning of §4.1. Consider again a system with a linearized dispersion relation $\omega(k, R)$, where R is a parameter. Instabilities can occur if ω_I becomes positive as R crosses a critical value R_c. This can happen in two very different ways: (1) ω_I changes sign from $-$ to $+$; (2) $\omega = \omega_1(k, R) \pm \sqrt{D(k, R)}$, where D changes sign from $-$ to $+$. These are termed *dissipative* and *dispersive* instabilities respectively.

Dissipative instabilities lead to the *Ginzburg-Landau* equation

$$(4.13) \qquad\qquad A_\tau = \pm\alpha A - \beta A|A|^2 + \gamma A_{\xi\xi},$$

where $\tau = \varepsilon^2 t$ and $\xi = \varepsilon(x - \lambda t)$, where λ is a constant, and A is, as in the NLS case, a complex amplitude. The reader is referred to the notes for details on the derivation of the Ginzburg-Landau equation.

Dispersive instabilities lead to a system which contains the sine-Gordon equation as a special case: There are now two amplitude functions, which satisfy (after a suitable scaling)

$$(4.14) \qquad\qquad (\partial_t + c_1\partial_x)(\partial_t + c_2\partial_x)A \;=\; \pm\alpha A - \beta AB$$

$$(4.15) \qquad\qquad (\partial_t + c_2\partial_x)B \;=\; (\partial_t + c_1\partial_x)|A|^2.$$

If $A = (2\beta)^{-1/2}(\partial_t + c_2\partial_x)\phi$ and $B = \pm(\alpha/\beta)(1 - \cos\phi)$, we recover a form of the SG equation:

$$(\partial_t + c_1\partial_x)(\partial_t + c_2\partial_x)\phi = \pm\alpha\sin\phi.$$

For details, see again the notes.

4.2 ISOSPECTRALITY

> The interpretation of a class of nonlinear evolution equations as isospectral deformations of linear eigenvalue problems is introduced, leading to the definition of Lax pairs, and, more generally, of commutator representations.

The first property of equations having solitons is that they can be written in a "commutator representation," so that they formally express that a certain eigenvalue problem is deformed in time in a simple way. The simplest situation of this kind is given in the following theorem, the proof of which is straightforward.

THEOREM 4.3 *Let $L(t)$, $A(t)$, $S(t)$ be three matrix-valued smooth functions of t. Assume $dS/dt = AS$ and $S(t) = I$. The following are then equivalent*

1. $L_t + [L, A] = 0$;

2. $L(t) = S(t)L(0)S(t)^{-1}$ for all t.

A related result is familiar in quantum mechanics, where A is a skew-adjoint operator, and $S(t)$ the group of unitary transformations it generates. It is this infinite-dimensional version which is relevant to the Korteweg-de Vries, and other PDE.

Since the operators $L(t)$ are unitarily equivalent, their eigenfunctions are mapped onto one another by $S(t)$, and their eigenvalues are independent of time. There should therefore be a large number of naturally defined *conserved quantities*.

More precisely, if $L(0)\psi_0 = \lambda\psi$, and $\psi(t) = S(t)\psi(0)$, so that

$$\psi_t = A\psi; \quad \psi(0) = \psi_0,$$

we have $L(t)\psi(t) = \lambda\psi(t)$ and we may write $d\lambda/dt = 0$.

A straightforward calculation shows that conversely, the equations $L\psi = \lambda\psi$, $\lambda_t = 0$, $\psi_t = A\psi$, imply

$$(L_t + [L, A])\psi = 0.$$

REMARK: Note however that one can construct matrices L and A such that $(L_t + [L, A])\psi = 0$ holds whenever ψ is an eigenvector of L, but such that the $L(t)$ are not similar, although they have the same (constant) eigenvalues.

A basic observation is that if A is a function of the entries of L, then the *Lax equation*

$$(4.16) \qquad\qquad L_t + [L, A] = 0$$

is a nonlinear evolution equation for the entries of L. In the infinite-dimensional case, L, A are operators, the coefficients of A are expressed in terms of the coefficients of L, and (4.13) is a PDE for these coefficients. If an equation can be cast in the form (4.13), one says that (L, A) is a *Lax pair*[4] for this problem.

Two basic examples are

EXAMPLE 1: THE TODA LATTICE. Take L and A to be $n \times n$ matrices, with L symmetric and A antisymmetric, $L_{ii} = b_i$, $L_{ij} = A_{ij} = a_i$ if $i = j + 1$, $L_{1n} = -A_{1n} = a_n$, and $L_{ij} = A_{ij} = 0$ if $|i - j| \neq 0, 1, n$. One then takes $u = (a_1 \ldots, a_n, b_1 \ldots, b_n)$, and obtains the nonlinear evolution

$$(4.17) \qquad\qquad \dot{b}_k = 2(a_k^2 - a_{k-1}^2) \qquad \dot{a}_k = a_k(b_{k+1} - b_k),$$

all indices being taken modulo n.

EXAMPLE 2: THE KDV EQUATION. We take $L = -\partial_x^2 + u(x, t)$ and take A to be a skew-adjoint operator of odd order $2m + 1$. Since we hope that the Lax equation is now an evolution equation for u, we determine the coefficients of A recursively by requiring that the operator $L_t + [L, A]$ be a multiplication operator; we therefore set to zero the coefficients of $\partial_x, \partial_x^2, \ldots$ One finds, upto a constant: for $m = 1$, $a = \partial_x$, for $m = 2$, $A = 4\partial_x^3 + (u\partial_x + \partial_x \circ u)$, and so on. It is customary to write ∂ for ∂_x, and to omit the $\circ$, so that $(\partial u)\psi$ equals $(u\psi)_x$ and *not* $u_x\psi$.

The evolution equations corresponding to these choices are

$$u_t = u_x$$

for $m = 1$,

$$u_t - 6uu_x + u_{xxx} = 0$$

for m=2, and so on. This sequence is called the *KdV hierarchy*; two explicit forms of this hierarchy are given in exercise 2.

To solve it, one may try to find $S(t)$ by solving $S_t = AS$ with $S(0) = I$, but this is generally awkward. However, in the infinite-dimensional case, *scattering theory* enables one to recover the coefficients of L using only asymptotic information on the eigenfunctions rather than the knowledge of the eigenfunctions themselves; this great simplification makes the theory effective. For this reason, one can make use of a Lax pair to generate special solutions without proving the solvability of the initial-value problem, or that the $L(t)$ are conjugate to $L(0)$.

---

[4]or *L-A* pair, in the Russian literature

An important generalization of the Lax pair formalism consists in introducing two equations of the form

$$\psi_x = X[u, x, t, \lambda]\psi, \quad \psi_t = T[u, x, t, \lambda]\psi,$$

where X and T are differential operators with coefficients depending on u and its derivatives, and λ is an "eigenvalue parameter." ψ and u may have several components. One then considers the compatibility condition

$$(4.18) \qquad\qquad [\partial_x - X, \partial_t - T] \equiv 0,$$

which reduces, in many interesting cases (see §4.3) to a nonlinear evolution equation for u. Eq. (4.18) is sometimes called a zero curvature representation, because on can interpret $\partial_x - X$ and $\partial_t - T$ as covariant differentiation operators.

4.3   THE KORTEWEG-DE VRIES EQUATION

We describe the solution of the KdV equation by inverse scattering. After showing the relation between the KdV equation and the one-dimensional Schrödinger equation, direct and inverse scattering for this equation are described, leading to a solution procedure for the KdV equation, and the construction of the multi-solitons.

We consider the solution of the initial-value problem for the KdV equation

$$(4.19) \qquad\qquad u_t - 6uu_x + u_{xxx} = 0$$

where $u(x, 0) = u_0(x) \in \mathcal{S}(\mathbf{R})$. We already know that the solution exists and remains in Schwartz space for all time. We are interested in a more precise description of the solution.

The starting-point is the Lax representation of (4.19), the origin of which is discussed in §4.7.

THEOREM 4.4 *For any constant c, the equations*

$$\begin{aligned}
L\psi := (-\partial_x^2 + u(x, t))\psi(x, t) &= \lambda\psi; \\
\partial\lambda/\partial t &= 0; \\
\psi_t = A\psi := -(4\partial_x^3 - 3(u\partial_x\psi + \partial_x(u\psi))) + c\psi &= 0
\end{aligned}$$

imply $(u_t - 6uu_x + u_{xxx})\psi = 0$.

The result follows by equating the two expressions of ψ_{xt} which are implied by the assumptions. The pair (L, A) is called a *Lax pair* for the KdV equation. In fact, the operator

$$L_t + [L, A]$$

is the operator of multiplication by the function $u_t - 6uu_x + u_{xxx}$.

We wish to apply this property to obtain information on the eigenfunctions of the operator L at time t from their properties at time $t = 0$. To this end, we must digress into the properties of these eigenfunctions and explain how information on L translates into information on u. This is the purpose of *inverse scattering*.

Eigenfunctions of L.

Since we are assuming that the potential u tends rapidly to zero at infinity, the essential spectrum of L is $[0, +\infty)$ and there are finitely many negative, simple eigenvalues. We are in the "limit-point case." (see Coddington and Levinson for basic material on Sturm-Liouville theory). Let us denote the eigenvalues by $\lambda_i = -\eta_i^2$, with $\eta_i^2 > 0$ and $1 \leq i \leq N$. There may of course be no point eigenvalue at all. It is also an elementary fact that the point eigenfunctions decay like $\exp(-\eta_i|x|)$ at infinity.

Corresponding to the essential spectrum, we now construct two families of eigenfunctions:

THEOREM 4.5 *There are two families of functions $\phi(x, k)$ and $\psi(x, k)$ defined for $Im\, k \geq 0$ which satisfy the following properties:*

1. $L\phi = k^2\phi$ *and* $L\psi = k^2\psi$;

2. *For every x, ϕ and ψ are continuous for $Im\, k \geq 0$ and are analytic in the open half-plane;*

3. $\Phi(x, k) = \phi(x, k)\exp(ikx)$ *(resp. $\Psi(x, k) = \psi(x, k)\exp(-ikx)$) is bounded for fixed x and tends to one as x tends to $+\infty$ (resp. $-\infty$.)*

These eigenfunctions generalize $\exp(\pm ikx)$. The reality of the potential u implies that $\psi(-k) = \bar\psi(k)$ for real k.

Proof: From the formula of variations of parameters, it immediately follows that Φ solves

$$(4.20) \qquad \Phi = 1 + \frac{1}{2ik}\int_{-\infty}^{x}[e^{2ik(x-y)} - 1]u(y)\Phi(y)\,dy.$$

This is an integral equation the kernel of which is dominated by $C(x - y)|u(y)|$ if k has nonnegative real part. Therefore, we can solve it by iterating in the space of continuous functions of x on some interval $[x_0, +\infty)$. If we start the iteration with $u = 0$, we obtain this way a uniformly convergent sequence of analytic functions of k, continuous in the upper half-plane. The existence part (1)-(2) is therefore proved. (3) now follows from the integral equation. Further properties are given in exercise 3. There are parallel properties for Ψ.

It is sometimes useful to have global bounds on the eigenfunctions. In this direction, we define

$$\sigma(x) = \int_x^\infty |u(t)| \, dt$$

and

$$\sigma_1(x) = \int_x^\infty (t-x)|u(t)| \, dt,$$

and we prove

THEOREM 4.6 *We have* $|\Psi(x,k) - 1| \le \sigma_1(x)e^{\sigma_1(x)}$.

THEOREM 4.7 *We have*

$$|\Psi(x,k) - 1| \le C\frac{1 + \max(0,-x)}{1 + |k|}.$$

A similar bound on Φ holds.

Proof of Th. 4.6: We begin with a Lemma.

Lemma. Let $Tf = \int_x^\infty CV(x,t)f(t) \, dt$, where $V \le V_0(x,t)$, and V_0 is non-increasing in x. Let $\rho(x) = \int_x^\infty V_0(x,t) \, dt$. Then

$$|T^n f(x)| \le \|f\|_\infty C^n \rho(x)^n/n!.$$

Note that a similar result trivially holds for integral operators involving integrals on $(-\infty, x)$.

Proof of Lemma: We have, after applying T successively n times,

$$
\begin{aligned}
|T^n f(x)| &\le C^n \|f\|_\infty \int_{x \le x_n \le \cdots \le x_1} \\
&\quad V_0(x,x_n)V_0(x_n,x_{n-1})\ldots V_0(x_2,x_1)\prod_i dx^i \\
&= C^n \|f\|_\infty \prod_j \int_{x \le x_j} V_0(x,x_j) \, dx^j \\
&= C^n \|f\|_\infty \rho(x)^n/n!.
\end{aligned}
$$

Here, we used the fact that (1) $V_0(x_{j+1}, x_j) \le V_0(x, x_j)$, and (2) the "octant" $\{x_1 \ge 0, \ldots, x_n \ge 0\}$ is the union of the $n!$ wedges obtained from $\{x \le x_n \le \cdots \le x_1\}$ by permutation of the variables.

This completes the proof of the Lemma.

We now use the integral equation for the eigenfunctions, noticing that

$$|[e^{2ik(t-x)} - 1]/2ik| \le (t-x)$$

for $t > x$. The conclusion of Th. 4.6 now follows from the Lemma with $V_0(x, t) = (t - x)|u(t)|$.

Proof of Th. 4.7: We prove that $|\Psi(x, k) - 1| \leq C(1 + \max(0, -x))$ for all (x, k), and $C/(1 + |k|)$ for $|k| \geq 1$. The result follows.

Since $[e^{2ik(t-x)} - 1]/2ik$ is bounded by $1/|k|$, the second estimate follows immediately. Indeed, we have the bound

$$|\Psi(x, k) - 1| \leq \sigma(x)|k|^{-1}e^{(\sigma(x)/|k|)},$$

using the Lemma with $V_0(x, t) = |u(t)|/|k|$. Since σ is bounded, we find the announced result.

For the first estimate, we bound $[e^{2ik(t-x)} - 1]/2ik$ by $(t - x)$, for $t > x$. We use it in two ways. First, using Th. 4.6,

$$
\begin{aligned}
|\Psi(x, k)| &\leq 1 + \int_x^\infty (t - x)|u(t)\Psi(t, k)|\, dt \\
&\leq 1 + \int_0^\infty t|u(t)\Psi(t, k)|\, dt - x\int_x^\infty |u(t)\Psi(t, k)|\, dt \\
&\leq 1 + \sigma_1(0)e^{\sigma_1(0)}\int_0^\infty t|u(t)| - Cx_-\int_x^\infty (1 + |t|)|u(t)|\, dt.
\end{aligned}
$$

This implies

$$|\Psi(x, k)| \leq C + \int_x^\infty (1 + |t|)|u(t)||\Psi(x, k)|/(1 + |t|)\, dt,$$

which, using the Lemma, proves that $|\Psi(x, k)| \leq (1 + |x|)$.

The desired result is slightly stronger. To obtain it, we write

$$
\begin{aligned}
|\Psi(x, k) - 1| &\leq \int_x^\infty (t - x)|u(t)\Psi(t, k)|\, dt \\
&\leq \int_0^\infty t|u(t)\Psi(t, k)|\, dt - x\int_x^\infty |u(t)\Psi(t, k)|\, dt \\
&\leq \sigma_1(0)e^{\sigma_1(0)}\int_0^\infty t|u(t)| - Cx_-\int_x^\infty (1 + |t|)|u(t)|\, dt \\
&\leq C(1 + |x|)\int_x^\infty (1 + |t|)|u(t)|\, dt.
\end{aligned}
$$

for $x < 0$, from which the result follows.

Another useful bound is

THEOREM 4.8 *We have the uniform bound*

$$|\Psi_k| \leq C(1 + x^2).$$

It enables one to prove the continuity of the transmission coefficient near zero. The method of proof is similar, but more involved (see Deift-Trubowitz (1979)).

Scattering data.

For real, nonzero values of k, it is easy to see that $\psi(k)$ and $\bar\psi(k) = \psi(-k)$ are linearly independent. In fact, define the Wronskian of two functions a and b to be

$$W(a, b) = ab_x - a_x b.$$

Then

$$W(\phi, \bar\phi) = 2ik,$$
$$W(\psi, \bar\psi) = -2ik.$$

It follows that there are functions $a(k)$ and $b(k)$ such that

(4.21) $$\phi(x, k) = a(k)\psi(x, -k) + b(k)\psi(x, k).$$

Using the above Wronskian relations, we find that

(4.22) $$W(\phi(k), \psi(k)) = 2ika(k); \quad W(\phi(k), \bar\psi(k)) = -2ikb(k).$$

In particular, a extends as an analytic function to the upper half-plane, and is continuous in its closure. On the other hand, b does not extend off the real axis[5]. We also see that a may have a pole at $k = 0$.

The *reflection* and *transmission* coefficients are defined by

$$R = b/a; \quad T = 1/a.$$

They correspond to the fact that a solution which behaves like a plane wave of frequency k traveling to the left is, as $t \to +\infty$, the sum of a "transmitted" part traveling in the same direction, and a "reflected" part in the opposite direction. Their amplitudes are related to the coefficients R and T; a precise formulation of this interpretation would require consideration of the hyperbolic equation obtained by Fourier transforming the eigenvalue problem with respect to k.

The functions a and b have a number of remarkable properties which we proceed to investigate.

THEOREM 4.9 *We have* $|a|^2 - |b|^2 = 1$ *on the real axis.*

Indeed, $\phi = a\bar\psi + b\psi$, from which we get

$$2ik = W(a\bar\psi + b\psi, \bar a\psi + \bar b\bar\psi) = 2ik(a\bar a - b\bar b).$$

QED

Thus, $|R|^2 + |T|^2 = 1$. It is sometimes useful to introduce the scattering matrix which connects the distinguished sets of eigenfunctions at $+\infty$ and $-\infty$. Th. 4.6 then expresses that *the scattering matrix is unitary*.

The next property shown that R is a nonlinear generalization of the Fourier transform of u.

---

[5]unless u has exponential decay.

THEOREM 4.10 *The functions a and b satisfy the relations*

$$2ik(1 - a(k)) = \int_{-\infty}^{+\infty} u(x)\phi(x, k)e^{ikx}\, dx,$$

$$2ikb(k) = \int_{-\infty}^{+\infty} u(x)\phi(x, k)e^{-ikx}\, dx.$$

This implies $R(0) = -1$ if $\int_{-\infty}^{+\infty} u(x)\phi(x, 0)\, dx \neq 0$. This also shows that, roughly speaking, R is essentially $\mathcal{F}u(2k)/2ik$ if u is "small." Since this function generally blows up for $k = 0$, while R is always bounded, one must be careful when using this idea.

Finally, we relate the poles of the transmission coefficient to the eigenvalues of L.

THEOREM 4.11 *If $Im(k_0) > 0$, then $\phi(k_0) = b\psi(k_0)$ if and only if k_0^2 is an eigenvalue of L. The functions $\phi(k_0)$ and $b\psi(k_0)$ are then normalizable, and we have $a(k_0) = 0$ and*

$$\int_{-\infty}^{+\infty} \psi(x, k_0)^2\, dx = ia'(k_0)/b.$$

Proof: The first part follows from the fact that ϕ and ψ have exponential decay at $-$ and $+\infty$ respectively. For the other statements, we differentiate the relation $W(\phi, \psi) = 2ika(k)$ with respect to k, and use the preceding Th. 4.11:

$$2i(a + ka'(k)) = W(\phi, \psi_k) + W(\phi_k, \psi),$$

where all the eigenfunctions are of course evaluated at some point x. For $k = k_0$, we have

$$W(\phi, \psi_k) = bW(\psi, \psi_k) = 2kb\int_x^{\infty} \psi^2\, dx$$

$$W(\phi_k, \psi) = -(1/b)W(\phi, \phi_k) = (2k/b)\int_{-\infty}^x \phi^2\, dx\, (2kb)\int_{-\infty}^x \psi^2\, dx.$$

Therefore, for $k = k_0$ (where a vanishes), we find

$$2ik_0a'(k_0) = 2k_0b\int_{-\infty}^{+\infty} \psi^2\, dx,$$

QED

Fourier transform of eigenfunctions.

Since the distinguished eigenfunctions Φ and Ψ are, for fixed x, of the form $1 + f(k)$ where f is analytic in the upper half-plane and satisfies $\int |f(k+ic)|^2\, dk \leq$

C for every $c > 0$, the Paley-Wiener theorem ensures the existence of functions K_+ and K_- such that

$$(4.23) \qquad (\mathcal{F}^{-1}\phi)(y) = \delta(x - y) + K_-(x, y)$$

$$(4.24) \qquad (\mathcal{F}^{-1}\psi)(y) = \delta(x + y) + K_+(x, -y);$$

furthermore, $K_+(x, y)$ (resp. $K_-(x, y)$) is zero for $y < x$ (resp. $y > x$). It is the function K_+ that will be reconstructed from the scattering data. Once it is known, we can recover u via

$$(4.25) \qquad K_+(x, x) = \frac{1}{2} \int_x^\infty u(s)\, ds.$$

The simplest way to prove this formula is to note first that ψ solves, in the sense of distributions,

$$[\partial_{yy} - \partial_{xx} + u(x)]\mathcal{F}\psi(x, y) = 0.$$

Now, since Ψ has an asymptotic expansion in inverse powers of k (to all orders if u in smooth), it can be written as sum of $\delta(x + y)$ and a piecewise smooth function (namely K_+). Writing the equation satisfied by K_+ in weak form, and using the fact that the equation is satisfied in the ordinary sense for $y > x$, (4.18) follows.

Reconstruction procedure.

THEOREM 4.12 *Let* $F(z) = \frac{1}{2\pi} \int_{-\infty}^{+\infty} e^{ikz} R(k)\, dk + \sum_j m_j^2 e^{-\eta_j z}$. *Then the Gel'fand-Levitan Marchenko (GLM) equation*

$$(4.26) \qquad K_+(x, y) + F(x + y) + \int_x^\infty K_+(x, t)F(t + y)\, dt = 0$$

holds for $y > x$.

Proof: The result follows by applying the inverse Fourier transform with respect to the eigenvalue parameter k to the equation

$$(4.27) \qquad (\frac{1}{a} - 1)\phi(k) = [\psi(-k) - \phi(k)] + R(k)\psi(k).$$

We recall that y is the variable conjugate to k; we assume $y > x$.

We write (4.27) as $I = II + III$ and compute the transform of I, II and III separately:

STEP 1. Since the transmission coefficient is continuous in the closed half-plane (including zero), we may apply Cauchy's theorem to the curve Γ formed by $[-R, R]$ and the counter-clockwise circle of radius R, large enough to enclose all the poles of T:

$$\int_\Gamma (\frac{1}{a} - 1)\phi(k)e^{iky}\frac{dy}{2\pi} = \sum_j \frac{i\phi(i\eta_j)e^{-\eta_j y}}{a'(i\eta_j)} = \sum_j \frac{ib_j}{a'(i\eta_j)}\psi(i\eta_j)e^{-\eta_j y}.$$

Since $\Phi \to 1$ and $a = 1 + O(1/k)$ as $k \to \infty$, we may bound the integrand by

$$(C/k)e^{-\operatorname{Im} k(y-x)},$$

which, integrated on the half-circle $|k| = R$, $\operatorname{Im} k \geq 0$, gives a quantity bounded by $\operatorname{const.}/R(y - x)$. Letting $R \to \infty$, and using (4.24), we find

(4.28) $\mathcal{F}^{-1}I = -\sum_j m_j^2[e^{-\eta_j(x+y)} + \int_x^\infty K_+(x,t)e^{-\eta_j(t+y)}\,dt].$

STEP 2. From (4.23-24), it easily follows that

(4.29) $\mathcal{F}^{-1}II = K_+(x,y)$ if $y > x$.

STEP 3. Let $r(y) = \mathcal{F}^{-1}R$. We have

$$
\begin{aligned}
\mathcal{F}^{-1}III &= (\mathcal{F}^{-1}R) *_y (\delta(x+y) + K_+(x,-y)) \\
&= \int_{-\infty}^{+\infty} r(z)[\delta(x+y-z) + K_+(x,-y+z)]\,dz \\
&= r(x+y) + \int_x^\infty K_+(x,t)r(t+y)\,dt.
\end{aligned}
$$

(4.30)

Putting these together, we find that the GLM equation holds.

The GLM equation is a Fredholm integral equation of the second kind, for each fixed x. Multiplying the equation by the conjugate of K, and using the property $|R| < 1$ for $k \neq 0$, one can prove that its homogeneous version has no non-trivial solution, hence that the GLM equation is uniquely solvable.

A noteworthy case in which the GLM equation can be solved in closed form is when the reflection coefficient vanishes. Such *reflectionless potentials* generate the multi-soliton solutions of KdV. If $R \equiv 0$, we seek $K = K_+$ in the form

$$K = \sum_k -m_k\alpha_k(x)e^{-\eta_k y}.$$

Substitution leads to the equation

(4.31) $(I + A(x))\alpha(x) = M(x),$

where

$$\alpha = \begin{pmatrix} \alpha_1 \\ \vdots \\ \alpha_N \end{pmatrix}; \quad M = \begin{pmatrix} m_1 e^{-\eta_1 x} \\ \vdots \\ m_N e^{-\eta_N x} \end{pmatrix}; \quad A = ((a_{kl})),$$

with $a_{kl} = \frac{m_k m_l}{\eta_k + \eta_l}e^{-(\eta_k + \eta_l)x}$.

THEOREM 4.13 *The matrix A is positive-definite for every x, and the corresponding potential is*

$$(4.32) \qquad u(x) = -2\frac{d^2}{dx^2}\ln\det(I + A(x)).$$

Proof: For any constant vector ξ, we compute

$$\sum_{k,l} a_{kl}\xi_k\xi_l = \int_x^\infty (\sum_k \xi_k m_k e^{-\eta_k x})(\sum_l \xi_l m_k e^{-\eta_l x})$$

which is manifestly positive for $\xi \neq 0$. Therefore, A is positive-definite. Letting $((b_{kl}))$ be the adjugate of A, and $\Delta(x) = \det(I + A(x))$, we now have

$$\alpha_k = \sum_l \frac{b_{kl}(x)}{\Delta(x)} m_l e^{-\eta_l x}.$$

In particular,

$$K(x,y) = -\frac{1}{\Delta}\sum_{k,l} m_k e^{-\eta_k y} b_{kl}(x) m_k e^{-\eta_l x}.$$

Now,

$$\begin{aligned}
\frac{d\Delta}{dx} &= \sum_{k,l} b_{kl}\frac{d}{dx}(\delta_{kl} + a_{kl}) \\
&= \sum_{k,l} -b_{kl}m_k m_l e^{-(\eta_k+\eta_l)x} \\
&= \Delta K(x,x),
\end{aligned}$$

and since $u(x) = -2(d/dx)K(x,x)$, we obtain the $u(x) = -2(d^2/dx^2)\ln\Delta$, QED.

Application to the KdV equation.

We now assume that u depends on a new variable t, in such a way as to satisfy the KdV equation (4.19). For every value of t we may construct distinguished eigenfunctions as before. Assume that $L\psi = k^2\psi$ and that $c = -4ik^3$ in the expression of A, so that $\psi_t - A\psi \to 0$ as $x \to +\infty$. Since $L_t + [A, L] = 0$, it easily follows that

$$(L - k^2)(\partial_t - A)\psi = 0.$$

but this equation has no decaying solutions, so we must have $\psi_t = A\psi$. One can similarly identify ϕ. The differentiability of ϕ and ψ in time can be proved from their construction (since u is now smooth in time), and it implies smoothness of a and b in t, since these are given by Wronskians of the eigenfunctions. As $x \to +\infty$, we therefore have now $\phi = ae^{-ikx} + be^{ikx}$, $\phi_t = a_t e^{-ikx} + b_t e^{ikx}$.

Substituting into the time evolution for ϕ and ψ, one derives the very simple relations

(4.33) $a_t = 0, \quad b_t = -8ik^3 b.$

Along similar lines, one derives

(4.34) $m_{kt} = 4\eta_k^3 m_k.$

We are now finally ready to give the inverse scattering solution of the KdV equation: starting with the initial data, compute a and b, as well as the eigenvalues and norming constants. Then form the quantity

$$F(z,t) = \frac{1}{2\pi}\int_{-\infty}^{+\infty} e^{-8ik^3 t}R(k)\,dk + \sum_k m_k^2 e^{8\eta_k^2 t - \eta_k z}.$$

Solve the associated GLM equation, and compute

$$u = -2\frac{d^2}{dx^2}\ln\Delta$$

as before. The result solves $u_t - 6uu_x + u_{xxx} = 0$.

The point is that the function F solves the equation $F_t + F_{zzz} = 0$. We have in this sense "linearized" the KdV equation.

Multi-solitons.

We must now describe the reflectionless solutions of the KdV equation, which are the simplest examples of pure soliton behavior. The main result is that the N-soliton solution defined here breaks up as $t \to \pm\infty$ into a sum of N one-solitons, and that each soliton undergoes a definite phase shift upon interacting with the others, while it remains otherwise unperturbed by the "interaction."

From what we already know, the issue is now to compute the determinant Δ for each N. This reduces to a determinant of the form

$$\det(\delta_{kl} + \frac{e^{\theta_k + \theta_l}}{\eta_k + \eta_l})$$

where $\theta_k = -\eta_k(x - x_k - 4\eta_k^2 t)$ (x_k being related to the norming constants). To study the large time behavior, we assume that the η_k are in their natural order, and we let x and t tend to $+\infty$ while keeping θ_s constant. The other θ_k will then tend to $+$ or $-\infty$.

The net result is that $\ln\Delta \sim e^{2\theta_s}D_{s+1} + D_s$, where

$$D_s = \prod_{k \geq s}\frac{1}{2\eta_k}\prod_{l < k}(\frac{\eta_l - \eta_k}{\eta_l + \eta_k})^2.$$

We therefore see that we are left with a one soliton with a translation in its phase. If we let x and t tend to $-\infty$ in the same way, we end up with a different phase. Therefore, one may put these two pieces of information together and say that the s-soliton has undergone a phase shift due to its interaction with the others. Remarkably enough, this shift is exactly the sum of those that would be due to pairwise interaction with each of the other solitons.

4.4   THE AKNS SYSTEMS

We discuss the family of (2×2) AKNS systems, with emphasis
on the sine-Gordon and nonlinear Schrödinger equations.

The AKNS systems (for Ablowitz-Kaup-Newell-Segur) are isospectral deformations of the two-component "eigenvalue" problem

$$(4.35) \qquad (\partial + ik)\psi_1 = q(x)\psi_2$$
$$(4.36) \qquad (\partial - ik)\psi_2 = r(x)\psi_1.$$

It follows that the vector $\psi = \begin{pmatrix} \psi_1 \\ \psi_2 \end{pmatrix}$ solves

$$\left(-\partial^2 + \begin{pmatrix} qr & q_x \\ r_x & qr \end{pmatrix}\right)\psi = k^2\psi,$$

so that this may be viewed as a two-component generalization of the Schrödinger eigenvalue problem. Note also that if $r \equiv -1$, we have

$$(-\partial^2 + q)\psi_2 = k^2\psi_2.$$

We will, however, be interested in the case when q and r are both in $\mathcal{S}$.

It turns out that this problem is associated with the SGE, the NLS and the mKdV equations.

There is in fact a systematic two-step procedure to derive families of evolution equations associated to this eigenvalue problem: (1) Postulate a time evolution for the eigenfunctions in the form

$$(4.37) \qquad \psi_{1t} = A\psi_1 + B\psi_2$$
$$(4.38) \qquad \psi_{2t} = C\psi_1 + D\psi_2;$$

Cross differentiation gives

$$(4.39) \qquad A_x = (-D)_x = qC - rB$$
$$(4.40) \qquad B_x + 2ikB = q_t - (A_D)q$$
$$(4.41) \qquad C_x - 2ikC = r_t + (A - D)r.$$

It is then natural to assume

$$A + D = 0.$$

(2) Now, assume that A, B, C are *polynomials* in k (with coefficients depending on x and t), and identify like powers of k. It turns out that for every choice of the degree of A, there is a choice of A, B, C such that (4.29-31) are equivalent to a nonlinear evolution equation for q and r. We give here the most important examples only, which are further reductions of these systems, under the assumption that r is proportional to q or $\bar{q}$:

Example 1: Assume that

(4.42) $r = \mp\bar{q},$

and that

$$\begin{aligned}
A &= 2ik^2 \pm i|q|^2 \\
B &= 2kq + iq_x \\
C &= \mp 2k\bar{q} \pm i\bar{q}_x.
\end{aligned}$$

Then cross-differentiation leads to the *nonlinear Schrödinger equation*

(4.43) $iq_t = q_{xx} \pm 2q|q|^2.$

In the case $r = \bar{q}$, the eigenvalue problem, which can be written

(4.44) $\begin{pmatrix} i\partial & -iq \\ ir & -i\partial \end{pmatrix} \begin{pmatrix} \psi_1 \\ \psi_2 \end{pmatrix} = k \begin{pmatrix} \psi_1 \\ \psi_2 \end{pmatrix}$

is easily seen to be self-adjoint. Using the symmetry properties of the spectrum, is is easy to prove that there are no point eigenvalues in this case.

The same eigenvalue problem, with a third degree A and real r and q, is associated with the modified KdV equation. If A is proportional to $1/k$ as in the next example, we are led to the SG and sinh-Gordon equations.

Example 2: Assume that

(4.45) $A = \frac{a}{k}; \quad B = \frac{b}{k}; \quad C = \frac{c}{k}.$

Then cross-differentiation leads to the following two choices:

(i) If $r = -q = u_x/2$, and

$$a = \frac{i}{4}\cos u; \quad b = c = \frac{i}{4}\sin u,$$

we find the single equation

$$u_{xt} = \sin u.$$

(ii) If $r = q = u_x/2$, and

$$a = \frac{i}{4}\cosh u; \quad b = -c = \frac{i}{4}\sinh u,$$

we find the equation

$$u_{xt} = \sinh u.$$

It is possible to find such a representation for the SGE in the usual form

$$u_{tt} - u_{xx} + \sin u = 0,$$

(exercise 12).

We now turn to the description of the scattering data for the AKNS systems. One could, from these data, set-up and solve an inverse problem; the procedure is entirely analogous to the KdV case is omitted (see Ablowitz et al. (1974) for details).

We define two sets of distinguished eigenfunctions $(\phi, \tilde{\phi})$ and $(\psi, \tilde{\psi})$ by the properties

$$\Phi = \phi e^{ikx} \to \begin{pmatrix} 1 \\ 0 \end{pmatrix}; \quad \tilde{\Phi} = \tilde{\phi} e^{-ikx} \to \begin{pmatrix} 0 \\ -1 \end{pmatrix}$$

as $x \to -\infty$ and

$$\Psi = \psi e^{-ikx} \to \begin{pmatrix} 0 \\ 1 \end{pmatrix}; \quad \tilde{\Psi} = \tilde{\psi} e^{ikx} \to \begin{pmatrix} 1 \\ 0 \end{pmatrix}$$

as $x \to +\infty$.

The scattering matrix is defined by

$$
\begin{align}
\psi &= a\tilde{\psi} + b\psi \tag{4.46} \\
\tilde{\phi} &= \tilde{b}\tilde{\psi} - \tilde{a}\psi. \tag{4.47}
\end{align}
$$

It is customary to define two "reflection coefficients" $R = b/a$ and $\tilde{R} = \tilde{a}/\tilde{b}$. One can also define a Wronskian by $W(u,v) = u_1 v_2 - u_2 v_1$, so that $W(\psi.\tilde{\phi}) = -1$, from which it easily follows that

$$a\tilde{a} + b\tilde{b} = 1.$$

The main properties of these functions are summarized in

THEOREM 4.14 *Φ and Ψ are analytic in the upper k-half-plane, while $\tilde{\Phi}$ and $\tilde{\Psi}$ are analytic in the lower k-half-plane. It follows that a and $\tilde{a}$ are analytic in the upper and lower half-planes respectively.*

Proof: We outline the argument for Φ; the other cases are similar. The idea
is to convert the equations into the form

$$\begin{aligned}
(\phi_1 e^{ikx})_x &= q\phi_2 e^{ikx} \\
(\phi_2 e^{-ikx})_x &= r\phi_1 e^{-ikx},
\end{aligned}$$

which are readily integrated into the integral equations

$$\begin{aligned}
\phi_1 e^{ikx} &= 1 + \int_{-\infty}^{x} dy \int_{-\infty}^{y} q(y)r(z)\phi_1(z,k)e^{ikz}e^{2ik(y-z)}\,dz \\
\phi_2 e^{ikx} &= \int_{-\infty}^{x} e^{2ik(x-y)}r(y)\phi_1(y,k)e^{iky}\,dy.
\end{aligned}$$

It is more convenient now to view these as equations on Φ rather than ϕ (with
k in the upper half-plane). Since they imply

$$|\Phi_1 - 1| \le \int_{-\infty}^{x} dy \int_{-\infty}^{y} |q(y)|\,|r(z)|\,|\Phi_1(z,k)|\,dz := T[\Phi_1],$$

an induction argument proves that

$$T^m[1] \le \frac{Q^m R^m}{(m!)^2}.$$

where $Q(x) = \int_{-\infty}^{x} |q(y)|\,dy$ and $R(x) = \int_{-\infty}^{x} |r(y)|\,dy$. It follows that

$$|\Phi_1 - 1| \le \sum_{m \ge 1} (QR)^m/(m!)^2$$

is uniformly bounded for all $x \in \mathbf{R}$ and all k in the upper half-plane, and is a
uniform limit of analytic functions. QED

Using these bounds and integration by parts, one can derive large k asymptotics for Φ and a, $\tilde{a}$:

$$(4.48) \qquad \phi_1 e^{ikx} = 1 - \frac{1}{2ik}\int_{-\infty}^{x} q(y)r(y)\,dy + O(1/k^2)$$

$$(4.49) \qquad \phi_2 e^{ikx} = -\frac{1}{2ik}r(x) + O(1/k^2)$$

$$(4.50) \qquad a(k) = 1 - \frac{1}{2ik}\int_{-\infty}^{\infty} q(y)r(y)\,dy + O(1/k^2)$$

$$(4.51) \qquad \tilde{a}(k) = 1 + \frac{1}{2ik}\int_{-\infty}^{\infty} q(y)r(y)\,dy + O(1/k^2).$$

Similar relations hold for ψ.

If $r = \pm\bar{q}$, we have

$$(4.52) \qquad \tilde{a}(k) = \bar{a}(\bar{k}); \quad \tilde{b}(k) = \mp\bar{b}(\bar{k}).$$

This follows from the fact that the "tilde" eigenfunctions are expressible in terms of ϕ and ψ:

$$(4.53) \qquad \tilde{\psi}(x,k) = \begin{pmatrix} \bar{\psi}_2(x,\bar{k}) \\ \pm\bar{\psi}_1(x,\bar{k}) \end{pmatrix}; \quad \tilde{\phi}(x,k) = \begin{pmatrix} \mp\bar{\phi}_2(x,\bar{k}) \\ \bar{\phi}_1(x,\bar{k}) \end{pmatrix}.$$

We have $a(k) = 0$ if and only if $\tilde{a}(\bar{k}) = 0$.

If $r = \pm q$, we find, by a similar argument, that

$$(4.54) \qquad\qquad \tilde{a}(k) = a(-k); \quad \tilde{b}(k) = \mp b(-k).$$

We have $a(k) = 0$ if and only if $\tilde{a}(-k) = 0$.

In both cases, the zeros of a in the upper half-plane can be computed from, those of $\tilde{a}$ in the lower half-plane. If $r = \pm q$ and is *real*, we are in the intersection of these cases, and

$$a(k) = 0 \Rightarrow \tilde{a}(\bar{k}) = \tilde{a}(-k) \Rightarrow a(-\bar{k}) = 0.$$

Therefore,

THEOREM 4.15 *If $r = \pm q$ and is real, the poles of $1/a$ come in pairs $\{k, -\bar{k}\}$.*

In particular, for the sine-Gordon equation, a simple pole on the imaginary axis is called a kink, and a pair of (simple) poles in the upper half-plane corresponds to a breather. The names correspond to the form of the solutions with these poles and with vanishing reflection coefficient(s).

Finally, we prove a property which will turn out to characterize our distinguished eigenfunctions.

THEOREM 4.16 *The distinguished eigenfunctions satisfy*

$$(4.55) \qquad\qquad \lim_{x\to+\infty} (\Phi/a, \Psi) \;=\; I \; \text{(the identity matrix)};$$

$$(4.56) \qquad\qquad \lim_{x\to-\infty} (\tilde{\Psi}, -\tilde{\Phi}/\tilde{a}) \;=\; I.$$

Note that the same limits obtain as $k \to \infty$ too, using the large k asymptotics we have derived.

Proof: We already know that Ψ tends to $\begin{pmatrix} 0 \\ 1 \end{pmatrix}$.

By the definition of the scattering coefficients, we know that, if k is real and x tends to infinity,

$$\Phi_1 = a + o(1); \quad \Phi_2 = be^{2ikx} + o(1).$$

Since $\Phi_1 = 1 + \int_{-\infty}^{x} q(y)\Phi_2(y,k)\,dy$, we conclude that

$$a(k) = 1 + \int_{-\infty}^{+\infty} q(y)\Phi_2(y,k)\,dy.$$

This relation can be continued to the upper half-plane (because the difference is
an analytic function which is zero on the boundary of the half-plane, as can be
seen using dominated convergence.) It now follows that $\Phi_1 \to a$. On the other
hand,

$$\Phi_2 = \int_{-\infty}^{x} r(y)\Phi_1(y,k)e^{2ik(x-y)} \, dy,$$

and therefore, since the exponential term tends pointwise to zero as $x \to \infty$ (for
fixed y), we conclude by dominated convergence that $\Phi_2 \to 0$, QED.

4.5 RIEMANN-HILBERT AND $\bar{\partial}$ FORMULATIONS

We discuss the Riemann-Hilbert formulation of inverse scattering
with application to the generalized AKNS systems and, in particular,
the three wave interaction.

We saw that the solution of the inverse problem consisted in (1) the identification of a set of distinguished eigenfunctions analytic in the lower/upper half-
plane and (2) the Fourier transformation of the linear relation that they satisfy.
We however had to multiply first this relation by the transmission coefficient,
which suggests that our choice of eigenfunction was not optimal. Furthermore,
we focused exclusively on the upper half-plane.

The Riemann-Hilbert formulation obviates both difficulties, and has a wider
range of applicability. The main point is that the basic relation (4.27) really
connects two distinguished sets of eigenfunctions, meromorphic on the upper
and lower half-planes respectively, on the boundary of their domain of definition.
They also satisfy bounds as $k \to \infty$. Now the theory of the Riemann-Hilbert (R-
H) problem suggests that a function which is meromorphic everywhere except
on a curve on which it has a jump should be determined by its asymptotic
properties at infinity and at its poles alone.

This is indeed the case.

We begin with the case of the 2×2 AKNS systems, and then turn to their
$n \times n$ generalizations.

Case of the AKNS systems.

We consider the matrix-valued function with columns equal to Φ/a and Ψ:

$$m_+(x,k) = (\frac{1}{a}\Phi(x,k), \Psi(x,k)).$$

We have seen that it satisfies $\lim_{x\to\infty} m_+(x,k) = I$. This suggests that it might
be possible to reach directly this matrix instead of computing Φ and Ψ separately. But there is more. If

$$m_-(x,k) = (\tilde{\Psi}(x,k), \frac{-1}{\tilde{a}}\tilde{\Phi}(x,k)),$$

then $\lim_{x\to\infty} m_-(x,k) = I$ as well, and, as a brief calculation shows, the two matrices are related by

$$m_+ = m_-(I + V),$$

where

$$V(x,k) = \begin{pmatrix} R\tilde{R} & \tilde{R}e^{-2ikx} \\ Re^{2ikx} & 0 \end{pmatrix}.$$

Actually, if $J = \begin{pmatrix} -1 & 0 \\ 0 & 1 \end{pmatrix}$ we find that

$$v(k) = e^{-ikxJ}V(x,k)e^{ikxJ}$$

is *independent of k*. Since both matrices are bounded at infinity, we expect that the $m_\pm$ are entirely determined by their jump condition and their limits at infinity.

We now present a generalization of these ideas to $n \times n$ AKNS-type systems.

Generalized AKNS systems.

We consider the "eigenvalue" problem

$$(4.57) \qquad \qquad \partial\psi = kJ\psi + q(x)\psi$$

where q, J and ψ are $n \times n$ complex-valued matrices, and k is a complex parameter. q is independent of k and J is a constant diagonal matrix with distinct eigenvalues:

$$J = \text{diag}\,(\lambda_1, \ldots, \lambda_n).$$

q is an off-diagonal matrix with entries in $\mathcal{S}$.

Since the solutions are proportional to $\exp(xkJ)$ when $q \equiv 0$, we let $\psi = m(x,k)e^{xkJ}$, so that

$$(4.58) \qquad \qquad \partial m = k[J,m] + qm.$$

If m_0 is a particular solution, any other solution has the form

$$m = m_0 e^{xkJ}v(k)e^{-xkJ}.$$

In fact, mm_0^{-1} solves (4.54) with $q = 0$; for this case, the general solution has the form $e^{xkJ}v(k)e^{-xkJ}$ if $q \equiv 0$.

Note also that if m has a simple pole at $k = k_j$, its residue at k_j also solves (4.54), as can be seen by multiplying the equation by $(k-k_j)$ and letting $k \to k_j$. We also introduce the set $\Sigma = \{k : \text{Re}\,(k\lambda_p) = \text{Re}\,(k\lambda_q) \text{ for some } p,q\}$.

The main results on the definition and properties of the scattering data are:

THEOREM 4.17 *There is a discrete set $Z \subset C \setminus \Sigma$, independent of x, such that for any $k \in C \setminus (\Sigma \cup Z)$, there is a unique solution m which is uniformly bounded, and tends to I as $x \to \infty$. It further satisfies*

1. *m is meromorphic in k on $C \setminus \Sigma$, with poles precisely on Z.*

2. *on each component of $C \setminus \Sigma$, $\lim_{k\to\infty} m(x,k) = I$.*

We denote by Σ_ν the rays of Σ, and by Ω_ν the connected components of $C \setminus \Sigma$. Also, $m_\nu^\pm$ is the limit of m for fixed x as k tends to Σ_ν from Ω_ν (resp. $\Omega_{\nu+1}$) (if it exists).

It is not possible to exclude multiple eigenvalues in general; we introduce however a class of "generic" potentials for which several conceivable degeneracies do not occur:

THEOREM 4.18 *There exists a dense open set $P \subset L^1$ of potentials for which*

1. *Z is finite.*

2. *m has only simple poles.*

3. *Distinct columns of m have distinct poles.*

4. *m has a continuous extension to each side of Σ and defines a continuous function on the closure of each Ω_ν.*

Assume from now on that $q \in P$ (i.e., is "generic"). We define the scattering data by

THEOREM 4.19 *For every ν, there is a unique matrix $v_\nu(k)$ (independent of x) such that for all x,*

$$m_\nu^+(x,k) = m_\nu^-(x,k)e^{xkJ}v_\nu(k)e^{-xkJ}.$$

Furthermore, if k_j is a pole of m, there are matrices $v(k_j)$ such that

$$\text{Res}_{k_j} m = \lim_{k\to k_j} m(x,k)e^{xk_jJ}v(k_j)e^{-xk_jJ}.$$

It turns out that q is entirely determined by the v_ν, k_j, $v(k_j)$. It is not quite true that one can prescribe these scattering data arbitrarily; the necessary additional constraints are known (exercise).

We now outline some of the technicalities involved in the construction of the scattering data, and in the reconstruction procedure.

We introduce the usual scalar product $(a, b) = \text{tr}\,(b^*a)$, where b^* is the Hermitian conjugate (i.e., the transpose of the conjugate matrix) of b. We let $|a| = (a,a)^{1/2}$ and $\mathcal{J}a = [J,a]$. One finds $\mathcal{J}^*a = [\bar{J},a]$, and a computation shows that $\mathcal{J}^*$ is normal and $(k\mathcal{J} + (k\mathcal{J})^*)/2$ is self-adjoint. Let us decompose $M_n(C)$ into the invariant spaces corresponding to the positive, negative, and zero eigenvalues for the latter operator. Let π_+, π_- and π_0 be the corresponding orthogonal projections. It will also be convenient to assume that in each sector Ω_ν, the eigenvalues are ordered in such a way that the sequence $\{\mathrm{Re}\,(k\lambda_j)\}$ be decreasing. We will refer to it as the "natural ordering" of the eigenvalues.

The eigenfunction is determined by the integral equation

$$m(x,k) = I + \int_{-\infty}^{x} e^{(x-y)k\mathcal{J}}$$

(4.59) $$(\pi_0 + \pi_-)(qm)(y)\,dy - \int_{x}^{\infty} e^{(x-y)k\mathcal{J}}\pi_+(qm)(y)\,dy$$

(4.60) $$:= I + T[m].$$

The limits of integration are motivated by the desire to find a solution which tends to I at $-\infty$, and remains bounded at $+\infty$.[6] In fact, both exponentials are uniformly bounded, and therefore m also is:

$$\|Tm\|_\infty \le \|q\|_1 \|m\|_\infty.$$

On the other hand, as $x \to \infty$, the integrands are dominated by $|q|$ and tend to zero pointwise. The property $\lim_{x\to\infty} m = I$ therefore follows as well.

CASE 1. If $\|q\|_1 < 1$, the existence of m (and its holomorphy in every sector) follows by Picard iteration. Since m is continuous in each closed sector, the existence of $m_\nu^\pm$ is assured. We have

$$\|m\|_\infty \le \frac{1}{1 - \|q\|_1}.$$

Letting $\psi = me^{xk\mathcal{J}}$, we find $\psi_x = (k\mathcal{J} + q)\psi$, so that

$$\frac{d}{dx}\det\psi = k\,\mathrm{tr}\,J\det\psi,$$

which proves $\det m = const. = 1$. We conclude that m^{-1} is uniformly bounded too.

Finally, if m_1 were another solution tending to I at $-\infty$ and bounded at $+\infty$, we would have $m_1 = me^{xk\mathcal{J}}a(k)$, and $e^{xk\mathcal{J}}a(k)$ would have to be bounded at $\pm\infty$. This forces a to be diagonal, and, since $e^{xk\mathcal{J}}a(k)$ tends to I at $-\infty$, $a = I$. We have proved the uniqueness of m.

Remark 1: If $\tilde{q} = -q^*$, we find that the corresponding eigenfunction $\tilde{m}$ satisfies $\tilde{m}(x,k)^{-1} = m(x,-\bar{k})^*$.

---

[6]Such a procedure is typical of proofs of the stable manifold theorem, see e.g. Coddington-Levinson.

Remark 2: Using the fact that $q_x \in L^1$, we can derive a more precise estimate on m. Noting that $\mathcal{J}^{-1} : Ran\, \mathcal{J} \to Ran\, \mathcal{J}$, and that $q \in Ran\, \mathcal{J}$, we let

$$n := I - k^{-1}\mathcal{J}^{-1}q(x).$$

Since $(\partial - k\mathcal{J})n = qn + f/k$ for some integrable f, we find that $(\partial - k\mathcal{J})(m^{-1}n)$ is an integrable function, and is $O(1/k)$. It follows easily that

$$|m - n| \leq C/|k|.$$

In other words, $m \sim I - \frac{1}{k}\mathcal{J}^{-1}q + O(1/k)$.

CASE 2. If $\|q\| \geq 1$, we argue by induction by assuming the result for $\|q\|_1 < 2^{N-1}$. The origin of the singularities of the eigenfunctions will become clear by this procedure. Let us assume that $\|q\|_1 < 2^N$. Assume also (by translating the potential if need be) that $\int_{-\infty}^0 |q|dx = (1/2)\|q\|_1 (< 2^{N-1})$. Write $q = q_1 + q_2$, where $q_1 = q\chi_{x<0}$. Let m_1 and m_2 be the corresponding eigenfunctions given by the induction assumption. The desired eigenfunction m must be have the form $m_1 e^{xk\mathcal{J}}a_1(k)$ for $x < 0$, and $m_2 e^{xk\mathcal{J}}a_2(k)$ for $x > 0$. Since it should be continuous, we need to have

$$m_1(0, k)a_1(k) = m_2(0, k)a_2(k).$$

Boundedness at infinity requires that $\pi_+ a_2 = 0$, while $m \to I$ at $-\infty$ requires $\pi_- a_1 = 0$ and $\pi_0 a_1 = I$. In the sector Ω_ν, if we assume that the eigenvalues are given their natural ordering, this amounts to the factorization problem

$$m_2(0, k)^{-1}m_1(0, k) = a_2 a_1^{-1},$$

with a_2 strictly lower triangular, and a_1 upper triangular with 1's on the diagonal. It is well-known that there is a unique choice of a_1 and a_2 which accomplishes this if and only if the leading minors of the left-hand side are non-zero. Since $m \to I$, this fails on a discrete set of the k plane. This is the origin of the set Z.

From the formulae for these matrices, one sees that both tend to I as $k \to \infty$.

Notions on the Cauchy integral and the $\bar{\partial}$ problem.

Let γ be a piecewise smooth curve in the complex plane. The Cauchy integral is the operator

$$f \mapsto Cf = C_\gamma(f)(x) = \frac{1}{2\pi i}\mathrm{pv} \int_\gamma \frac{f(z)\, dz}{x - z}$$

for $x \in \gamma$. We may assume, for the time being, that f is smooth and compactly supported. If $\gamma = \mathbf{R}$ (with the usual orientation), this is the convolution by $\frac{1}{2\pi i}\mathrm{pv}\,(1/x)$. We also define

(4.61) $$C^\pm f = \frac{1}{2\pi i}\lim_{\varepsilon \to \pm 0}\int_\gamma \frac{f(z)\, dz}{x - (z + i\varepsilon)}.$$

Since $(x \pm i0)^{-1} = \mp i\pi\delta(x) + \mathrm{pv}\,(1/x)$, we recover *Plemelj's formulae:*

$$(4.62) \qquad\qquad C^{\pm}f = \pm\frac{1}{2}f + Cf.$$

These formulae hold for more general curves, provided that we define the "plus" side of γ to be "on the left" when γ is traveled in its positive direction; the proof is easy for smooth γ and f. Since $C^+f - C^-f = f$, we see that we immediately have a solution to the "additive Riemann-Hilbert problem" which consists in expressing f as a difference of two functions analytic in the lower and the upper half-plane respectively.

Since $\mathcal{F}[(x \pm i0)^{-1}] = \mp 2\pi i H(\pm\xi)$, where H is Heaviside's function, the operators $C^{\pm}$ are complementary orthogonal projections in $L^2(\mathbf{R})$.[7]

To apply the preceding, we note that the eigenfunctions of the AKNS problem satisfy jump conditions of the form

$$m_\nu^+ - m_\nu^- = m_\nu^-[e^{xkJ}v_\nu(k) - I].$$

on Σ_ν. This can be now turned into an integral equation for the eigenfunctions. The details are extremely lengthy and can be found in Beals-Coifman (1985).

4.6   CRITERIA FOR INTEGRABILITY

We discuss a few algorithmic procedures to decide whether a given equation possesses solitons. In particular, the Painlevé test for PDE, the symmetry test, the Wahlquist-Eastabrook method, and Hirota's method are presented, with focus on their application to the KdV equation. Another criterion based on the domain of analyticity of multi-solitons is also described, with application to the breather problem.

There is at the present time no comprehensive criterion to decide whether a given PDE is integrable by inverse scattering. There are however a few procedures of an algorithmic nature which single out equations which are believed to have special structure. In a limited number of cases, these do produce either a Lax pair formulation or, for ODEs, the maximum possible number of "regular" integrals. We describe first some of the important algorithmic criteria for integrability; in each case the achievements of that particular method will be outlined. Some aspects of the problem of defining what one may mean by the word "integrable" are then outlined,

---

[7]$\mathcal{F}[(x \pm i0)^a] = 2\pi e^{\pm i\pi a/2}\chi_{\mp}^{-a-1}$, where $\chi_+^a = x_+^a/\Gamma(a+1)$ and a is any complex number.

The WTC test.

The WTC test (for Weiss-Tabor-Carnevale) or generalized Painlevé test, tests for the existence of meromorphic solutions. Its local form is extremely easy to implement and generally provides a large number of constraints.

Historical background. The WTC test has very deep roots in the analytic theory of PDE and ODE, and its significance extends well beyond the limited context of integrability. It is part of a very comprehensive program to study the singular behavior of solutions of PDE by postulating and justifying formal expansions. This line of thought is a very natural outgrowth of the historical development of the subject, which we therefore proceed to recall in some detail.

a. From Cauchy to Briot-Bouquet. The discovery of local existence theorems for ODEs with analytic nonlinearities and coefficients (Cauchy, 1835) presented as the next problem the study of the behavior of solutions near singularities. These singularities could be due to the coefficients, in which case they were termed *fixed*, since they did not change with the initial data. Singularities that did depend on data or constants of integration were termed *movable*.

The case of the hypergeometric equation (Gauss (1812), Riemann (1857)) suggested that the solution have a series representation near fixed singularities, and that the singularity structure might actually characterize the equation completely.

The first nonlinear example studied in detail was

$$(4.63) \qquad z\frac{du}{dz} = \lambda u + f(z, u),$$

where f is analytic near $(0,0)$, and vanishes for $z = u = 0$. Such equations were studied by Briot and Bouquet (1854, 1856). They proved that (i) if λ is not a positive integer, there is exactly one analytic solution which vanishes for $z = 0$; (ii) if λ is a positive integer, there is no analytic solution unless the coefficients of f satisfy a constraint; if the constraint holds, there are infinitely many solutions which vanish at 0. They also realized that if the constraint fails, there is still a solution, but it has an expansion involving powers of $z \ln z$, and is therefore not analytic near 0. From this work and work of Poincaré (1866), it follows that there is a reduction procedure, based on the construction of a Newton polygon, to reduce rather general equations of the form

$$\frac{du}{dz} = \frac{g(z, u)}{h(z, u)}$$

to $z^m u' = \lambda u + f(z, u)$.

By the end of the nineteenth century, it had become clear that similar statements hold for linear equations of *Fuchsian type*, which are nowadays written as first-order systems:

$$(4.64) \qquad z\frac{du}{dz} = Az + zf(z, u),$$

where u is a vector, and A is constant. There is a fundamental matrix of solutions which has the form $z^A v(z)$, where v is analytic. Therefore, all solutions have expansions in powers of z or $z \ln z$. For such an equation, or the associated system, the origin is said to be a regular singular point.

Most special functions of mathematical physics were found to be obtained from a second-order ODE with five regular singular points, allowing for coalescence.

b. The Painlevé transcendents. The general problem of classification of possible singularities solutions of a single ODE was found to be intractable; in fact, it is still unsolved. Attention was therefore directed towards characterizing those cases in which the singularities were simplest possible. A simple restriction is that the solution should be *uniform*, i.e., single-valued. A rather complete theory is available in the case of first-order equations (see Hukuhara *et al.*). But for the second-order case, even this problem was found too complicated and attention was further limited to the case where the solution only exhibits poles as movable singularities. In this case, it follows from the work of Painlevé and Gambier that an equation $u'' = f(z, u, u')$ with this property must, if f is rational in u', algebraic in u and analytic in z, be reducible to one of a list of sixty canonical equations, the list of which has been recently reestablished by Cosgrove. Of these, 54 led to known functions, but the remaining six defined new functions: the *Painlevé transcendents*. We do not discuss here the issue of how one proves that a transcendent is indeed "new;" it is closely related to Galois theory.

A useful tool in such investigations is Painlevé's "α-method:" it turns out that if one inserts a parameter α into the equation in such a way as to generate, by a theorem of Poincaré, an analytic family $u(z, \alpha)$ of solutions, if the general solution of the equation for $\alpha \neq 0$ is uniform, then the same must hold true for $\alpha = 0$. One then applies this idea to the family of equations obtained from the change of variables $z = z_0 + \alpha \zeta$, obtaining in this way a simplified equation for $\alpha = 0$.

Chazy obtained further results for third order equations, discovering in particular an equation having solutions with a movable natural boundary, within which it remains single-valued.

c. Beyond analytic solutions? Research gradually moved to non-analytic nonlinearities. It seems that the work of Levinson (1949) is representative of this change: he extended the work of Briot-Bouquet to the case of a non-analytic f, by setting up an iteration which enabled him to perform an "asymptotic integration" of the ODE from the singularity. Of course, the solutions obtained in this way behave precisely like the first term in the expansions of Briot-Bouquet. On the other hand, the interpretation of multiple solutions as related to the arbitrariness of some coefficient in these expansions is lost, since the solutions are not smooth enough for these terms to exist.

The realization of the need for a good spectral theory of the one-dimensional

Schrdinger operator $-d^2/dx^2 + q(x)$, for the needs of quantum mechanics, also shifted interests in other directions. Also, the explosion of qualitative methods in dynamical systems suggested that it might be possible to obtain robust results on the asymptotic behavior of general classes of dynamical systems under weak regularity assumptions, in a non-perturbative context.

d. Solitons and Painlevé transcendents. After the discovery of the IST, and its extension to large classes of equations, it was realized that the ODE reductions of these equations could always be reduced to one of the Painlevé-Gambier list. The ARS conjecture (Ablowitz-Ramani-Segur) suggested that any equation integrable by IST would have this property. Olver and McLeod proved that the existence of a solution procedure using a GLM-type integral equation would always satisfy the ARS conjecture, under a reasonable assumption on the reflection coefficient. The idea is that resolvents of Fredholm operators can be shown to be meromorphic in the eigenvalue parameter.

To check the ARS conjecture on any specific equation, one needs (i) to determine all possible ODE reductions; (ii) check that the only type of singularity at a "generic" point is a pole. To achieve (ii) for an equation of order m, the usual procedure is to show that there is a formal solution

$$u = (z - z_0)^{-\nu}(a_0 + a_1(z - z_0) + \ldots)$$

where ν is an integer (ensuring u is single-valued) and where $m - 1$ of the coefficients a_j are arbitrarily specifiable. If this series converges, it appears that we have here a "general" solution, containing m constants of integration, namely z_0 and the arbitrary a_j. Unfortunately, many equations have *several* series of this form for the same z_0; they are called *branches*. For instance, the equation

$$u^{(4)} + 30uu^{(3)} + 30u'u'' + 180u^2u' + cu' = 0$$

has *two* such series for every z_0, one having four, and the other three arbitrary coefficients.

Adler and van Moerbeke have shown that for classes of systems of *homogeneous* ODE, one can interpret some of the branches as limiting cases of the solutions represented by the branch with the maximum number of arbitrary coefficients.

Seeking to derive a test of the ARS conjecture which would not require the computation of all ODE reductions, WTC suggested testing (ii) directly by seeking a solution in the form of a formal series

$$u = \phi^\nu \sum_k u_k \phi^k,$$

where ν is again a negative integer, and ϕ and u_k depend on all independent variables. The series is therefore not uniquely determined by u even if it exists; ϕ generalizes the expansion variable $z - z_0$ which was used in the ODE case.

They then found on several examples, see Th. 4.20 below, that a Lax pair and a Bäcklund transformation could be derived by careful consideration of the coefficients of this series. It therefore appears that the analytic structure of the singularities of the solutions of a PDE carries qualitative information related to its integrability, and not just smoothness information.

The next three questions are: how general is this procedure? do these formal solutions converge? can one identify the "general solution" with such an expansion?

e. What about nonintegrable equations? An easy criticism of the theory of solitons is that it applies only to approximate equations, and that detailed models are usually not accessible to it. This requires us to consider the singularity structure of solutions of general equations, which therefore intersects with the topic of Chapter 3.

Kichenassamy and Littman have shown that very general semi-linear systems in $\mathbf{R}^{n+1}$ actually have singular solutions of the form

$$u = \phi^\nu F(x^1, \ldots, x^n, \phi, \phi \ln \phi, \ldots, \phi(\ln \phi)^l)$$

for some integer l, where F is analytic in its arguments. The WTC obtains if ν is a negative integer and F does not contain any logarithmic terms. Applied to linear equations, this yields the usual theory of regular singular points. If $\nu = 0$, we recover the Cauchy-Kowalewska theorem. Finally, the results of Briot-Bouquet and all previous convergence results for ODEs are included as well.

That the general solution is indeed given by this singular expansion has been recently established for the equation $\Box u = \exp(u)$ with high enough Sobolev class regularity.

In conclusion, the WTC algorithm is a natural continuation of a continuous line of research originating in the last century, and has pointed to a connection between analyticity properties and integrability. At the same time, the expansion it suggests are defined and convergent for a wide range of equations well beyond the range of soliton theory. It provides precise models for blow-up.

The WTC algorithm. We now turn to a more detailed exposition of the WTC method. The algorithm may be described as follows. Consider a PDE

$$F[u] = 0$$

of order m. All considerations will be made in the vicinity of one point to begin with. Given any analytic function ϕ such that $\phi = 0$ is a non-characteristic hypersurface, seek a solution of the form

$$(4.65) \qquad\qquad u = \phi^\nu \sum_{k \geq 0} u_k(x,t) \phi^k.$$

where ν is a constant to be determined. We say that the equation passes the test if

1. there is a negative integer value of ν such that such a formal expansion exists;

2. this formal expansion has the further property that exactly $m - 1$ of the coefficients u_k can be prescribed arbitrarily. Counting ϕ, the expansion therefore involves m arbitrary functions in all.

We say that the equation has the *weak Painlevé property* if the above statements are true with an expansion of the form

$$u = \phi^{\nu/r} \sum_{k \geq 0} u_k(x, t) \phi^{k/r},$$

for some integer r.[8]

This algorithm has its origin in the "Painlevé test" based on the observation that an inordinate number of ODE reductions of integrable equations had no movable critical points; thus, the six Painlevé transcendents (see Ince, Ablowitz-Clarkson, Cosgrove) can all be obtained in this fashion. The WTC method obviates to some extent the need to consider *all* ODE reductions of the equation under consideration.

Since the gradient of ϕ is non-zero by assumption, it is advantageous to use the "reduced Ansatz" where $\phi = x_n - \psi(x_1, \ldots, x_{n-1})$ and the coefficients u_k are independent of x_n. It is clearly equivalent to the previous formulation, upto a permutation of the x_k. More structure is sometimes unraveled by the original formulation, as we will see presently.

The test provides some information even when it does not succeed. Indeed, as soon as the first term has been computed, it is possible, as we show, to recast the equation in Fuchsian form, in which case we know that there always exist convergent series solutions involving logarithms. The absence of logarithms is equivalent to the passing of the test. Before giving a general result in this direction, let us work out an example in some detail.

EXAMPLE: Consider the KdV equation written in the form

(4.66) $$u_{xxx} = u_t + uu_x.$$

Let us assume the putative singularity surface is given by $x = \psi(t)$, where ψ is analytic near $t = 0$, and assume $\psi(0) = 0$ to fix ideas. Let us use as new coordinates $(x - \psi(t), t)$, which we still denote by (x, t) for simplicity. We find

$$u_{xxx} = u_t - \psi' u_x + uu_x.$$

---

[8]Some authors require that the solution be globally meromorphic. There is, except in limited cases, no algorithmic way to check this at the present time. Also, one usually requires $\gcd(-\nu, r) = 1$.

Substituting $u = x^\nu(a_0(t) + a_1(t)x + \ldots)$, we find that the terms of the equation have leading orders $a_0\nu(\nu-1)(\nu-2)x^{\nu-3}$, $a_{0t}x^\nu$, $-\psi'a_0\nu x^{\nu-1}$ and $a_0^2\nu x^{2\nu-1}$. We would like to have $a_0 \neq 0$ (otherwise the leading exponent would have to be greater than ν). Also, ψ' is assumed to be nonzero since we are interested in general singularity surfaces. This requires that at least two of these four leading terms be equal. If $\nu \geq 0$, we need either $\nu - 1 = \nu - 3$, which is absurd, or $\nu(\nu-1)(\nu-2) = 0$. But in the latter case, we are not dealing with a singular expansion, but actually with a Cauchy problem, covered by the Cauchy-Kowalewska theorem. We therefore focus on $\nu < 0$. We now require $\nu - 3 = 2\nu - 1$, these being the lowest exponents. The desired leading order is therefore $\nu = -2$, and $a_0 = 12$.

To proceed, we convert the equation into a system by letting $u_j = (x\partial_x)^j u$ for $j = 0, 1, 2$, which results into the system

$$\begin{aligned}
x\partial_x u_0 &= u_1; \\
x\partial_x u_1 &= u_2; \\
x\partial_x u_2 &= 3u_2 - 2u_1 + x^3 u_{0t} + x^2 u_0 u_1.
\end{aligned}$$

Since $\nu = -2$, we let $u_j = x^{-2}v_j$, which produces the system

$$\begin{aligned}
(x\partial_x - 2)v_0 &= v_1; \\
(x\partial_x - 2)v_1 &= v_2; \\
(x\partial_x - 2)v_2 &= 3v_2 - 2v_1 + v_0 v_1 - \psi' x^2 v_1 + x^3 v_{0t}.
\end{aligned}$$

For $x = 0$, we find that (since $v(0) \neq 0$), we need $v = (12, -24, 48)$. Letting $v_0 = 12 + xw_0$, $v_1 = -24 + xw_1$, $v_2 = 48 + xw_2$, we obtain

$$\begin{aligned}
(x\partial_x - 1)w_0 &= w_1; \\
(x\partial_x - 1)w_1 &= w_2; \\
(x\partial_x - 4)w_2 &= 10w_1 - 24w_0 + xw_0 w_1 - \psi' x^2 w_1 + 24x\psi' + x^3 w_{0t}.
\end{aligned}$$

This is a Fuchsian system of the form

$$xw_x + Aw = xf(t, v, v_x),$$

with

$$A = \begin{pmatrix} -1 & -1 & 0 \\ 0 & -1 & -1 \\ 24 & -10 & -4 \end{pmatrix}.$$

Its eigenvalues are 2, -3 and -5, which implies that there is a convergent solution which is analytic in $(x, \ldots, (x \ln x)^l)$, and which involves (at least) two arbitrary functions corresponding to the two negative eigenvalues. However, an explicit calculation upto order five reveals that the coefficients of the logarithmic terms are all zero, so that the KdV equation does pass the WTC test.

There are also points of contact between the structure of the singular expansion constructed above, and the Lax pairs and Bäcklund transformations. Indeed, let us come back to the redundant form of the expansion. It may be hoped that the complexity in the function ϕ can be used to simplify the form of the expansion. In fact, the expansion truncates: Let us, to simplify the algebra, write the KdV equation in the form $u_t + uu_x + u_{xxx} = 0$. We find, (after substitution and identification of like powers of ϕ), a sequence of equations, the first few being

$$(4.67) \qquad u_0 = -12\phi_x$$

$$(4.68) \qquad u_1 = 12\phi_{xx}$$

$$(4.69) \qquad \phi_x\phi_t + 4\phi_x\phi_{xxx} - 3\phi_{xx}^2 + \phi_x^2 u_2 = 0$$

$$(4.70) \qquad \phi_{xt} + \phi_{xx}u_2 + \phi_{xxxx} - \phi_x^2 u_3 = 0$$

$$(4.71) \qquad \partial_x[\phi_{xt} + \phi_{xx}u_2 + \phi_{xxxx} - \phi_x^2 u_3] = 0.$$

The coefficients u_4 and u_6 are arbitrary; they correspond to the presence of two (complicated) identities in the above sequence. However, it is possible to *truncate* the series by setting $u_k = 0$ for $k \geq 3$, provided that u_0, u_1 adn u_2 are given by the above formulae, and

$$\phi_{xt} + \phi_{xx}u_2 + \phi_{xxxx} = 0$$

$$u_{2t} + u_2 u_{2x} + u_{2xxx} = 0.$$

In that case, since we have truncated the singular series *at its constant term*, it is natural that u_2 should solve the same equation as u. Note that

$$u - u_2 = 12\partial_x^2 \ln \phi.$$

Elimination of u_2 leads to a constraint on ϕ of the form

$$\frac{\phi_t}{\phi_x} + \{\phi; x\} = \lambda,$$

where $\{\phi; x\} = \phi_{xxx}/\phi_x - (3/2)(\phi_{xx}/\phi_x)^2$ is the Schwarzian derivative of ϕ with respect to x.

The form of (4.54) suggests the change of unknown $\phi_x = v_2$.

THEOREM 4.20 *If u_2 is any given solution of KdV and*

$$(4.72) \qquad \phi_x = v^2$$

$$(4.73) \qquad 6v_{xx} + u_2 v = \lambda v;$$

$$(4.74) \qquad 2v_t + u_2 v_x + 2v_{xxx} + \lambda v_x = 0,$$

where λ is a constant, then $u = u_2 + 12\partial_x^2 \ln \phi$ also solves KdV.

We therefore find that both the Lax pair and a Bäcklund transformation are embedded in the WTC expansion, when suitably interpreted. Note however that reductions of the self-dual Yang-Mills equations can exhibit a movable essential boundary, so that one must be careful in any general claim on the validity of the test. A number of proofs of the Painlevé conjecture are indicated in the Notes.

The symmetry test.

This procedure checks for the existence of higher symmetries as defined below.

Consider a one-dimensional PDE of the form

$$u_t = F[u].$$

We write u_i for the successive x-derivatives of u and define the total derivative operators

$$
\begin{aligned}
D = D_x &= \partial_x + \sum_{i \geq 0} u_{i+1} \partial/\partial u_i; \\
D_t &= \partial_t + \sum_{i \geq 0} D^i F \partial/\partial u_i.
\end{aligned}
$$

We say that a function $G(x, t, u, u_1, \dots)$ is a symmetry of (4) if

$$D_t G = F_* G := \sum_{i \geq 0} \frac{\partial F}{\partial u_i} D^i G.$$

In other words, G solves the linearization of (4).

One connection with integrable systems is that for any symmetry, one has the operator equation

$$D_t G_* - D_\tau F_* + [F_*, G_*] = 0,$$

where $D_\tau = \sum_{i \geq 0} D^i G \partial/\partial u_i$. This expresses the compatibility of two flows: $v_t = F_* v$ and $v_\tau = G_* v$. As an example, the r.h.s. of the fifth KdV equation provides a symmetry for the third order KdV equation.

The algorithm relies on the construction of canonical densities which can be computed from the equation. It then remains to investigate the possibility of finding, for any such density ρ, a function σ such that $D_t \rho = D_x \sigma$. (See the paper by Mikhailov-Shabat-Sokolov in Zakharov (1991).)

Analytic characterizations.

Most of the criteria we have described aim at characterizing integrability by algebraic properties. It is possible in some cases to single out integrable equations by the analyticity properties of their solutions. The WTC test is one such case; we present another.

The sine-Gordon equation $u_{tt} - u_{xx} + \sin u = 0$ has a family of periodic solutions, called breathers

$$(4.75) \qquad u_{SG} = 4\arctan\left(\frac{\varepsilon}{\sqrt{1-\varepsilon^2}}\frac{\cos(t\sqrt{1-\varepsilon^2})}{\cosh(\varepsilon x)}\right),$$

where ε is related to the period T by

$$\varepsilon^2 + (2\pi/T)^2 = 1.$$

It turns out that these are characteristic of the sine nonlinearity in the following sense: let g be entire with $g(0) = 0$ and $g'(0) = 1$; we write

$$g(u) = \sum_{m=1}^{\infty} g_m u^m.$$

We now consider the problem

$$(4.76) \qquad \begin{cases} u_{tt} - u_{xx} + g(u) = 0 \\ u(x, t+T) = u(x,t);\ u(x,t) \to 0 \text{ as } |x| \to \infty;\ u_t \not\equiv 0. \end{cases}$$

Any solution of this problem will be called a breather.

The sine-Gordon breather was discovered in the sixties and investigated as a model of elementary particle which, due to its internal "pulsation" would have more structure than a traveling wave. The question then arises whether there are such solutions for more general equations.

Breathers were sought numerically mostly for the double sine-Gordon and ϕ^4 cases ($g(u) = \sin u + k \sin 2u$ and $-(1+u) + (1+u)^3$ respectively). For the latter in particular, one can easily find kink-type traveling waves, which tend to different limits at infinity. By launching a solution which consists initially of a kink and its reflection (anti-kink), well-separated, one may experiment on the "collision" of such waves. It was found that if the relative velocity of the kink and the anti-kink fell in certain intervals, a long-lived breather-like oscillation took place, and then was destroyed. Recent work of this type is by Campbell and Peyrard, in which references to some earlier work can be found; see also Dodd, Eilbeck, Gibbon and Morris and section 7 of Kichenassamy (1991). The existence of ϕ^4 breathers was also suggested by formal calculations due to Dashen, Hasslacher and Neveu.

One can in fact construct, under quite general conditions, series representing approximate breathers to all orders; however, they have singularities as functions of the period, except if the equation is equivalent to the sine-Gordon equation. Furthermore, a rearrangement of this divergent series can be identified, if g is odd, with a series which converges for large x. In other words, if the breather is to exist and resemble the sine-Gordon breather, it must be the analytic continuation of this second series. More precisely,

1. There are formal breathers:

    (4.77) $u = \varepsilon u_1(\varepsilon x, t\sqrt{1-\varepsilon^2}) + \varepsilon^2 u_2(\varepsilon x, t\sqrt{1-\varepsilon^2}) + \cdots$

    for (3), characterized by periodicity, decay and parity conditions, if and only if

    (4.78) $\dfrac{5}{6}g_2^2 - \dfrac{3}{4}g_3 > 0.$

2. If series (4) defines a solution of (3) analytic in ε, εx and $t\sqrt{1-\varepsilon^2}$ for $|\varepsilon| < \frac{1}{\sqrt{2}} + \delta$ for any $\delta > 0$ and, say, x, t close to the real axis, and if this function has an expansion in powers of ε, $e^{-\varepsilon x}$ and $\cos(t\sqrt{1-\varepsilon^2}$ convergent for large x, then $u + g(u) = \alpha^{-1}\sin(\alpha u)$. (In this case, the series does have the asserted analyticity property.)

3. If g is *odd*, and
    $$|g_m| \le \alpha^m/m!$$
    for some $\alpha > 0$, then eq. (3) does have a time periodic solution on some interval $(A, +\infty)$ given by a power series in $e^{-\varepsilon x}$, convergent for large x and small ε. This solution is the only one which is odd in $\cos(t\sqrt{1-\varepsilon^2})$ and tends to zero as x tends to $+\infty$.

This result has been extended to fifth order perturbations of the KdV equation, and has been cast as a general procedure for the testing of complete integrability.

A useful consequence of the existence of formal breathers is that truncating the series at level N produces a solution of an equation of the form

$$u_{tt} - u_{xx} + g(u) = \varepsilon^{N+1}k(\varepsilon, \varepsilon x, t\sqrt{1-\varepsilon^2}),$$

where k is smooth, and exponentially decaying in its second argument. Its H^s norm is therefore uniformly bounded by $c(N)\varepsilon^{2N+1}$. It easily follows that if we take as Cauchy data those of the truncated series, and solve the equation $u_{tt} - u_{xx} + g(u) = 0$, we will obtain a solution which is close to the truncated series for times of order $1/\varepsilon$ (and more, if N is large). This gives a candidate for the long-lived breathers that have been observed numerically.

There is another way to convince oneself that the sine-Gordon equation may be singled out by the existence of breather solutions. (Birnir-McKean-Weinstein, Denzler) It is however restricted to nonlinearities with coefficients close to those of the sine. Indeed, if we assume that the breather (4.77) exists for a set of ε, and depends smoothly on the g_k, one could differentiate the equation with respect to the g_k and conclude that the variation of the breather must solve the linearization of the sine-Gordon equation at the sine-Gordon breather. It turns out that the kernel of this linearization can be computed explicitly, and we can therefore derive infinitely many orthogonality conditions, which yield necessary constraints on the g_k for a breather depending smoothly on the coefficients of g to exist. It turns out that these constraints seem to be none other than the linearization of the residues of the poles of (4.77). Requiring that they all vanish can be shown to exclude again all equations except the sine-Gordon equation.

The Wahlquist-Eastabrook method.

This method enables one to decide whether or not a given equation has a commutator representation of a specified form, by constructing a related Lie algebra. We illustrate it on the case of KdV.

Let u satisfy $u_t = 6uu_x - u_{xxx}$, and let us seek two equations

$$\begin{aligned} \psi_x &= F(u, \lambda)\psi, \\ \psi_t &= G(u, u_x, u_{xx}, \lambda)\psi. \end{aligned}$$

One could of course start by allowing F and G to depend on more derivatives of u. In an unfamiliar example, one would need to make an educated guess.

We express that the compatibility of these equations is equivalent to KdV:

$$F_u u_t - D_x G + [F, G] = 0,$$

where D stands for a total derivative. Substituting u_t and identifying terms, one finds first, from the coefficient of $u_3 = u_{xxx}$ that

$$G_{u_2} = -F_u,$$

or $G = -u_2 F + a(u, u_x)$. The u_2 terms now give

$$a_{u_x} = u_x F_u + [F_u, F],$$

hence $a = u_x^2 F_u / 2 + u_x[F_u, F] + b(u)$. Finally, we arrive at the equations

$$\begin{aligned} F_{uuu} &= 0; \\ [b, F] &= 0; \\ \partial_u[F_u, F] + \frac{1}{2}[F_{uu}, F] &= 0; \\ b_u &= 6uF_u + [[F_u, F], F]. \end{aligned}$$

From the first equation, we find that there are constant matrices X_0, X_1, X_2 such that

$$F = X_0 + X_1 u + X_2 u^2.$$

If we introduce $X_6 = [X_1, X_0]$, $X_4 = [X_1, X_6]$, $X_5 = [X_2, X_6]$, we find that $\{X_0, \dots, X_6\}$ form a Lie algebra with explicitly known structure constants. It remains to find a representation of this algebra to work one's way backwards and find a commutator representation for KdV as desired.

Hirota's method.

This procedure consists in casting a given equation into a "bilinear form" defined below, from which it is possible to investigate in a simple way the existence of rational solutions and of solutions which involve rational functions of a finite number of exponentials. Since the multi-soliton solutions are usually of the latter form, the procedure has a direct point of contact with inverse scattering. It is believed that there are deeper connections between Hirota's method and IST (see Newell (1984)).

Let us present the procedure for the case of KdV, and make brief comments on other equations. Some motivation for these steps is given in §4.7.

If $u = v_x$ solves the KdV equation, and vanishes with all is derivatives at spatial infinity, we find

$$v_t + 3v_x^2 + v_{xxx} = 0.$$

If $v = 2\tau_x/\tau$, we find

$$(4.79) \qquad \tau\tau_{xt} - \tau_x\tau_t + \tau\tau_{xxxx} - 4\tau_x\tau_{xxx} + 3\tau_{xx}^2 = 0.$$

Introduce *Hirota's bilinear operator* by

$$D_x\sigma \cdot \tau := \sigma_x\tau - \sigma\tau_x,$$

and more generally

$$\prod_i D_{x_i}^{\alpha_i}\sigma \cdot \tau := \prod_{r=1}^n \sum_{\varepsilon_r \to 0} (\frac{\partial}{\partial\varepsilon_r})^{\alpha_r}\sigma(x_r + \varepsilon_r)\tau(x_r - \varepsilon_r),$$

for functions σ and τ involving n variables. (4.79) then takes the *bilinear form*

$$(4.80) \qquad (D_xD_t + D_x^4)\tau \cdot \tau = 0.$$

Since 1 is a trivial solution of this equation, one then tries a perturbation series

$$\tau = 1 + \varepsilon f_1 + \varepsilon^2 f_2 + \dots$$

This produces a recurrence relation for the f_j. In particular, we need

$$\begin{aligned}
2\partial_x(\partial_t + \partial_x^3)f_1 &= 0; \\
2\partial_x(\partial_t + \partial_x^3)f_2 &= (D_xD_t + D_x^4)f_1 \cdot f_1; \\
2\partial_x(\partial_t + \partial_x^3)f_3 &= (D_xD_t + D_x^4)(f_1 \cdot f_2 + f_2 \cdot f_1).
\end{aligned}$$

We first take f_1 to be a polynomial. This produces rational solutions. From the first equation, we must take $f_1 = a_0 + a_1x + a_2x^2 + a_3x^3 + a_4(x^4 - 24xt) + bt,$

and we may truncate by taking $f_j = 0$ for $j \geq 2$ is $a_4 = b - 12a_3 = 3a_1a_3 - a_2^2$. As an example, we recover (for $a_1 = 0$)the solution $u = -6x(x^3 - 24t)/(x^3 + 12t)^2$.

We can also take $f_1 = \sum_{i=1}^{N} a_i \exp(k_i x + \omega_i t)$, where $\omega_i + k_i^3 = 0$. We now use the (easily proved) property

$$(4.81) \qquad P(D_t, D_x)e^{\eta_1} \cdot e^{\eta_2} = \frac{P(\omega_1 - \omega_2, k_1 - k_2)}{P(\omega_1 + \omega_2, k_1 + k_2)} P(D_t, D_x)e^{\eta_1 + \eta_2} \cdot 1.$$

We find

$$f_2 = \sum_{i>j} e^{\eta_i + \eta_j + A_{ij}},$$

where $\exp(A_{ij}) = (\frac{k_i - k_j}{k_i + k_j})^2$. We recognize the phase shift for the KdV equation.

Defining "integrability".

The word "integrability" usually refers to solvability in closed form, to solvability via a variant of IST, or to the existence of (a Hamiltonian structure and) the existence of the maximum number of conserved quantities (hopefully in involution with respect to some symplectic structure). Criteria for integrability usually address one of these aspects. From a practical standpoint, the issue is really to know whether it is possible to obtain detailed information on the long-time behavior of solutions analytically. From a more general standpoint, the problem is to explain what the above definitions have in common.

(1) An integrable system is, in mechanics, a system of ODE which is Hamiltonian with respect to a symplectic structure in a phase space of dimension $2n$, and which has n first integrals in involution *which are globally smooth*. In infinite dimensions, one would like to have a natural Hamiltonian structure, and a statement of "completeness" of the first integrals of the flow under consideration. Such a set-up can be constructed from the IST for KdV and many other equations integrable by IST; essentially, the values of (the logarithm of) the transmission coefficient for different values of k are the relevant first integrals.

We saw that the WTC algorithm does produce such integrals in a limited number of cases.

A related requirement is to demand an infinite set of polynomial conserved densities. This can be checked in an algorithmic fashion.

(2) If one wishes to solve a given equation by an inverse method, one must look for an eigenvalue problem that would be relevant. The method of Wahlquist and Eastabrook (and the WTC method to some extent) address this aspect of the issue. The construction of families of equations from a given eigenvalue problem does provide lists of "integrable" equations.

(3) If an approach via IST is possible, it is hoped that it is also possible to derive a Bäcklund transformation. Similarly, there should be analogues of multi-solitons and rational solutions; one can therefore test for the existence of those, to detect equations with special properties.

4.7 FURTHER RESULTS AND PROBLEMS

1. (a) In the notation of §4.3, prove that (for smooth potentials), there is an asymptotic expansion

$$m \sim \sum_{j=0}^{\infty} m_j(x) k^{-j}$$

with $m_0 = I$, $m_j \to 0$ as $x \to \infty$, and $(\partial - q)m_j = [J, m_{j+1}]$. Conclude in particular that $q = \lim_{k \to \infty} -[J, k(m - I)]$.

(b) Let A be a constant matrix. Define F_j by the expansion $mAm^{-1} \sim \sum_{j=0}^{\infty} F_j(x, A)k^{-j}$ as $k \to \infty$, where $F_0 = A$ and $F_j \to 0$ as $x \to -\infty$ for $j > 0$. For any nonnegative integer r, show that the evolution equation

$$\frac{\partial q}{\partial t} = -[J, F_{r+1}(x, A)]$$

corresponds to the scattering data evolution

$$v_t = [k^r A, v].$$

(c) Similarly, for $r < 0$, G_j by the Taylor expansion of mAm^{-1} at $k = 0$ (which exists if v vanishes to infinite order an $k = 0$): $mAm^{-1} \sim \sum_{j=0}^{\infty} G_j(x, A)k^j$ and discuss the corresponding flows (see Beals-Coifman (1985)).

2. (a) Compute explicitly the fifth member of the KdV hierarchy. What equation corresponds to the choice $A = A_5 + cA_3$?

(b) Take $A = a + b\partial_x$, and $a = -b_x/2$. Postulate the relation $b = b_n \lambda^n + \cdots + b_0$ by analogy with the AKNS case. Take $b_n = -1$ to fix ideas. Derive the relations $\partial_x b_k = -(1/4)Mb_{k+1}$ for $k \geq 1$, and show that the corresponding evolution equation is

$$u_t = -\frac{1}{2}Mb_0,$$

where $M = \partial_x^3 + 2(u\partial_x + \partial_x u)$.

(c) Writing $M = \partial_x R$ with $R = \partial^2 + 4u - 2I_x \circ u_x$ (where $I_x = \int_{-\infty}^{x}$, conclude that we may take B_0 to be a multiple of $\partial_x N^n$. This is one explicit form of the KdV hierarchy. Despite appearances, the nth flow *is* local.

(d) Define the square root of $-L$ to be a formal series $(-L)^{1/2} = \partial + \sum_{j \leq 0} a_j \partial_x^j$, where the composition rule, which generalizes Leibniz' rule) is given by:

$$a\partial^k \circ b\partial^l = a[\sum_{j \geq 0} \binom{k}{j} (\partial^j b)\partial^{k+l-j}],$$

with $\binom{k}{j} := k(k-1)\ldots(k-j+1)/j!$; note that k and l can be negative. Let also $(-L)^{m/2} = ((-L)^{1/2})^m$. Prove that the mth flow is given by $u_t = [(-L)^{m/2}]_+$, where the $+$ indicates that only positive powers of ∂_x are retained. This construction extends to higher-order operators (see I. M. Gel'fand and L.

A. Dikii, Fractional powers of operators and Hamiltonian systems (1976) *Funct. Anal. and Appl.*, *10*: 259–273, and, for the solution of the resulting equation by inverse scattering, Beals-Deift-Tomei (1988)).

3. (a) By iterating twice the integral equation (4.20), prove that

$$\Phi = 1 - \frac{1}{2ik}\int_{-\infty}^{x} u(y)\,dy - \frac{u(x)}{(2ik)^2} + \frac{1}{2}(\frac{1}{2ik}\int_{-\infty}^{x} u(y)\,dy)^2 + o(1/k^2)$$

as $k \to \infty$ in the upper half-plane. Prove the corresponding statement for Ψ. Specify the decay and differentiability assumptions this requires.

(b) Iterating the formula in Th. (4.10) in the same way, prove that

$$\frac{1}{a(k)} = 1 + \frac{1}{2ik}\int_{-\infty}^{\infty} u(x)\,dx - \frac{1}{2}(\frac{1}{2ik}\int_{-\infty}^{\infty} u(y)\,dy)^2 + o(1/k^2).$$

(c) Conclude that

$$\lim_{k\to\infty} (2ik)^2[\frac{\Phi\Psi}{a} - 1] = -2u(x).$$

4. (a) Let ϕ and ψ be two smooth functions of (x,y,z) subject to the condition $\phi_y + \phi_z\psi = \psi_x + \psi_z\phi$. Show that the system

$$\begin{aligned} F_x &= \phi(x,y,F), \\ F_y &= \psi(x,y,F), \\ F(x_0,y_0) &= F_0 \end{aligned}$$

has a unique solution near (x_0, y_0).

(b) Let $u'(x,t)$ satisfy $u'_{xt} = 0$. Show that the equations

$$\begin{aligned} u_x &= u'_x - a\exp\frac{u+u'}{2}; \\ u_t &= -u'_t - \frac{2}{a}\exp\frac{u-u'}{2}, \end{aligned}$$

provides a solution of $u_{xt} = e^u$. Show conversely that one can solve for u' is u is any solution of $u_{xt} = e^u$. Show that if $u' = f(x) + g(t)$,

$$d(\exp(-\frac{1}{2}(u - f + g))) = \frac{a}{2}e^f\,dx + \frac{1}{a}e^{-g}\,dt.$$

Conclude that there are, locally, two increasing functions $F(x)$ and $G(t)$ such that $u = \ln[2F'G'/(F+G)^2]$.

(c) Show that the relations

$$\begin{aligned} u_x &= u'_x - 2a\sin\frac{u+u'}{2}; \\ u_t &= -u'_t - \frac{2}{a}\sin\frac{u-u'}{2}, \end{aligned}$$

are solvable for u if and only if $u'_{xt} = \sin u$, and that one then has $u_{xt} = \sin u$ too. Conversely, discuss the possibility of solving for u' is u is given. [This is an *auto-Bäcklund transformation* for the SG equation.] Find the kink solution starting from $u' = 0$. See also problem 5g below.

5. Let u_i, $i = 0$, 1, be two functions of x, in $\mathcal{S}$. Let $L_i = -\partial_x^2 + u_i$. Let finally ψ_0 satisfy $L_0\psi_0 = \lambda\psi_0$.

(a) If A is a function and μ is a constant, show that the relations

$$\begin{aligned}
\psi_1 &= A\psi_0 + \psi_{0x}, \\
u_1 - u_0 &= 2A_x \\
A^2 - A_x - u_0 + \mu &= 0
\end{aligned}$$

imply $L_1\psi_1 = \lambda\psi_1$. [Thus, one can transform any eigenfunction of L_0 into an eigenfunction of L_1.]

(b) Let y satisfy $L_0 y = \mu y$. Show that one can take $A = -y_x/y$.

(c) Assume that $y \sim \alpha_\pm e^{\eta|x|}$ as $x \to \pm\infty$, with $\eta > 0$. (This implies that $-\eta^2$ is not and eigenvalue of L_0. Compute the scattering functions a_1 and b_1 of u_1 in terms of those of u_0. Show in particular that a_1 will have a zero at $i\eta$ which a_0 does not share. Interpret the result. [*Remark:* If $-\eta^2$ is an eigenvalue of $L - 0$, one can use (a-b) to *remove*, rather than add, an eigenvalue. It suffices in fact to exchange the roles of L_0 and L_1 to make this observation precise.]

(d) *Example 1.* Take in (c) $u_0 = 0, y = \cosh(\eta x)$ for some $\eta > 0$. What is μ? Find u_1 and, using property (P) for $\psi_0 = \exp(\pm i\xi x)$, find the reflection and transmission coefficients of u_1. Could the result be expected? Find the (L^2-) eigenfunctions of u_1.

(e) *Example 2.* Take in (c) $u_0 = -2/\cosh^2 x, y = \cosh^2 x$. What is μ? Find u_1, the reflection and transmission coefficients, and the L^2-eigenfunctions.

(f) *Example 3.* Take in (c) $u_0 = -2/\cosh^2 x, y = 1/\cosh x$. What is u_1? Can you explain the result?

(g) Coming back to the situation of (a-c), let $\mu = \mu_1$, and introduce a third potential u_2 corresponding to the value $\mu = \mu_2$. Prove that, if we let $v_{ix} = u_i$ for $i = 1, 2$, with the convention that $v_i - v_0 = 2A_i$, one has

$$(v_i + v_0)_x = 2\mu_i + \frac{1}{2}(v_i - v_0)^2$$

for $i = 1$ or 2. Show that there is a function u_3 related to both u_1 and u_2 by transformations of parameters μ_2 and μ_1 respectively, so that

$$(v_3 + v_i)_x = 2\mu_{2-i} + \frac{1}{2}(v_3 - v_i)^2.$$

(Why are the normalizations for v_3 compatible?) [*Hint:* Compute the scattering functions of u_3.] Conclude that

$$(v_3 - v_0)(v_1 - v_2) = 4(\mu_2 - \mu_1).$$

(This is an example of *permutability* of Bäcklund transformations. Another classical example is the case of SG, which goes back to Bianchi (see Lamb, (1971) *Rev. Modern Phys.*, *43*: 99, and *Physica 66*: 298 (1973). The SG case may be interpreted as a 2×2 generalization of the result of (a).)

6. Apply the (weak) Painlevé test to the following equations

(a) $y'y''' - 2(y'')^2 + 18y' = xy(y')^2$. (Test for singularities at $x = 0$; the case $y'(0) = 0$ must be treated separately.)

(b) $u_{xx}u_{tt} - u_{xt}^2 = 0$.

(c) $u_t = u^3 u_{xxx}$. (This equation (H. Dym equation) does have a commutator representation).

(d) $x' = ax + cxy; \quad y' = by + dxy$.

Discuss in particular whether or not -1 is a resonance. (See the discussion in Kichenassamy and Srinivasan (1995)).

(e) $u_t = u_{xx} + uu_x$.

7. Show that the one-soliton for the AKNS systems may blow-up in finite time if no symmetry condition on (q, r) is imposed.

8. Show that the coefficient a in the scattering theory of KdV has, if u is smooth, an asymptotic expansion in inverse powers of k. Its coefficients are all constants of motion for the KdV equation; they are furthermore given by integrals of local expression in u and its derivatives. Same question for the NLS equation.

9. Prove that if $u = -2\alpha v_x/v$,

$$u_t + uu_x - \alpha u_{xx} = -\frac{2}{v}(\partial_x - \frac{v_x}{v})(v_t - \alpha v_{xx}).$$

This is the Cole-Hopf transformation, which played an important role in the early history of shock waves.

10. There is a combination of the third and fifth order KdV equations which has a two-soliton solution corresponding to two solitons traveling at exactly the same speed.

11. Apply the reductive perturbation method to obtain the indicated model equations.

(a) $u_{tt} - u_{xx} + u + g_2 u^2 + \cdots = 0$ (NLS).

(b) $u_{tt} - c^2 u_{xx} - a^2 u_{xxxx} = 0$, with $c^2 = (1 + u_x^2)^{-3/2}$ (mKdV).

12. Construct a 2×2 eigenvalue problem associated to the sine-Gordon equation in the form $u_{tt} - u_{xx} + \sin u = 0$. (See D. J. Kaup and A. Newell (1978) Solitons as particles, oscillators, and in slowly varying media: a singular perturbation theory, *Proc. Roy. Soc. London A361*: 413–446, and Goursat and Cauchy problems for the sine-Gordon equation, *SIAM J. Appl. Math., 34*: 37–54.)

NOTES

The property of elastic interaction was observed numerically by Zabusky and Kruskal in their study of the Fermi-Pasta-Ulam paradox; their results correspond to the KdV equation with periodic boundary conditions. The inverse scattering solution of the KdV equation on the whole line was introduced in Greene-Gardner-Kruskal-Miura (1968) and immediately interpreted in terms of isospectral deformations in Lax (1969) where Lax pairs and the KdV hierarchy are introduced. A few technical properties related to Bäcklund transformations had been observed much earlier, and have sometimes a differential-geometric interpretation.

The universality of certain envelope equations was recognized gradually from the sixties by Gardner-Su, Newell-Whitehead, Eckhaus, to name a few. We have followed the systematic derivation of Taniuti and co-workers (see Taniuti-Wei (1968)) for the NLS, KdV, and mKdV equations. For the treatment of instabilities, we have followed Gibbon (1980), which contains many references to earlier work. This topic has been recently reinvestigated by Calogero and Eckhaus. Note that there are recent precise estimates of the time of validity of the Ginzburg-Landau equation for important cases (van Harten (1990), Collet-Eckmann (1990)).

The technicalities on the inverse scattering via the GLM equation are taken from Faddeev (1963) and Deift-Trubowitz (1979) (see also Melin). For further results, see also Calogero-Degasperis. The AKNS formalism was introduced in Ablowitz-Kaup-Newell-Segur (1974), following the results of Zakharov-Shabat and Lamb on the NLS and SG equations respectively. The Toda lattice was studied in Flaschka (1974) and Manakov (1975). A list of integrable equations with recent references can be found in Ablowitz and Clarkson (1990). It has been suggested that all integrable equations may be reductions of the self-dual Yang-Mills equation; a large number have been proved to be (see Chakravarty-Ablowitz-Clarkson (1990)).

The Riemann-Hilbert formulation of the generalized AKNS systems is taken from Beals-Coifman (1984) and (1985); see also Novikov et al. (1984); the latter contains information on the periodic problem, which has not been touched upon here.

The formulation of the KdV equation as a completely integrable Hamiltonian system can be found in the monograph of Faddeev and Takhtajan. This issue has been investigated recently by Beals and Sattinger (1991) for the three-wave interaction.

The WTC test has been introduced in Weiss-Tabor-Carnevale (1983), following the Painlevé conjecture for ODE; see Ablowitz-Segur, Olver-McLeod. The latter paper contains one case where one can prove the Painlevé conjecture. A large number of equations have been tested (see the monograph by Euler and Steeb); two recent discussions of this test are Clarkson and Kruskal and

Flaschka-Tabor. A related construction is the method of Kowalewski exponents (see e.g. Yoshida *et al.* (1985)) for ODE with homogeneous nonlinearities. The convergence results are due to Kichenassamy and Littman (1994), and Kichenassamy and Srinivasan (1995) contains a detailed treatment of the application of these ideas in practical situations, and shows the role of the invariant theory of binary forms in streamlining the computations involved. Details on the symmetry test can be found in Sokolov (1979) and Mikhailov-Shabat-Sokolov (1983) (see Zakharov (1991)); the latter paper contains several lists of third and fifth order equations which pass the test. Some seventh-order equations were investigated by Bilge (1993). On the Wahlquist-Eastabrook method, see Wahlquist and Eastabrook (1975) and Dodd and Fordy (1983). On Hirota's method, see Hirota, Jimbo and Miwa. For rational solutions of KdV, see Adler and Moser (1978); many families of rational solutions have also been obtained by WTC as a by-product of their method. The two results on the breather problem are due to Kichenassamy (1991) (which contains references to earlier work), and Birnir-McKean-Weinstein and Denzler (1994). The first method has been applied to other equations in Kichenassamy-Olver (1993).

REFERENCES

ABLOWITZ, M. AND CLARKSON, P. A. (1991) *Solitons, Nonlinear Evolution Equations and Inverse Scattering,* Cambridge University Press.

ABLOWITZ, M., KAUP, D. J., NEWELL, A. C., SEGUR, H. (1974) The inverse scattering transform — Fourier analysis for nonlinear problems, *Studies in Appl. Math., 53*: 249–315.

ADLER, M. AND MOSER, J. (1978) On a class of polynomials connected with the Korteweg-de Vries equation, *Comm. Math. Phys. 61*: 1–30.

BEALS, R. AND COIFMAN, R. R. (1984) Scattering and inverse scattering for first order systems, *Comm. Pure Appl. Math, 37*: 39–90.

BEALS, R. AND COIFMAN, R. R. (1985) Inverse scattering and evolution equations, *Comm. Pure Appl. Math, 38*: 29–42.

BEALS, R., DEIFT, P. AND TOMEI, C. (1989) *Direct and Inverse Scattering on the Line,* Math. Surveys and Monographs, AMS, Providence, R.I., vol. 28.

BEALS, R. AND SATTINGER, D. (1991) On the complete integrability of completely integrable systems, *Comm. Math. Phys., 138*: 409–436.

BIRNIR, B., McKEAN, H. AND WEINSTEIN, A. (1994) The rigidity of sine-Gordon breathers, *Comm. Pure Appl. Math, 47*: 1043–1051.

CALOGERO, F. AND ECKHAUS, W. (1987) Nonlinear evolution equations, rescalings, model PDEs and their integrability, *Inverse Pbs. 3*: 229–262 (see also pages L27-L32) and *4*: 11–33 (1988).

CALOGERO, F. (1989) Universality and integrability of the nonlinear PDE describing N-wave interactions, *J. Math. Phys. 30*: 28–40, see also pages 639–654 in the same volume.

CHAKRAVARTY, S., ABLOWITZ, M. J., AND CLARKSON,. P. A. (1990) *Phys. Rev. Lett. 65*: 1085.

COLLET, J.-P. AND ECKMANN, P. (1990) The time dependent amplitude equation for the Swift-Hohenberg problem, *Comm. Math. Phys. 132*.

DEIFT, P., AND TRUBOWITZ, E. (1979) Inverse scattering, on the line *Comm. Pure Appl. Math, 32*: 121-251.

DENZLER, J. (1993) Nonpersistence of breather families for the perturbed sine-Gordon equation, *Comm. Math. Phys. 158*: 397–430, see also a forthcoming paper in *Ann. IHP. Anal. Non Linéaire*.

DODD, R. K. AND FORDY, A. P. (1983) *Proc. Roy. Soc. London A385*: 389–429.

GARDNER, C. S., GREENE J. M., KRUSKAL, M. D., MIURA, R. M. (1967) Method for solving the Korteweg-de Vries equation, *Phys. Rev. Lett. 19*: 1095–1097.

GARDNER, C. S., GREENE J. M., KRUSKAL, M. D., MIURA, R. M. (1974) Korteweg-de Vries equation and generalizations, *Comm. Pure Appl. Math, 27*: 97–133.

FADDEEV, L. D. (1959) The inverse problem in the quantum theory of scattering, *Uspekhi Mat. Nauk 14*: 57–119, translated as *J. Math. Phys. 4*: 72–104 (1963). See also *AMS Tansl. 65*: 139–166 (1965).

FLASCHKA, H. (1974) *Phys. Rev. B*: 1924–5.

FORDY, A. P. AND WOOD, J. C. (Eds.) (1994) *Harmonic Maps and Integrable Systems,* Aspects of Mathematics, vol. E 23, Vieweg, Wiesbaden

GIBBON, J. D., JAMES, I. N. AND MOROZ, I. (1979) *Proc. Roy. Soc. London A367*: 219–237.

GIBBON, J. D. AND McGUINNESS, M. J. (1981) Amplitude equations at the critical points of unstable dispersive physical systems, *Proc. Roy. Soc. London A377*: 185–219, (see also Akylas and Benney (1980) *Stud. Appl. Math. 63*: 209–226).

VAN HARTEN, A. (1991) On the validity of the Ginzburg-Landau equation, *J. of Nonlinear Sci. 1*: 397–422.

HIROTA, R. (1980) Direct methods in soliton theory, in *Topics in Current Physics, 17,* R. Bullough and P. Caudrey eds., Springer, 157–175; see also E. Date, M. Jimbo, M. Kashiwara and T. Miwa's article in *Proc. RIMS Symp. on Nonlinear Integrable Systems,* M. Jimbo and T. Miwa eds., World Scientific, 1983.

KICHENASSAMY, S. (1991) Breather Solutions of the Nonlinear Wave Equation, *Comm. Pure and Appl. Math. 44*: 789–818, see also *Contemporary Mathematics, 122* 73–76 (1991).

KICHENASSAMY, S. AND LITTMAN, W. (1993) Blow-up Surfaces for Nonlinear Wave Equations, Part I, *Comm. in P. D. E., 18 (3&4)*: 431–452, Part II, *Comm. in P. D. E., 18 (11)*: 1869–1899.

KICHENASSAMY, S. AND OLVER, P. J., (1992) Existence and Non-existence of Solitary Wave Solutions to Higher Order Model Evolution Equations, *SIAM J. Math. Anal. 23, No. 5*: 1141–1166.

KICHENASSAMY, S., AND SRINIVASAN, G. K., (1995) The structure of WTC expansions and applications, *J. Phys. A,: Math. Gen., 28*: 1977-2004.

KRUSKAL, M. D. AND CLARKSON, P. A. (1992) The Painlevé-Kowalevski and Poly-Painlevé tests, *Stud. Appl. Math., 86*: 87–165.

LAX, P. D. (1976) Almost periodic solutions of the KdV equation, *SIAM Rev., 18*: 351–375.

LAX AND LEVERMORE, C. D. (1983) *Comm. Pure Appl. Math. 36*: 253–290, 471–495, 809–830, VENAKIDES, S. (1984) *Comm. Pure Appl. Math. 37.*

MANAKOV, S. (1975) *Sov. Phys. JETP 40*: 269–74.

MCLEOD, J. B. AND OLVER, P. J. (1983) The connection between partial differential equations soluble by inverse scattering and ordinary differential equations of Painlevé type, *SIAM J. Math. Anal. 14*: 488–506.

MELIN, A. (1985) Operator methods for inverse scattering on the line, *Comm. PDE, 10*: 677–766.

NEWELL, A. C. AND WHITEHEAD, J. (1969) Finite bandwidth, finite amplitude convection, *J. Fluid Mech. 38*: 279–303.

NOVIKOV, S., MANAKOV, S., PITAEVSKII, L. AND ZAKHAROV, V. (1984) *Theory of Solitons,* Consultants Bureau, Plenum, New York.

SOKOLOV, V. V. (1988) On the symmetries of evolution equations, *Russian Math. Surveys 43 (5)*: 165–204.

TANIUTI, T. AND WEI, C. C. (1978) The reductive perturbation method, *J. Phys. Soc. Japan 24*: 941.

WAHLQUIST, H. AND EASTABROOK, F. (1975) *J. Math. Phys. 16*: 1–7.

WEISS, J., TABOR, M. AND CARNEVALE, C. (1983) The Painlevé property for PDE, *J. Math. Phys. 24 (3)*: 522–526; see also, in the same journal, the series of papers by Weiss: (1983) *24 (6)*: 1405–1413; (1984) *25 (1)*: 13–24; *24 (7)*: 2226–2235; (1985) *26 (2)*: 258–269; *26 (9)*: 2174–2180.

YOSHIDA, H., GRAMMATICOS, B. AND RAMANI, A. (1987) Painlevé analysis versus Kowalewski exponents, *Acta Appl. Math., 8*: 75–103.

ZAKHAROV, V. E. (ed.) (1991) *What is Integrability?* Springer.

Chapter 5

Perturbation Methods

A perturbation problem consists in finding a function $u = u(x, \varepsilon)$, which solves a PDE of the form

$$(5.1) \qquad\qquad F[u, \varepsilon] = 0,$$

for small values of ε. Thus, the initial-value problem is a perturbation problem, with time playing the role of ε. However, (5.1) usually does not contain any derivatives with respect to ε, and is therefore extremely degenerate from a PDE view-point. The established strategy in such cases is to find a *formal* solution, which is usually a power series in ε and $\ln \varepsilon$, but may be more complicated. One then seeks to prove that this solution converges, or is asymptotic, to a genuine solution. In some cases, the series does not correspond to any genuine solution of the problem. Finally, one would like to continue the solution to large values of ε, a step which is sometimes considered a part of *global bifurcation theory*.

This chapter describes some salient issues which may arise in the realization of this program for nonlinear waves.

We begin with a brief review of perturbation theory for linear operators, which already contains many difficulties (§5.1). The perturbation of spectra is in fact an ingredient of most perturbative treatments of nonlinear waves, in which it arises in the analysis of linearized equations. We then discuss a generalization of the implicit function theorem in some detail in §5.2. The rigorous justification of expansions of solutions of nonlinear equations is often accomplished by reducing the problem to the application of the implicit function theorem, the existence of formal solutions being equivalent to the invertibility of the linearization for $\varepsilon = 0$. For PDE, the inverse of the linearization is often *unbounded* for $\varepsilon \neq 0$, and requires this extension of the IFT.

§5.3 is devoted to the method of geometrical optics (or WKB method) for nonlinear hyperbolic equations. This method aims at constructing highly oscillatory solutions of nonlinear equations. As is shown in §5.4, closely related results can be obtained by a formal averaging of a Lagrangian formulation, over one period of the oscillation. This procedure, due to Whitham, enables one to define a local frequency and wave number for nonlinear problems.

Finally, §5.5 describe the formal perturbation theory of solitons using the inverse scattering formalism; very few rigorous results are known in this area, but this formalism is encountered very frequently in the literature, and has had considerable heuristic value.

A few further directions are mentioned in the Notes and the problems.

5.1   PERTURBATION OF SPECTRA

We recall a few difficulties which occur in the perturbation theory of linear eigenvalue problems. In addition to indicating typical difficulties, these consideration have implications for nonlinear problems since they apply to their linearizations.

A possible pathology.

We begin by giving an example of a perturbation series which leads to an incorrect conclusion, as a warning of what can go wrong.

Consider the eigenvalue problem

$$(5.2) \qquad (\phi, f)\phi + \varepsilon x f(x) = \lambda f(x),$$

where $f(x)$ is an unknown function in $L^2(\mathbf{R})$, while $\phi \in \mathcal{S}(\mathbf{R})$ is given, and has norm 1. (ϕ, f) is the L^2 scalar product.

THEOREM 5.1 *Problem (1) has a formal solution*

$$f = f_0 + \varepsilon f_1 + \ldots; \quad \lambda = \lambda_0 + \varepsilon \lambda_1 + \ldots,$$

with $\lambda_0 = 1$, but has no exact solution if ϕ is positive everywhere. The formal solution is unique if we normalize it by $(f, f) = 1$ and $f_0 = \phi$.

Proof: Substitution of the formal series and identification of like powers of ε results in

$$(5.3) \qquad (\phi, f_0)\phi = \lambda_0 f_0,$$

and more generally,

$$(5.4) \qquad (\phi, f_k)\phi + x f_{k-1} = f_k + \lambda_1 f_{k-1} + \cdots + \lambda_k f_0$$

for $k \geq 1$. We also know that since $(f, f) = 1$ at all orders,

$$(5.5) \qquad \sum_{j=0}^{k} (f_j, f_{k-j}) = 0.$$

Now, (5.3) forces f_0 to be proportional to ϕ, and we choose $f_0 = \phi$ to satisfy the condition $(f_0, f_0) = 1$. Multiplying (5.4) by ϕ, we compute λ_k in terms of $(f_0, \ldots, f_{k-1}, \lambda_0, \ldots, \lambda_{k-1})$. Finally, (5.4) and (5.5) determine respectively $f_k - (\phi, f_k)\phi$ and $(\phi, f_k)\phi$. Therefore f_k and λ_k can be computed by induction.

On the other hand, if there is an exact solution, it must satisfy

$$f = C\phi/(\lambda(\varepsilon) - \varepsilon x)$$

for some constant C. But if ϕ is everywhere positive, this function cannot be square-integrable, and there is no solution.

Let us now turn to positive results.

"Type (A)" methods.

The main point is that isolated, simple eigenvalues can be perturbed, as long as the perturbation is not only smaller but also weaker than the reference operator for $\varepsilon = 0$. In general, the perturbation of eigenvalues of finite multiplicity leads to perturbation series involving fractional powers of the perturbation parameter; however, fractional powers are not needed for *self-adjoint* operators.

Any perturbation of continuous spectra or of eigenvalues embedded in a continuous spectrum can have pathological features, some of which are given in the exercises.

Let us now turn to specific criteria where perturbation series give the correct result.

Let us consider operators in a Hilbert space with norm $\| \quad \|$. We say that a family of operators $A(\varepsilon)$ depending on a parameter is a *type (A)* family if

1. $D(A(\varepsilon))$ is independent of ε;

2. $\psi \mapsto A(\varepsilon)\psi$ is an analytic function of ε for each ψ in this domain.

Type (A) families can be found thanks to the following result:

THEOREM 5.2 *If A_0 is closed and has non-empty resolvent set, then $A(\varepsilon) = A_0 + \varepsilon A_1$ is of type (A) if and only if $D(A_1) \supset D(A_0)$ and*

$$\|A_1\psi\| \leq a\|A_0\psi\| + b\|\psi\|$$

for some a and b.

We say in this case that A_1 is a *relatively bounded perturbation* of A_1.

The basic perturbation theorem for type (A) families is

THEOREM 5.3 *If $\{A(\varepsilon)\}$ is a type (A) family with $A(0)$ self-adjoint, and if λ_0 is a simple isolated eigenvalue of $A(0)$, then there exist analytic functions $\lambda(\varepsilon)$ and $\phi(\varepsilon)$ defined for small ε, such that*

1. $\sigma(A(\varepsilon)) \cap (\lambda_0 - \alpha, \lambda_0 + \alpha) = \lambda(\varepsilon)$ for α small enough;

2. $A(\varepsilon)\phi(\varepsilon) = \lambda(\varepsilon)\phi(\varepsilon)$, and $(\phi(\varepsilon), \phi(\varepsilon)) = 1$.

"Type (B)" methods deal with the perturbation of quadratic forms associated to the familiy of operators.

Bounds on the number of eigenvalues.

It is in practice necessary, as a preliminary step in a perturbation method, to have information on the location of the spectrum of a concrete eigenvalue problem. We list therefore a few results on operators of the type $-\Delta + V(x)$. Many other specialized results can be found in Reed-Simon, vol. 4.

(1) In one space dimension, if $xV(x) \in L^1(\mathbf{R})$, the number of eigenvalues of $-\partial_x^2 + V(x)$ does not exceed $1 + \int_{-\infty}^{\infty} |xV(x)|$.

(2) In three space dimensions, the number of eigenvalues of $-\Delta + V(x)$ can be estimated by $c \iint |V(x)V(y)| \, dx \, dy / |x - y|^2$ (Rollnik norm).

(3) A potential with $O(1/r)$ fall-off may have positive eigenvalues (embedded in the continuous spectrum), the best-known example being the Wigner-von Neumann potential, defined in such a way that $u(r) = \sin r / [r(1 + g(r))^2]$ be an eigenfunction with eigenvalue 1, where $g(r) = 2r - \sin(2r)$. There are however no positive eigenvalues if $V = V_1 + V_2$ where V_1 and V_2 are bounded near infinity, $V_1 = o(1/r)$, $v_2 = o(1)$, and V_2 is C^1 for large r and satisfies $r\partial V_2 / \partial r \leq 0$.

(4) In any number of space dimensions, if V is locally integrable, bounded below and tends to infinity at infinity, then $-\Delta + V(x)$ has purely discrete spectrum.

(5) If $V \leq 0$, tends to infinity at infinity, and if $V \in L^{n/2} + L^{\infty}$, then any eigenfunction satisfies decay estimates of the form $|u(x)| \leq c(a)\exp(-a|x|)$ for *any $a > 0$.*

Such theorems are important in deciding reasonable function spaces in which one may hope to set-up a perturbation argument.

5.2   IMPLICIT FUNCTION THEOREMS

We present a few forms of the Nash-Moser or "hard" IFT. Both are based on Newton-type iteration in a scale of Banach spaces. There are essentially two varieties: one is typically used in spaces of analytic functions, another in Sobolev or C^k spaces.

Nash-Moser theorem without smoothing.

Let $\{X_s\}$ and $\{Y_s\}$ be two scales of Banach spaces, with respective norms $\|\ \|_s$ and $|\ |_s$, where $s \in [0,1]$. We assume that $X_{s'} \supset X_s$ and $Y_{s'} \supset Y_s$ if $s' < s$. A typical example is the case when X_s consists of functions analytic in a strip $|\text{Im } z| \leq s$, with the norm being the uniform norm.

We wish to solve the equation

$$F(u) = 0,$$

where F satisfies the following assumptions, where s, C and R are given:

(H1) F is defined for $\|u\|_s < R$, and sends X_s to $Y_{s'}$ for every $s' < s$.

(H2) There is a mapping $L(u) : X_\sigma \to Y_{\sigma'}$, for any $\sigma' < \sigma \leq s$, such that

$$|F(v) - F(u) - L(u)(v-u)|_{\sigma'} \leq C(\sigma - \sigma')^{-p}\|u - v\|_\sigma^{1+\delta},$$

for some $\delta \in (0,1]$.

(H3) For any $f \in Y_{\sigma'}$, one find $w \in Y_{\sigma'}$ such that $L(u)w = f$, and

$$\|w\|_\sigma \leq c(\sigma - \sigma')^{-q}|f|_{\sigma'}.$$

By abuse of notation, we write $w = L(u)^{-1}f$.

The result can now be stated.

THEOREM 5.4 *There is a function $\varepsilon_0(s)$ such that is $|F(0)|_1 < \varepsilon_0(s)$, there is a solution of $F(u) = 0$ in X_s.*

Proof: We define an iteration by $u_0 = 0$, and

(5.6) $$u_{k+1} = u_k - L(u_k)^{-1}F(u_k).$$

We recognize Newton's method. Since we will assume, $F(0) \in Y_1$, we can find $u_1 \in X_\sigma$, $\sigma < 1$, *but not* in X_1 in general. Therefore, we cannot iterate in a fixed space. To tackle this problem, we define a sequence $\{s_k\}$ with $s_k \downarrow s$, and estimate $\|u_k\|_{s_k}$. It will be convenient to define another sequence $\{t_k\}$ and to track $|F(u_k)|_{t_k}$.

More precisely, we let

$$s_0 = 0; \quad s_{k-1} - s_k = \rho k^{-2}$$

for $k \geq 2$, and

$$t_k = s_k - \frac{1}{2}\rho(k+1)^{-2},$$

for $k \geq 0$. For s_k to tend to s, we must take $\rho = (1 - s) \sum_{k=1}^{\infty} k^{-2}$. We also let $\varepsilon = |F(0)|_1$, and $a_k = |F(u_k)|_{t_k}$. Note that $s_k > t_k > s_{k+1}$.

In the following, the letter C denotes various positive constants independent of k.

We prove by induction on k that $\|u_k\| < R/2$,

$$(5.7) \qquad\qquad a_k \leq \varepsilon(k + 1)^{-r}$$

and

$$(5.8) \qquad\qquad \|u_{k+1} - u_k\|_{s_{k+1}} \leq C\varepsilon(k + 1)^{-(1+2(p+q)/\delta)},$$

if ε is small enough.

Using (H2), the definition of u_{k+1}, and (H3), we find, for any $\sigma \in (t_{k+1}, t_k)$,

$$(5.9) \qquad a_{k+1} \ \leq \ c\|L(u_k)F(u_k)\|_{\sigma}^{1+\delta}(\sigma - t_{k+1})^{-p}$$
$$(5.10) \qquad\qquad \leq \ C[a_k(t_k - \sigma)^{-q}]^{1+\delta}(\sigma - t_{k+1})^{-p}$$
$$(5.11) \qquad\qquad = \ Ca_k^{1+\delta}(t_k - \sigma)^{-q(1+\delta)}(\sigma - t_{k+1})^{-p}.$$

We minimize this quantity by taking

$$\sigma = t_{k+1} = \frac{p(t_k - t_{k+1})}{p + q(1 + \delta)}.$$

We find

$$a_{k+1} \leq Ca_k^{1+\delta}(t_k - t_{k+1})^{-p-q(1+\delta)}.$$

It now follows from the induction hypothesis that

$$a_k \leq \varepsilon(k + 1)^{-r},$$

where $r = 1 + 1(p + q(1 + \delta))/\delta$, provided that ε is small enough.

As for the last estimate, we have, using the recurrence relation and (H3),

$$\|u_{k+1} - u_k\|_{s_{k+1}} \ \leq \ Ca_k(t_k - s_{k+1})^{-q}$$
$$= \ Ca_k[2(k + 1)^2/\rho]^q$$
$$\leq \ C\varepsilon(k + 1)^{-(1+2(p+q)/\delta)}.$$

Finally,

$$\|u_{k+1}\|_{s_{k+1}} \leq C\varepsilon\sum_{1}^{\infty}(k + 1)^{-(1+2(p+q)/\delta)},$$

is less than $R < 2$ if ε has been chosen small enough.

This proves the desired estimates.

We see that the sequence $\{u_k\}$ certainly converges in X_s, and since $a_k \to 0$, its limit is a solution, QED.

Nash-Moser theorem with smoothing.

The preceding result assumes that the inverse of the linearization of F sends X_σ to $X_{\sigma'}$ for any $\sigma' < \sigma$. This mimics the behavior of the operator d/dz on $X_s = \{u : |u(x + iy)| \text{ bounded for } |y| < s\}$. It is clearly inadequate in Sobolev-type spaces, where differential operators induce a definite *loss of derivatives*. This loss forces us to modify the iteration (5.6) into:

$$(5.12) \qquad u_{k+1} = u_k - S_k L(u_k)^{-1} F(u_k).$$

where S_k is a smoothing operator. This new iteration procedure will be shown to converge, under a new set of hypotheses.

It is convenient not to restrict the index s to lie between 0 and 1.

There are two sets of hypotheses: one pertaining to the smoothing, the other to F. In all, $M \geq 1$ is a fixed constant and $\sigma < s$.

Let us assume that there is a family $S(t)$ of smoothing operators such that

- (S1) $\|S(t)u\|_{s+\sigma} \leq Mt^\sigma \|u\|_s$ if $u \in X^s$.

- (S2) $\lim_{t\to\infty} \|(I - S(t))u\|_s = 0$ if $u \in X^s$.

- (S3) $\leq \|(I - S(t))u\|_{s-\sigma} \leq Mt^{-\sigma}\|u\|_s$ if $u \in X^\sigma$.

- (S4) $\|(d/dt)S(t)u\|_s \leq Mt^{s-\sigma-1}\|u\|_s$ if $u \in X^\sigma$.

We will not make use of (S4), but it usuallly hols, and does enter in other proofs.

We also assume that the map F has the following properties:

- (F1) F is of class C^2 from X^s to Y^s, with first and second derivatives bounded by $M \geq 1$.

- (F2) $F'(u)$ has a right inverse $L(u)$ which sends Y^s to X^{s-a}, and which satisfies the estimate

$$\|L(u)F(u)\|_{s+b} \leq M(1 + \|u\|_{s+a+b})$$

  for some $b > 8a$.

In short, if s is identified with a degree of differentiability, F' fails to be invertible because $L(u)$ lands in X^{s-a} instead of X^s.

THEOREM 5.5 *Under assumptions (S1)-(S4) and (F1)-(F2), the equation $F(u) = 0$ has a solution if $F(0)$ is small enough in X^s.*

Remarks: 1) The assumptions are stronger than requiring that F' have an unbounded inverse. In fact, this latter assumption alone would lead to an incorrect result, as the case of the mapping $F(u) = u - u_0$ from H^s to H^{s-1} shows, if $u_0 \notin H^s$. What happens here is that (F2) requires $L(u)F(u)$ to be smoother and smoother if u is. This excludes the counter-example we just gave. Also, in practice, it is much easier to find $L(0)$ than it is to construct $L(u)$ for $u \neq 0$. At the formal level, the existence of $L(0)$ suffices to construct a formal series to all orders. This series may however be completely meaningless.

2) We can construct a family of smoothing operators by choosing a function $\phi(\xi)$ which equals 1 for $|\xi| < 1$ and 0 for $|\xi| > 2$, with $0 \leq \phi \leq 1$, and by letting $S(t)u = \mathcal{F}^{-1}\phi(\xi/t)\mathcal{F}u$, where $\mathcal{F}$ denotes the Fourier transform. Indeed, we then have $\phi(\xi/t)(1 + |\xi|^2)^\sigma \leq (1 + 4t^2)^\sigma$ and $(1 - \phi(\xi/t))(1 + |\xi|^2)^{-\sigma} \leq (1 + t^2)^{-\sigma}$

Proof: Let $q_k = \exp[\lambda\rho^k]$ and $S_k = S(q_k)$, where $\rho = 3/2$ (the reader is invited to examine the rest of the proof for $\rho \in (1,2)$).

We define $\{u_k\}_{k \geq 0}$ by $u_0 = 0$ and

$$u_{k+1} = u_k - S_k L(u_k)F(u_k).$$

We prove by induction that there exist positive constants μ and ν such that

(5.13) $$\|u_k - u_{k-1}\|_s \leq q_k^{-\mu a}; \quad 1 + \|u_{k-1}\|_{s+a+b} \leq q_k^{\nu a}.$$

These estimates will ensure that the iteration is well-defined, and converges in the s-norm.

We first estimate the difference of two consecutive approximations:

$$\begin{aligned}
\|u_{k+1} - u_k\|_s \\
\leq \ & M q_k^a \|L(u_k)F(u_k)\|_{s-a} \\
\leq \ & M^2 q_k^a |F(u_k)|_s \\
= \ & M^2 q_k^a \Big[|F(u_{k-1}) - F'(u_{k-1})S_{k-1}L(u_{k-1})F(u_{k-1})|_s \\
& + |\int_0^1 (1 - \sigma)F''((1 - \sigma)u_{k-1} + \sigma u_k) \cdot (u_k - u_{k-1})^2|_s\Big] \\
\leq \ & M^2 q_k^a \left\{|F'(u_{k-1})(I - S_{k-1})L(u_{k-1})F(u_{k-1})|_s + M q_k^{-2a\mu}\right\}.
\end{aligned}$$

Now, using (S3) and (F2), we find

$$\|L(u_{k-1})F(u_{k-1})\|_{s+b} \leq M(1 + \|u_{k-1}\|_{s+b+a}) \leq M q_{k-1}^{a\nu},$$

while (F1) and (S3) imply

$$|F'(u_{k-1})(I - S_{k-1})v|_s \leq M^2 q_{k-1}^{-b}\|v\|_{s+b}.$$

Therefore

$$\|u_{k+1} - u_k\|_s \leq M^5 q_k^a q_{k-1}^{a\nu-b} + M^3 q_k^{a-2\mu}.$$

Since $M \geq 1$, it suffices to have

(5.14) $$M^5(q_k^a q_{k-1}^{a\nu-b} + q_k^{a(1-2\mu)}) \leq q_{k+1}^{-a\mu}$$

to ensure the announced estimate.

As for the other bound, we estimate

$$
\begin{aligned}
1 + \|u_{k+1}\|_{s+b+a} &\leq 1 + \sum_{j=0}^{k} \|S_k L(u_k)F(u_k)\|_{s+b+a} \\
&\leq 1 + M \sum_j q_j^a \|L(u_k)F(u_k)\|_{s+b} \\
&\leq 1 + M^2 \sum_j q_j^a (1 + \|u_k\|_{s+b+a}) \\
&\leq 1 + M^2 \sum_j q_j^{a(1+\nu)}.
\end{aligned}
$$

We therefore need

$$(5.15) \qquad\qquad 1 + M^2 \sum_j q_j^{a(1+\nu)} \leq q_{k+1}^{a\nu}.$$

This amounts to requiring, since $q_{k+1} \geq q_{j+1}$,

$$1 \geq q_{k+1}^{-a\nu}(1 + M^2 e^{\lambda a(1+\nu)}) + M^2 \sum_{j=1}^{k} e^{\lambda \rho^j a[(1+\nu) - \rho\nu]},$$

which, since $\rho = 3/2$, holds if $\nu > 2$ and λ is large enough.

As for (5.14), we must require

$$\lambda \rho^{k-1}(a\nu - b + a\rho + \mu\rho^2) \leq -5\ln M - \ln 2,$$

and

$$\lambda \rho^k (a(1 - 2\mu) + a\mu\rho) \leq -5\ln M - \ln 2.$$

These hold as well for λ large, provided that

$$\mu > 2 \quad \text{and} \quad b \geq a\nu + 3a/2 + 9\mu/4.$$

This estimate also ensures that $\|u_k\|_s$ remains les than $R/2$, so that the iterations are well-defined.

To summarize, all we need is to be able to choose $\mu > 2$, $\nu > 2$, and $b > a\nu + 3a/2 + 9a\mu/4$. This is certainly possible if b is greater than $2a + 3a/2 + 18a/4 = 8a$, which is the case by assumption.

To start the induction, we need to consider $u_1 = -S_0 L(0)F(0)$. Since $\|S_0 L(0)F(0)\|_{s+a+b} \leq M q_0^a \|L(0)F(0)\|_{s+b} \leq M^2 q_0^a$ (from (S2)), we require

$$(5.16) \qquad\qquad M^2 q_0^a \leq q_1^{\nu a}.$$

For the first part of (5.13), we write

$$\|u_1\|_s \leq M q_0^a |L(0)F(0)\|_{s+a} \leq M^2 q_0^a |F(0)|_s,$$

which leads to the condition

(5.17) $$|F(0)|_s \le q_0^{-a} q_1^{-\mu a}/M^2.$$

We therefore choose λ large enough to satisfy (5.15), and then check the smallness condition on $F(0)$.

If $F(0)$ is small enough, we see that the iterations remain in a small neighborhood of 0 (and are therefore well-defined, and converge in X^s norm. It follows from the continuity of F and te existence of a uniform bound on $L(u)$ that $L(u_k)F(u_k)$ converges in X^{s-a}, and since the smoothing operators approximate the identity, we may write

$$\|(I - S_k)L(u_k)F(u_k)\|_{s-a-1} \le C q_k^{-1} \to 0$$

as $k \to \infty$. We conclude that $u\infty = \lim_{k \to \infty} u_k$ solves

$$L(u_\infty)F(u_\infty) = 0,$$

in the space X^{s-a-1}. Applying $F'(u_\infty)$, we conclude that

$$F(u_\infty) = 0,$$

QED.

An application.

We now give a typical application of the analytic Nash-Moser theorem, which illustrates the amount of work needed to apply it in concrete situations.

The question is the construction of solution of

(5.18) $$u_{xx} + u_{yy} + \lambda a(y)u = f(\lambda, y, u)$$

where $0 < y < 1$. We require that $u(x,0) = u(x,1) = 0$ for all x, and that u decay exponentially as $x \to \pm\infty$. The function a is nonnegative, and positive somewhere. The function f satisfies

$$f = a_k(\lambda, y)u^k + o(|u|^k)$$

as $|u| \to \infty$, uniformly in y, when λ remains bounded; $k > 1$ is an integer.

This is a model problem for solitary water waves in a channel. A similar method can be applied to water waves, but we chose this example for its relative simplicity.

The first nontrivial step is to cast this question as a perturbation problem. If we think of u as having small amplitude, we are tempted to linearize the equation, which results in discarding the nonlinear term. Now it is easy to see, by elementary methods that the linearization at the zero solution has *no* solution which decays as $x \to \infty$ in both directions. We therefore scale variables in such a way that u_{xx} and u^k have the same order:

$$u' = u\varepsilon^{1/(k-1)}, \quad x' = x\varepsilon^{1/2}.$$

It turns out that this is not enough, because seeking $u = \sum_m u_m \varepsilon^m$ generates one condition for the calculation of each term; we need in fact to allow λ to vary with ε, just as we did in the case of eigenvalue perturbation;

After performing these changes of coordinates and unknown, we obtain the following equation, where all primes have been dropped for convenience:

$$F(u, \varepsilon) := u_{yy} + \lambda e^{\alpha\varepsilon} a(y)u + \varepsilon u_{xx} - \varepsilon^{-1/(k-1)} f(\lambda e^{\alpha\varepsilon}, y, \varepsilon^{1/(k-1)}u) = 0.$$

Let us briefly investigate the existence of a formal solution. If we substitute formally $u = \sum_m u_m \varepsilon^m$, we find

$$u_{0yy} + \lambda a(y)u_0 = 0,$$

so that $u_0 = S(x)\phi(y)$, where ϕ is an eigenfunction of $\partial_{yy} + \lambda a$ with Dirichlet boundary conditions. We assume this eigenvalue to be simple. This of course sets restrictions on λ if we want to ensure that our solution is nontrivial.

At the next stage, we find

$$(\partial_{yy} + \lambda a)u_1 + \lambda\alpha a(y)u_0 + u_{0xx} - a_k u_0^k = 0.$$

Multiplying by ϕ and integrating, we find a condition on S:

$$(5.19) \qquad S_{xx} - c_1 S + c_2 S^k = 0,$$

where $c_1 = -\alpha\lambda \int a\phi^2 dy$ and $c_2 = -\int a_k\phi^2 dy$. This equation defines the profile of the wave at lowest order. It is easy to see that S decays at infinity precisely when c_1 and c_2 are both positive, in which case

$$S = \left[\frac{(k+1)c_1}{2c_2}\text{sech}^2(\frac{2x}{(k-1)\sqrt{c_1}})\right]^{1/(k-1)}.$$

Note that u_1 is now determined upto a function of the form $S_1(x)\phi(y)$.

The higher-order approximations lead to equations of the form

$$(\partial_{yy} + \lambda a)u_m + \phi(y)(\partial_{xx} - c_1 + kc_2 S^{k-1})u_{m-1} = \text{terms involving } u_0, \dots u_{m-2}.$$

At this point, the terms $u_0, \dots u_{m-1}$ will have been determined, except for a term $S_{m-1}(x)\phi(y)$. Multiplying the equation by ϕ and integrating, we again obtain a single condition, which has the form

$$(\partial_{xx} - c_1 + kc_2 S^{k-1})S_{m-1} = \text{known}.$$

It is easy to see that $(\partial_{xx} - c_1 + kc_2 S^{k-1})$ is invertible on the space of even, squre-integrable functions, and therefore the inductive construction of the formal solution con be continued.

To prove that there actually is such a solution, we normalize ϕ to have L^2 norm one, and introduce the projector Q by $Qv = (v, \phi)\phi$. The formal process we just described determines u_k by solving

$$(I - Q)F = 0$$

at level k and

$$QF = 0$$

at level $k + 1$ in ε. It is therefore natural to define a new mapping which puts these two calculations at the same level:[1]

$$G(u, \varepsilon) := \varepsilon^{-1} Q F(u, \varepsilon) + (I - Q)F(u, \varepsilon).$$

We must now apply the Nash-Moser IFT for fixed ε. Let us define the scales of spaces $\{X_s\}$ and $\{Y_s\}$ as follows: X_s and Y_s consist of functions $v(x, y)$ analytic in x on the strip $|\text{Im}\, x| \leq s$, for $0 \leq y \leq 1$, and completed with respect to the norms:

$$\|v\|_s := \sup \left\{ e^{c|\text{Re}\, z|} |v(x, y)| : |\text{Im}\, x| \leq s \text{ and } 0 \leq y \leq 1 \right\}$$

and

$$|v|_s := \sup \left\{ e^{c|\text{Re}\, z|} |\partial_x^j \partial_y^k v(x, y)| : j + k \leq 2, |\text{Im}\, x| \leq s \text{ and } 0 \leq y \leq 1 \right\}$$

respectively.

The only difficulty here is in the inversion of the linearization of G. The argument works because the linearization is, for $\varepsilon \neq 0$, invertible with loss for $\varepsilon \neq 0$. To see this, we argue in three steps:

STEP 1: There is a w_1 such that $QG'_u Qw_1 = z_1$. Indeed, this means that the linearization of (5.19), on the space of square-integrable even functions of x, is invertible.

STEP 2: We define z_2 by solving

$$(\varepsilon \partial_{xx} + \partial_{yy})w_2 + \lambda a w_2 = z_2,$$

which is easily accomplished by Fourier transform.

STEP 3: Compute $G'_u(w_1 + w_2) - z$ and show that it is small. This proves that the operator $z \mapsto w_1 + w_2 \mapsto G_u(w_1 + w_2)$ is invertible on Y_s. The other assumptions of the theorem follow.

---

[1] This idea is extremely old and useful. A form already appears in Poincaré's work in celestial mechanics, see exercise 6.

Nash-Moser theorem and paradifferential operators.

We saw that the Nash-Moser theorem was made necessary because the inverse of the map $F_u(u, \varepsilon)$ was not necessarily bounded, so that the usual IFT was not applicable. We consider here a situation where the difficulty is obviated by making use of the smoothness properties of paramultiplication (Ch. 3). The technique is illustrated on a form of the Nash embedding theorem, and a more general set-up can be found from the Notes.

We consider the problem of embedding isometrically a torus into a Euclidean space of sufficiently high dimension. We will limit ourselves to the question of showing that given a C^∞ free embedding (i.e., an embedding $x \mapsto u_0(x)$ such that $\partial u_0/\partial x^j$, $\partial^2 u_0/\partial x^j \partial x^k$, $1 \le j, k \le n$ are linearly independent), one can perturb u_0 into an isometric embedding provided that the metric on the torus is sufficiently close to the pull-back of the Euclidean metric by u_0.

We must therefore solve the equation

$$(5.20) \qquad \qquad \Phi(u) = g,$$

where $g = (g_{jk})$ is the given metric on the torus, and

$$\Phi(u)_{jk} = \delta_{\mu\nu} \frac{\partial u^\mu}{\partial x^j} \frac{\partial u^\nu}{\partial x^k}.$$

A right inverse for $\Phi'(u)$ can be computed as follows: Given h, solve the system

$$\left(\frac{\partial u}{\partial x^j}, v \right) = 0; \quad \left(\frac{\partial^2 u}{\partial x^j \partial x^k}, v \right) = -\frac{1}{2} h_{jk}.$$

This is possible if u is close to u_0, since it will be also a free embedding. We then have identically $(du, v) = 0$, hence $\Phi'(u)v = 2(du, dv) = -(d^2 u, v) = h$, as desired. We write $v = \Psi(u)h$. Note that v has the same smoothness as h, while u is smoother than $\Phi(u)$ in general: the inverse is unbounded.

We nevertheless wish to solve

$$\Phi(u_0 + u) = \Phi(u_0) + g$$

if g is small. We assume $g \in C_*^s$ (Zygmund spaces) for $s > 2$. The solution will be found in C_*^{s+1}.

STEP 1: We construct a modification of the inverse Ψ by applying the usual IFT to

$$u \mapsto u - \Psi(u_0)h - T_{\Psi(u+u_0)-\Psi(u_0)}h,$$

where $h \in C_*^{s+1}$ is given. It maps C_*^{s+1} to itself, and its differential with respect to u for $u = 0$ is the identity. Therefore there is a unique $u = U(h)$ with the property that

$$u = \Psi(u_0)h + T_{\Psi(u+u_0)-\Psi(u_0)}h.$$

STEP 2: Now, let us consider the mapping

(5.21) $F : h \mapsto \Phi(u_0 + U(h)),$

which, at face value, maps C_*^{s+1} to C_*^s. If we could prove that it is locally surjective, the proof would be finished. It turns out that this can be proved using the ordinary IFT only, because F is, despite appearances, a map from C_*^{s+1} *to itself.*

To see this, let us first note that

$$\Phi(u + u_0) = \Phi(u) + T_{\Phi'(u+u_0)} + R(u),$$

where the remainder is a smooth map from C_*^{s+1} to C_*^{2s}. Therefore $h \mapsto R(U(h))$ has the correct regularity. $\Phi(u_0)$ is certainly smooth, so we are left with the paraproduct term:

$$T_{\Phi'(u+u_0)}U(h) = T_{\Phi'(u+u_0)}T_{\Psi(u+u_0)}h + T_{\Phi'(u+u_0)}(\Psi(u_0) - T_{\Psi(u_0)})h.$$

The factor $(\Psi(u_0) - T_{\Psi(u_0)})$ is infinitely smoothing. Since $\Psi(u + u_0)$ takes its values in C_*^{s-1}, we see that since $\Phi'(u + u_0)\Psi(u + u_0) = \mathrm{Id}$, the operator

$$T_{\Phi'(u+u_0)}T_{\Psi(u+u_0)} - \mathrm{Id}$$

depends smoothly on $u \in C_*^{s+1}$ and sends C_*^σ to $C_*^{\sigma+s-1}$ for any σ, if $s - 1 > 1$. In particular, if $\sigma = s$, since $\sigma + s - 1 > s + 1$ for $s > 2$, we conclude that F is a smooth map from C_*^{s+1} to itself, as announced.

STEP 3: It now remains to note that the differential of h at the origin is the identity, and to apply the (ordinary) IFT.

5.3 NONLINEAR GEOMETRICAL OPTICS

We describe nonlinear versions of the method of geometrical optics presented in Ch. 1. The main departures form the linear theory are that the transport equations are nonlinear ("distortion of signals") and that different frequencies can interact ("resonate"). The formal aspects have much in common with the derivation of universal model equations in Ch. 4.

We consider for definiteness a vector $u = (u^1, \ldots, u^N)$ which solves

(5.22) $A(u)Du + B(u) := A^0(x,u)\partial_t u + \sum_j A^j(x,u)\partial_j u + B(x,u) = 0.$

The summation sign will often be omitted. All nonlinearities will be assumed to be smooth. We assume the equation to be strictly hyperbolic with respect ot t.

We are interested in solutions of the form

$$(5.23) \qquad u = u(t, x, \theta_1, \ldots, \theta_m, \varepsilon),$$

where $\theta_k = \phi_k(x, t)/\varepsilon$. We therefore need a function of $(t, x, \theta_1, \ldots, \theta_m, \varepsilon)$, which solves

$$A(u)Du + B(u) + \frac{1}{\varepsilon}\sum_k A(u)D_x\theta_k D_{\theta_k}u = 0.$$

This is now a perturbation problem.

By analogy with the WKB method, where the phases θ_k enter the solution through complex exponentials, we require that the solution be *periodic* in the θ_k, with period 2π. The solution has therefore the form

$$u = \sum_{j \geq 0}\varepsilon^j u_j = \sum_{j,a}u_{ja}(x, t)\varepsilon^j e^{ia\cdot\theta},$$

where $ia \cdot \theta = a_1\theta_1 + \cdots + a_m\theta_m$, and the a_k are integers.

FIRST TERM: At lowest order, we find

$$\sum_k A(u_0)D\theta_k D_{\theta_k}u = 0.$$

HIGHER ORDER TERMS: The j-th coefficient u_j is determined by an equation of the form

$$Eu_j := \sum_k A(u_0)D\theta_k D_{\theta_k}u_j = L_j u_{j-1},$$

where the coefficients of L_j depend on $u_0, \ldots, u_{j-2}$. We therefore must express that $L_j u_{j-1}$ lies in the range of E, and then determine u_j upto the addition of an element of the kernel of E. To this end, we construct an operator F which enables us to characterize the range and kernel of E.

Strict hyperbolicity enables us to define M smooth functions $\lambda_p(\xi)$ by

$$\det(A^j(u_0)\xi_j - \lambda_p(\xi)A^0(u_0)) = 0.$$

Let l_p and r_p be a set of left and right eigenvectors corresponding to these eigenvalues. Now

$$E\sum_a v_a e^{ia\cdot\theta} = A(u_0)\sum_k[ia_k D\theta_k]e^{ia\cdot\theta}v_a$$

so we really need to analyze the range and kernel of $A(u_0)\sum_k[ia_k D\theta_k]$. Now, from the knowledge of the λ_p, we see that $A(u_0)(\xi, -\lambda)v = w$ is uniquely solvable for v, except if $\lambda = \lambda_p(\xi)$ for some p. In that case, we must require $l_p w = 0$ for a solution to exist, and the solution is then determined upto a multiple of r_p.

5.4 WHITHAM'S THEORY

Whitham's theory aims at describing the modulation of nonlinear waves by a system of equations for amplitude and phase functions which depend on space and time. This system can be derived formally by an averaging procedure if the problem possesses a variational formulation. The averaged system may however be elliptic or hyperbolic, and may form shocks.

Whitham's theory is concerned with the asymptotic description of slow modulations of uniform periodic wavetrains. It is therefore similar to the derivation of the nonlinear Schrödinger equation in Ch. 2; however, the reference solution is not a plane wave solution anymore, but rather a function $U(\theta)$ where θ is a nonlinear function of space and time variables. The first order perturbation equations take, in the simplest examples, the form of a nonlinear first-order system for the local frequency, wave number, and "energy." These equations can be directly derived from a variational formulation of the full equations, by averaging over one period of the phase θ. This feature makes it plausible that we have here a genuinely nonlinear Ansatz, which goes essentially further than the other methods described in this chapter.

We discuss below one example in some detail; see exercise 5 for a more general theory.

We consider the equation

$$(5.24) \qquad u_{tt} - u_{xx} + V'(u) = 0,$$

in one space dimension. To fix ideas, we will also take $V'(u) = u - u^3$.

We seek solutions of the form $u = u(X,T,\theta)$, periodic of period 2π in $\theta = \theta(x,t)$, where $X = \varepsilon x$, and $T = \varepsilon t$. These are therefore slowly varying perturbations superimposed on a periodic wavetrain. We define κ and ω by

$$(5.25) \qquad \partial_x \theta = \kappa(X,T), \quad \partial_t \theta = \omega(X,T).$$

The functions κ and ω are part of the unknown, and are at this point only constrained by the relation

$$(5.26) \qquad \kappa_T + \omega_X = 0.$$

The chain rule gives

$$\partial_x = \kappa \partial_\theta + \varepsilon \partial_X, \quad \partial_t = -\omega \partial_\theta + \varepsilon \partial_T.$$

We therefore find that u should satisfy:

(5.27)
$$(\omega^2 - \kappa^2)u_{\theta\theta} + V'(u) - \varepsilon\left[2\kappa u_X + 2\omega u_T + (\kappa_X + \omega_T)u\right] + \varepsilon^2[u_{TT} - u_{XX}] = 0.$$

We then substitute a formal solution in powers of ε.

At lowest order, we find

$$(\omega^2 - \kappa^2)u_{0\theta\theta} + V'(u) = 0.$$

Integrating, we find that there is a function $E(X, T)$ such that

$$(\omega^2 - \kappa^2)u_{0\theta}^2 + V(u) = E(X, T).$$

Expressing that the period is 2π, we obtain

(5.28)
$$(\omega^2 - \kappa^2)^{1/2}\int \sqrt{2(E - V(u))}\, du = 1.$$

At first order, we find

$$(\omega^2 - \kappa^2)\frac{\partial}{\partial\theta}(u_{1\theta}u_\theta - u_1 u_{\theta\theta}) = F_1,$$

where

$$F_1 = u_\theta[2\omega u_{\theta T} + 2\kappa u_{\theta X} + \omega_T u_\theta + \kappa u_\theta].$$

Periodicity of the u_k in θ requires that F_1 have mean zero. Writing out this condition, we find

$$\frac{\partial}{\partial T}\left[\omega\int_0^1 u_\theta^2\, d\theta\right] + \frac{\partial}{\partial X}\left[\kappa\int_0^1 u_\theta^2\, d\theta\right] = 0.$$

Since we express u in terms of the variable θ, we find

(5.29)
$$\frac{\partial}{\partial T}\left[\omega(\omega^2 - \kappa^2)^{-1/2}\int \sqrt{2(E - V(u))}\, du\right]$$
$$+ \frac{\partial}{\partial T}\left[\kappa(\omega^2 - \kappa^2)^{-1/2}\int \sqrt{2(E - V(u))}\, du\right] = 0.$$

(5.26), (5.28) and (5.29) form three equations for E, ω and κ.

It turns out that the same equations can be directly obtained from the Lagrangian formulation of the problem, as follows:

Define $L(u, u_x, u_t) = (1/2)[u_t^2 - u_x^2] - V(u)$ and

$$\mathcal{L}(\kappa, \omega, E) = (\omega^2 - \kappa^2)^{1/2}\int \sqrt{2(E - V(u))}\, du - E,$$

where κ and ω are constrained to be the components of the gradient of θ.

Then the above equations can be obtained by considering the variation of $\mathcal{L}$ with respect to E and θ respectively.

5.5 SOLITON PERTURBATION

When considering a perturbation of an equation integrable by inverse scattering, it is possible to derive formally equations for the evolution of soliton parameters (eigenvalues, norming constants, scattering functions). This theory in presented for the AKNS and KdV cases, and some predictions found in the literature in this connection are given.

It would be desirable to be able to use the detailed description provided by the inverse scattering transform to describe solutions of nearly-integrable equations. This entails the computation of the variation of scattering data for an eigenvalue problem such as the AKNS system (4.35)–(4.36), in the event that the potentials vary with time. We proceed to describe the formal aspects of this theory, which is often used without comment in the specialized literature. Let us assume $r = -\bar{q}$ since this is the case for the sine-Gordon equation.

Case of the 2×2 AKNS system.

We first express the variation of the scattering data in terms of the time derivatives of the potential. The second issue is to use a given PDE on the potentials to simplify the resulting expression. It is assumed that the number of eigenvalues does not change. Let the eigenvalues in the upper half-plane be $\{k_1, \ldots, k_p\}$. The norming constants are denoted by $\{c_1, \ldots, c_p\}$.

The first question leads, by computing the time derivative of the eigenvalue problem, to an inhomogeneous system for ψ_t. The formula of variation of parameters gives a particular solution involving products of eigenfunction components. From the expression of the scattering data in the proof od Th. 4.16, one can, after some algebra, derive the following general relations:

$$(5.30) \qquad k_{it} = -c_k \int_{-\infty}^{\infty} (\bar{q}_t \psi_{i1}^2 + q_t \psi_{i2}^2)\, dx,$$

$$(5.31) \qquad c_{it} + \frac{a_k''}{a_k'} k_{it} c_i = -\frac{1}{a_k'^2} \int_{-\infty}^{\infty} \left(\bar{q}_t \frac{\partial \psi_{i1}^2}{\partial k} + q_t \frac{\partial \psi_{i2}^2}{\partial k}\right) dx,$$

and

$$(5.32) \qquad (\bar{b}/a)_t = -\frac{1}{a^2} \int_{-\infty}^{\infty} (\bar{q}_t \psi_1^2 + q_t \psi_2^2)\, dx$$

for any real k. The $\psi_i = \psi(k_i)$ are the eigenfunctions at k_i with the normalization of Ch. 4.

A completely integrable AKNS system with dispersion relation $\omega = \Omega(k)$, has the form

$$\partial_t \begin{pmatrix} \bar{q} \\ q \end{pmatrix} = 2\Omega(L) \begin{pmatrix} \bar{q} \\ q \end{pmatrix},$$

where

$$L = \frac{1}{2i} \begin{pmatrix} \partial_x + 2\bar{q}I_x \circ q & 2\bar{q}I_x \circ \bar{q} \\ -2qI_x \circ q & -\partial_x - 2qI_x \circ \bar{q} \end{pmatrix},$$

where $I_x \circ qu := \int_{-\infty}^{\infty} q(x)u(x)\,dx$. Now we must note that the squared eigenfunctions

$$\begin{pmatrix} \psi_1^2 \\ \psi_2^2 \end{pmatrix} \quad \text{and} \quad \begin{pmatrix} \tilde{\psi}_1^2 \\ \tilde{\psi}_2^2 \end{pmatrix}$$

are eigenfunctions of L. Using this, one finds

$$k_{it} = 0, c_{it} + \frac{a_k''}{a_k'}k_{it}c_k = 2\Omega(k_i)c_i, (\bar{b}/a)_t = 2\Omega(k)\bar{b}/a.$$

If we add to such an equation a perturbation $\varepsilon(F_1, F_2)$, the same calculation gives the equations

$$k_{it} = -\varepsilon c_k \int_{-\infty}^{\infty} (F_1\psi_1^2 + F_2\psi_2^2)\,dx,$$

$$c_{it} + \frac{a_k''}{a_k'}k_{it}c_i = 2\Omega(k_i)c_i - \frac{\varepsilon}{a_k'^2} \int_{-\infty}^{\infty} (F_1\frac{\partial\psi_{i1}^2}{\partial k} + F_2\frac{\partial\psi_{i2}^2}{\partial k})\,dx,$$

and

$$(\bar{b}/a)_t = 2\Omega(k)\bar{b}/a - \frac{\varepsilon}{a^2} \int_{-\infty}^{\infty} (F_1\psi_1^2 + F_2\psi_2^2)\,dx.$$

These formulae are often simplified using the relation $|a|^2 + |b|^2 = 1$.

These formulae are in practice useful for the perturbation of pure soliton solutions, since in that case the eigenvalues and eigenfunctions are known exactly (see exercise 9). While the above formulae are "exact" (in the sense that no terms were discarded in the formal process), it becomes necessary to start approximating, and in particular replacing the eigenfunctions by their value at $t = 0$.

For instance, for

$$u_{xt} = \sin u + \varepsilon g(x + t),$$

and a kink solution (one purely imaginary eigenvalue k in the upper half-plane), one finds, since $\psi_1^2 + \psi_2^2$ is initially $q/(2i)$, we find

$$k_t = -\frac{i\varepsilon}{8} \int_{-\infty}^{\infty} g(x + t)u_x\,dx.$$

It is also noteworthy that one can estimate the variation of a and b (variously referred to as "radiation" or "excitation of the continuous spectrum") in this manner. However, it is sometimes difficult to see why the terms discarded are smaller than those retained.

Case of KdV.

We turn to the relevant formulae for the case of the KdV equation. The application of the result to the "shelf problem" will illustrate some of the merits of the procedure.

The formal part is fairly similar to the AKNS case, and we therefore give the result without comment: if

$$(5.33) \qquad\qquad q_t + \partial_x[P(M)u] = \varepsilon F,$$

where $M = -\frac{1}{4}\partial_x^2 - q - \frac{1}{2}I_x \circ q$, one finds

$$(5.34) \qquad\qquad m_j k_{jt} \;=\; \frac{\varepsilon}{2jk_j a''(k_j)} \int_{-\infty}^{\infty} F\psi_j^2 \, dx,$$

$$(5.35)\; m_{jt} + \Big(\frac{a''(k_j)}{a'(k_j)} + \frac{1}{k_j}\Big)m_j k_{jt} \;=\; 2iP(k_i^2)m_k$$

$$(5.36) \qquad\qquad\qquad + \frac{\varepsilon}{2ik_j a'(k_j)^2} \int_{-\infty}^{\infty} \frac{\partial}{\partial k}(\psi^2)(k_j) F \, dx,$$

$$(5.37) \qquad\qquad \frac{\bar{b}}{a}\Big|_t \;=\; 2iP(k^2)\frac{\bar{b}}{a} + \frac{\varepsilon}{2ika^2} \int_{-\infty}^{\infty} F\psi^2 \, dx.$$

Consider now the following perturbation of the KdV equation:

$$q_t + 6qq_x + q_{xxx} = -g(t)q.$$

This is a model for water waves in a channel of slowly varying depth. The unperturbed problem $(g = 0)$ has the one-soliton solution

$$q_0 = 2k^2 \mathrm{sech}^2[k(x - x_0)].$$

Note that the parameter x_0 determines the norming constant, and that k is the eigenvalue parameter. Approximating g by a non-zero constant, one finds that

$$k_t \approx -\frac{2}{3}gk; \quad x_{0y} \approx 4k^2 + \frac{1}{3}g/k.$$

However, one should be able to verify the exact conservation law

$$\frac{\partial}{\partial t}\int_{-\infty}^{\infty} q \, dx = -g \int_{-\infty}^{\infty} q \, dx,$$

which follows by integrating the equation. However, substitution of the first order perturbation parameters shows that this is false at first order. It has been suggested that the discrepancy is due to the fact that the part of the solution corresponding to the continuous spectrum causes the appearance, behind the soliton, of a region in which the solution is close to a non-zero constant (a "shelf"). Its length would grow in time. However, this makes the solution resemble a "square well" potential, which suggests that several new eigenvalues should appear. A rigorous estimate of the number of eigenvalues involved has not been worked out.

The preceding discussion should illustrate the difficulties in the interpretation of the result of soliton perturbation in the absence of a preliminary study of the actual number of eigenvalues in the perturbed solution.

We refer to the Notes and the exercises for further results in this area.

5.6 APPLICATION OF INVARIANT MANIFOLD THEORY

We discuss a general technique for reducing nonlinear elliptic problems on a strip to a finite-dimensional ODE. This approach is relevant to the construction of traveling waves, and has proved successful for waves over shallow water. Its main advantage is that it enables one to avoid a small-amplitude expansion and to deal directly with the full equation of interest.

It is well-known that the dynamics of an ODE $\dot{x} = f(u)$ in N dimensions, for small x, can be described in terms of three manifolds intersecting at zero: the stable, unstable and center manifolds respectively (W^s, W^u and W^c). Only the stable and unstable manifold are uniquely determined. They have the property of being invariant, and tangent at the origin to the stable, unstable and central spaces associated to $f'(0)$, that is, those spaces spanned by the eigenvectors corresponding to eigenvalues with negative, positive, or vanishing real part. It can be shown that any solution which is bounded for all positive and negative time must live on W^s.

The relevance of this picture for us is that if one is interested in waves in an infinite channel (say, in the x direction), it is possible to view the governing equation as an infinite-dimensional dynamical system with x as "time" variable, so that the solutions of interest are bounded for all x. If there exists a finite-dimensional center manifold, it may be possible to classify those waves in some detail. This is actually the case for systems governed by *elliptic* equations, such as those arising in shallow-water theory. This is valuable in view of the difficulties in estimating the error made in approximating the full water wave equations by the KdV or Boussinesq equations.

Abstract results.

Consider an abstract ODE

$$du/dx = Ku + f(x, u),$$

where u lives in a Hilbert space X. K is linear, and f is smooth and quadratic near $u = 0$. The nonlinearity may involve additional parameters. Assume that $K = K_1 \oplus K_2$, where K_1 and K_2 are unbounded operators acting on complementary spaces X_1 and X_2 respectively. Corresponding to this splitting of X, we write $u = u_1 + u_2$. The basic assumptions on K are

1. The spectrum of K_1 is contained in the imaginary axis and K_1 generates a continuous group $(\exp(tK_1))_{t \in \mathbf{R}}$ with polynomial growth: $\| \exp(tK_1) \| \leq C(1 + |t|)^m$. [Thus, K_1 corresponds to the central part.]

2. The resolvent set of K_1 is contained in the imaginary axis and one has

$$\|(K_2 - iy)^{-1}\| \leq C(1 + |y|)^{-1}$$

for any real y.

One then has

THEOREM 5.6 *There is a smooth function $h : U \to X_2$, where U is a neighborhood of 0 in X_1, such that $h(x, 0) = 0$, $D_{u_1} h(x, 0) = 0$, for which the set*

$$M = \{(x, u_1 + h(x, u_1)) : x \in \mathbf{R} \, ; u_1 \in U\}$$

is locally invariant and contains all small bounded solutions of the original system.

It turns out that the manifold depends smoothly on any parameters f depends smoothly on, and that it usually inherits invariance properties of f (in particular, reversibility). For proofs, see the Notes. The idea is the same as the proof of the finite-dimensional version: one casts the equation into the form of an integral equation by writing $u_x - Ku = f$, and applying an inverse of $\partial_x - K$ (i.e., a Green's function) which is adapted to the condition that the solution should remain bounded at infinity.

Unfortunately, if the original equation is derived from a variational principle, the same may not hold for the reduced flow on the center manifold. (Example: Consider the flow on the cotangent space of $\mathbf{R}^4$ derived from the Lagrangian $L(q, \dot{q}) = \frac{1}{2}(\dot{q}_1^2 - \dot{q}_2^2) + q_1 \dot{q}_2$.) It is therefore preferable to use a reduction theorem for Hamiltonian systems instead.

Examples.

To illustrate the concrete applicability of the above considerations, we give below their results in two cases.

Internal waves in a 2D channel. One seeks a (pseudo-)stream function $\psi(x,y)$. x varies from $-\infty$ to $+\infty$, while $g_1(x) < y < g_2(y)$. By letting

$$y = g_1(x) + \eta(g_2 - g_1),$$

one reduces the problem to an equation on a strip in the (x, η) plane. One also assumes $g_2 \equiv 1$, and $g_1 = \varepsilon g(x)$, where g is compactly supported. This means that we have a small bump at the bottom of the channel. Study of the operator K in this situation reveals that there is, if the Froude number is close to a critical value, the center manifold is two-dimensional and corresponds to a flow with a Hamiltonian which can be computed to all orders near the origin. It follows that all small amplitude solitary waves can be analyzed by a phase plane analysis.

These results should be compared with the predictions of the previous section.

Surface waves. The water-wave equations can again be put in a Lagrangian form for the velocity potential and the surface elevation and their x derivatives. One is interested in solitary waves, so the time dependence can be removed by a change of variables. It is possible to use center-manifold theory near the uniform flow solution (surface elevation $=1$). The actual infinite-dimensional ODE is obtained after some manipulation, and it can be found in Iooss and Kichgässner (1992). Two parameters are the Bond number b, which is $T/(\rho h c^2)$, and the Froude number $\lambda^{-2} := c/\sqrt{gh}$, where h, g, c and T are respectively the undisturbed depth, the acceleration of gravity, the wave speed, and the surface tension coefficient.

If the Bond number b is greater than $1/3$, the solitary waves are either periodic waves (in x) or waves of depression depending on whether λ is smaller or greater than unity. If $b < 1/3$ and λ is in a suitable region, one can find decaying solitary waves (qualitatively similar to the KdV one-soliton). For $b < 1/3$, there are periodic waves, but also asymptotically periodic solutions as well as quasiperiodic solutions, depending on the parameter range. There are approximate solitary waves to all orders, which decay as $x \to \pm\infty$.

5.7   FURTHER RESULTS AND PROBLEMS

1. (The Lagrange expansion) This problem gives one case where a perturbation expansion can be computed in closed form.

Let g be holomorphic in $U \subset \mathbf{C}$, bounded by a smooth simple curve. Let $a \in U$. Then if $|\varepsilon g(z)| < |z - a|$ for every $z \in C$, there is a unique solution $u = u(a)$ of the equation

$$u = a + \varepsilon g(u).$$

(a) If f is holomorphic on an open domain $V \supset \bar{U}$, one has

$$f(u) = f(a) + \sum_{n=1}^{\infty} \frac{\varepsilon^n}{n!} \left(\frac{d}{da} \right)^{n-1} [f'(a)g(a)^n].$$

(b) What well-known expansion do you recover by taking $g(u) = u^2 - 1$ and $f(u) = u$? $g(u) = u$ and $f(u) = e^{-u}$?

(c) Generalize the statement of (a) to the C^∞ case. How about several variables?

2. (a) Show that $-\Delta + \varepsilon V(x)$ is a type (A) family in three space dimensions, if $V \in L^2 + L^\infty$.

(b) Compute the eigenvalues of $-\Delta + (\varepsilon - 1)/|x|$ explicitly, in three space dimensions. Show that there are eigenvalues which have an analytic continuation to all values of ε, but which are actual eigenvalues for ε small enough only.

3. Show that the perturbation theorem for type (A) families works for

$$|\varepsilon| < (a + \frac{1}{\delta}(b + a(|\lambda_0| + \delta)))^{-1},$$

where $\delta = (1/2)\mathrm{dist}(\lambda_0, \sigma(A_0) \setminus \{\lambda_0\})$.

4. (a) Discuss the form of the first term of the geometric optics expansion in the case when a multiple characteristic speed exists.

(b) Use this information to show that a Cauchy problem with a single double characteristic may be ill-posed in C^1. [*Remark:* It is still possible to solve problems with multiple characteristics in Gevrey classes.]

5. Compute the Whitham averaged equations for the Euler-Lagrange equations associated to a Lagrangian $L(u(x), \nabla u(x)) \, dx$, in any number of dimensions. Show that they can be derived from the averaged Lagrangian

$$\mathcal{L}(\kappa, E) = \int L(u, H\kappa) \, du + E[\int \frac{du}{H} - 1],$$

where $u = u(\theta)$ is periodic in θ and is defined implicitly by

$$\sum_i \kappa_i L_{u_i} \partial u / \partial_\theta - L(u, \kappa_i U_\theta) = E,$$

and $H = \partial u / \partial \theta$. (See Luke (1966))

6. (a) We consider the search for periodic solutions of the ODE

$$u'' + u = \varepsilon g(u).$$

For any a, solve the initial-value problem with $u(0) = a$ and $u'(0) = 0$. Define the map F by $F(a, \tau, \varepsilon) = u(\tau, \varepsilon)/\varepsilon$ for $\varepsilon \neq 0$. Show that if $F = 0$, it follows that the solution u is periodic with period 4τ (use reflection).

(b) Show that F is indeed well-defined, and smooth in an interval containing $\varepsilon = 0$.

(c) Give a sufficient condition on g which ensures that $\partial F/\partial \tau \neq 0$ for $\varepsilon = 0$ and conclude using the implicit function theorem.

(d) (Non-reversible case.) Consider now

$$u'' + u = \varepsilon g(u, u').$$

Solve again with data $u(0) = a$ and $u'(0) = 0$, and consider now the map $(a, T, \varepsilon) \mapsto (u(T) - 1, u'(T))/\varepsilon$. Show that periodic solutions can still be found by the implicit function theorem, but that the amplitude a is not in general arbitrary anymore. Apply the procedure to the van der Pol equation ($g = (1 - u^2)u'$). What happens if $g = g(u)$?

(e) For both (c) and (d), compute the first two terms of the expansion of the period as a function of ε, near $\varepsilon = 0$.

7. It is possible to use the Nash-Moser iteration even if only an *approximate inverse* for the inverse is defined. This is essential for conjugacy problems.

(a) More precisely, if F acts on scales of Banach spaces indexed by $s \geq 0$. Assume that for any Q, and any u, g, there exists a vector v such that

1. $|F'(u)v - g|_0 \leq KQ^{-\mu}$;

2. $|v|_r \leq KQ$;

3. $|F'(u)v|_0 \geq |v|_0$;

4. $|F(u + v) - F(u) - F'(u)v|_0 \leq M|v|_0^{2-\beta}|v|_r^\beta$.

Assume further that the parameters λ, μ and β satisfy

$$0 < \lambda + 1 < (\mu + 1)/2$$

and

$$0 < \beta < \frac{\lambda\mu}{(\lambda + 1)(\mu + 1)}(1 - 2\frac{\lambda + 1}{\mu + 1}).$$

Fix u_0 and let $f_0 = F(u_0)$. Then there is a number $K_0 > 1$ such that the inequalities $|u_0|_r < K_0$, $|f|_s \leq MK_0$ and $|f - f_0| < K_0^{-\lambda}$ ensure the existence of a solution u of $F(u) = f$. (This is Moser's original form of the result, see Moser (1966)).

(b) Let f be an analytic function of z, with $f(0) = 0$ and $f'(0) = \lambda$. Assume that for $q = 1, 2, \ldots$, one has $|\lambda^q - 1| \geq c_0 q^2$ for some $c_0 > 0$. Show that there is a function $u(z) = z + O(z^2)$ such that $u^{-1} \circ f \circ u(z) = \lambda z$.

8. The perturbation series of §5.1 (Rayleigh-Schrödinger series) can be useful even when they diverge, because they can be summed by Borel's method, or via

Padé approximants. An example of the latter is discussed in this problem (see Reed-Simon, §XII.4, for details).

(a) Assume that $f(\varepsilon) \sim \sum_n a_n \varepsilon^n$ in the following strong sense: there is a sector $\{\varepsilon : 0 < \varepsilon < R, |\arg \varepsilon| \le \frac{1}{2}\pi + \alpha\}$, for some $\alpha > 0$, in which

$$|f(\varepsilon) - \sum_n \le N a_n \varepsilon^n| \le C a^{N+1}(N+1)! |\varepsilon|^{N+1},$$

for some constants C and a. Let

$$g(\varepsilon) = \sum_n a_n \frac{\varepsilon^n}{n!}$$

be the *Borel transform* of f. Show that

$$f(\varepsilon) = \int_0^\infty g(\varepsilon t) e^{-t}\, dt$$

for $|\arg \varepsilon| < \alpha$.

(b) Apply this result to the ground state (= lowest eigenvalue) of $-d^2/dx^2 + x^2 + \varepsilon x^4$. [*Remark:* The analyticity of the ground state also implies that some higher eigenvalues are in fact continuations of each other under analytic continuation around the origin, if the eigenvalues are defined in a sector of aperture greater than 2π.] Incidentally, the case ε real and positive is accessible to type (B) methods (see Kato).

9. (Weakly nonlinear asymptotics) Consider a hyperbolic conservation law in one space variable:
$$u_t + (f(u))_x = 0.$$
Compute, as in §4.1, a formal solution of the form $u(\varepsilon) = u_0 + \varepsilon u_1(x - \lambda t, \varepsilon t) + \varepsilon^2 u_2 + \dots$. Show in particular that $u_1 = \phi R_1$ where ϕ solves a scalar Burgers equation. Show that one can arrange for the error made after truncating at level ε^2 to be independent of time. This suggests that the formal solution may be valid after the function u_1 develops singularities.

10. (Incompressible limit for fluid flows) The equations of isentropic incompressible fluid flow are

$$\frac{D\rho}{Dt} + \rho \operatorname{div} v = 0; \quad \frac{Dv}{Dt} + \frac{1}{\rho}\operatorname{grad} p = 0,$$

with $p = a\rho^\gamma$, $\gamma > 1$. $\frac{D}{Dt} = \frac{d}{dt} + v \cdot \operatorname{grad}$. Define a parameter $\varepsilon = [M(\gamma a)^{1/2}]$, where M is the Mach number. After rescaling of the variables and unknowns, one finds
$$(\gamma p)^{-1}\frac{Dp}{Dt} + \operatorname{div} v = 0; \quad \rho(p)\frac{Dv}{Dt} + \varepsilon^{-2}\operatorname{grad} p = 0.$$

(a) Take initial data of the form $v(x, 0) = v_0(x) + \varepsilon v_1(x)$, $p = p_0 + \varepsilon^2 p_1(x)$, where $\operatorname{div} v_0 = 0$, in some H^s. Show that the solution exists in a time domain independent of ε on which $v \to v_0$ as $\varepsilon \to 0$, in $C([0,T]; H^{s'})$, for some $s' < s$, if s is large enough.

(b) grad p tends to grad p_0 weak* in $L^\infty([0,T]; H^{s-1})$.

(c) Derive a set of equations for v_1 and p_1 (linearized acoustics).

(d) In particular, show that if $v_i = w_i + \nabla\psi_i$ with div$w_i = 0$ for $i = 0, 1$, then $(\partial_{tt} - c_0^2\Delta)\psi_1 = 0$ for a particular constant c_0.

NOTES

There are many subtleties in the choice and construction of formal solutions, which can be found in particular in the monographs by Kevorkian-Cole, Nayfeh, van Dyke among others. The September 1994 issue of SIAM review contains several historical papers on perturbation theory with emphasis on matched asymptotic expansions, and is a good source for relevant references.

Perturbation theory remains active to this day because it is an indispensable complement to computational work.

On perturbation theory for linear operators, see the classics by Rellich, Kato, Reed and Simon. It is remarkable that even finite-dimensional perturbation theory of spectra was treated in detail only in the forties.

An early version of the introduction of strained coordinates can be found in Poincaré, in his explanation of Lindstedt's results in particular. This type of method is now known as the Poincaré-Lighthill-Kuo (PLK) method. Multiple scales were discussed in the previous chapter in the discussion of universality issues.

For the Nash-Moser implicit function theorem, we described the approaches in Schwartz (1969), Moser (1966), and Nirenberg (1972). Our presentation has been simplified by refraining from describing the version when the inverse of the linearization of the equation does not exist. This problem occurs essentially in conjugacy problems, which was one of the first areas where the technique proved its value. In many other applications, however, this is not the issue. The first use of this type of argument was by Nash in his proof of the isometric embedding theorem. The next important application was by Moser. This versatile tool has been useful in water waves in particular. Hörmander (1986) develops a convenient version in the case of scales with compact embeddings (such as C^r spaces, or H^s spaces of periodic functions. Hamilton (1983) presents a very general framework which leads to slightly longer proofs, but yields the smoothness of the inverse, in case the mapping F is smooth; in the spirit of some earlier proofs, this work avoids the use of Newton-type iteration. Hamilton (1983) also contains interesting applications in differential geometry. It is instructive to compare the results of Kirchgässner (1982) with the (slightly different) results of Sachs who uses the Nash-Moser IFT. Our treatment of the Nash embedding theorem follows Hörmander (1990), which contains a general version of the

IFT using the paraproduct. The reduction of the Nash embedding theorem to a perturbation problem is discussed in Schwartz (1969) for instance. For this particular application, another reduction to the usual IFT is due to M. Günther.

Since a form of the Nash embedding theorem can be reduced to the application of the ordinary IFT, and the existence of progressive solitary waves can be handled without the Nash-Moser IFT, one may wonder about the importance one should give to this technique. However, in practice, the Nash-Moser theorem appears to be remarkably flexible and easy to use. Its proofs are not as complicated as they may seem at first. It is therefore likely to remain an invaluable tool. For situations where it is applicable, Hörmander's substitution of a "para-inverse" for the unbounded inverse is likely to further streamline the application of the "hard" IFT.

The method of geometrical optics was extended to nonlinear problems by Choquet-Bruhat (1969), motivated by the results of Lax (1957), Leray, and Garding-Kotake-Leray among others. She studied the case of a single phase, corresponding to a simple or a multiple characteristic. The subject was taken up more recently by Hunter and Keller (1983), Hunter, Majda and Rosales (1984) and Majda and Rosales (1988), where several phases, possibly resonant, were considered. This led to the "coherence" assumption. Difficulties with the case when the solution is only quasi-periodic in the phase functions were recently studied in detail by Joly, Métivier and Rauch (1994), where earlier references can be found. There has been a recent renewal of interest in this area, with more detailed results in the case of $\Box u = f(u_t)$, where blow-up may precede caustic formation. A complete picture is yet to emerge, but see Joly *et al.* (1995) (to appear in *Trans. AMS* for references.

The justification of the Whitham averaged system is in its infancy. Interesting rigorous results have been obtained for equations integrable by inverse scattering, see Venakides and references therein. The treatment in the text follows Whitham, Lighthill and Luke.

Our treatment of soliton perturbation follows mostly Kaup and Newell (1978) and Kaup (1976). There are very few rigorous results in this area, apart from the completeness theorem by Sachs (1983), which provides a solid framework for the solution of the linearization of the KdV equation. The formal arguments described in the text are however so widely found in the literature that it was impossible not to include them. On the shelf problem, one should mention the work of Mielke (1986) which does study rigorously traveling waves for this type of problem.

The material in §5.6 follows Mielke (1991) and Iooss and Kirchgässner (1992), where the reader can find several other results in that direction. The results for water waves are still among the most useful non-trivial results in this area, even though they pertain only to the case of traveling wave solutions. Indeed, soliton perturbation by inverse scattering has still not been completely justified. Note that even when it does not enable one to conclude, the center manifold approach can sometimes recover approximate solutions to all orders, which may explain numerical results. Thus, Iooss and Kirchgässner (1992) recover approximate solitons which are similar to those found in Kichenassamy (1990) and Kichenassamy and Olver (1992) [references in the previous chapter] for nonlinear Klein-Gordon and perturbations of the fifth order KdV equations respectively.

REFERENCES

CHOQUET-BRUHAT, Y. (1969) Ondes asymptotiques et approchées pour des systèmes d'équations aux dérivées partielles non linéaires, *J. Math. Pures et Appl., 48*: 117–158.

VAN DYKE, M. (1964) *Perturbation Methods in Fluid Mechanics,* Academic Press, New York.

HAMILTON, R. S. (1983) The Inverse Function Theorem of Nash and Moser, *Bull. AMS, 7*: 65–222.

HÖRMANDER, L. (1985) On the Nash-Moser implicit function theorem, *Ann. Acad. Sci. Fenn., 10*: 255–259.

HÖRMANDER, L. (1990) The Nash-Moser theorem and paradifferential operators, in *Analysis, et cetera*, Research papers published in honor of Jürgen Moser's 60th birthday, P. H. Rabinowitz and E. Zehnder Eds., Academic Press.

HUNTER, J., AND KELLER, J. (1983) Weakly nonlinear high frequency waves, *Comm. Pure Appl. Math., 36*: 547–569. See also *Wave motion, 6*: 79–89.

IOOSS, G. AND KIRCHGÄSSNER, K. (1992) Water waves for small surface tension: an approach via normal form, *Proc. Roy. Soc. Ediburgh, 122A*: 267–299.

KATO, T. (1980) *Perturbation Theory for Linear Operators,* 2nd ed., corrected printing, Springer, Berlin, New York.

KAUP, D. J. (1976) Closure of the squared Zakharov-Shabat eigenstates, *J. Math. Anal. Appl., 54*: 849–864.

KAUP, D. J. AND A. NEWELL, A. (1978) Solitons as particles, oscillators, and in slowly changing media: a singular perturbation theory, *Proc. Roy. Soc. London, A361*: 413–446.

KEVORKIAN, J. AND COLE, J. D. (1981) *Perturbation Methods in Applied Mathematics,* Springer.

KIRCHGÄSSNER, K. (1982) Wave solutions of reversible systems and applications, *J. Diff. Eq., 45*: 113–127.

KOGELMAN, S. AND KELLER, J. (1973) *SIAM J. Appl. Math., 24*: 352–361.

LAX, P. D. (1957) Asymptotic solutions of oscillatory initial value problems, *Duke Math. J., 24*: 627–646.

LUKE, J. C. (1966) A perturbation method for nonlinear dispersive wave problems, *Proc. Roy. Soc. London, A292*: 403–412.

MAJDA, A. AND ROSALES, R. (1986)

MIELKE, A. (1986) A reduction principle for nonautonomous systems in infinite-dimensional spaces, *J. Diff. Eq., 65*: 68–88 (see also the following paper in this journal).

MIELKE, A. (1991) *Hamiltonian and Lagrangian flows on Center Manifolds*, Lect. Notes in Math., vol. 1489, Springer.

MOSER, J. K. (1966) A rapidly convergent iteration method and non-linear equations, I & II, *Ann. Sc. Norm. Sup. Pisa, 20*: 265–315 and 499–535.

NAYFEH (1993) *Introduction to Perturbation Techniques*, Wiley Classics Library Ed.

O'MALLEY JR., R. E., VAN DYKE, M., COLE, J. D. AND ECKHAUS, W. (1994) Four papers on the history of singular perturbations and matched asymptotic expansions, in *SIAM Review, 36*: 413–439.

SACHS, R. L. (1983) Completeness of derivatives of squared eigenfunctions and explicit solutions of the linearized KdV equation, *SIAM J. Appl. Math., 14*: 674–683

SACHS, R. L. (1990) Bifurcation for semi-linear elliptic problems via the Nash-Moser technique, in *Analysis, et cetera*, Research papers published in honor of Jürgen Moser's 60th birthday, P. H. Rabinowitz and E. Zehnder Eds., Academic Press.

SCHWARTZ, J. T. (1969) *Lectures on Nonlinear Functional Analysis*, Gordon and Breach, New York.

VENAKIDES, S. (1984) The generation of modulated wavetrains in the solution of the KdV equations, *Comm. Pure Appl. Math.*

WHITHAM, G. B. (1965) *Proc. Roy. Soc. London, A283*: 238–261.

Chapter 6

General Relativity

General Relativity generalizes the Newtonian theory of gravitation. It also supersedes prerelativistic models of continuum mechanics and electromagnetism, and in this sense includes most of the equations studied in this book as approximations. Also, the development of General Relativity is intimately related to the emergence of the modern theory of the hyperbolic Cauchy problem.

After some background on differential geometry (§6.1), we introduce the basic equations of General Relativity, namely *Einstein's equations*, which read, in local coordinates

$$(6.1) \qquad\qquad G_{ab} = \chi T_{ab},$$

where G_{ab} involves the derivatives of the components of the space-time metric upto second order, as explained in §6.1. T_{ab} is the energy-momentum tensor, and $\chi = 8\pi$. Our main concern will be to set-up and solve a Cauchy problem for this system (§6.3). The unknowns include the 10 independent components of the symmetric tensor (g_{ab}), together with some components of T_{ab}, given some information on the structure of the latter (§6.2). This problem differs essentially from those of Ch. 2 because (6.1) is under-determined: the 10 equations satisfy four identities

$$\nabla^a(G_{ab} - \chi T_{ab}) = 0.$$

Furthermore, the solutions are not unique because the tensorial character of (6.1) allows for arbitrary changes of coordinates. Finally, Cauchy data cannot be prescribed arbitrarily; they are determined by solving *constraint equations*. These difficulties are due to the fact that the unknown is the space-time itself, rather than the particular representations of its metric in a given coordinate patch.

Section 6.4 is devoted to the problem of linearization stability of (6.1). It is shown that the presence of Killing vectors leads to obstructions. A few other perturbation problems in General Relativity are also discussed, including linearized gravity. It turns out that the non-relativistic limit of (6.1) is a singular limit, which leads to special difficulties.

Section 6.5 discusses the analogue for (6.1) of "blow-up" and "global existence," namely singularity theorems and cosmic censorship. It turns out that the maximal Cauchy development given in §6.3 may sometimes give only part of the space-time. This feature has no equivalent for the equations of Ch. 2. Once a manifold has been maximally extended, it may still be geodesically incomplete in a sense made precise in §6.5. This is interpreted by saying that spacetime is *singular*. We will see that other definitions of singular spacetimes, for instance based on curvature scalars becoming infinite, are not general enough. Here again, due to the geometric nature of the problem, the notion of "global existence" of Ch. 2 is wholly inadequate.

Section 6.6 reviews some important exact solutions. Exact solutions are of course useful examples; but they also form the basis of the most important experimental verifications of the theory. Three systematic methods of generation of solutions will be outlined. One class of spaces leads to equations possessing Bäcklund transformations in the sense of Ch. 4.

Thus, the mathematical study of Einstein's equations displays most of the difficulties we have encountered so far in the study of nonlinear waves, compounded by the role of coordinate independence. It therefore appears as the culmination of the theory of nonlinear waves.

6.1 PRELIMINARIES

This section contains background material on Lorentzian spacetimes. After reviewing basic properties of metric connections and their curvature, we briefly discuss the conformal curvature tensor and the Bel-Robinson tensor. The conservation of the latter has been proposed as a substitute for the energy identity for the wave equations. Some properties of congruences of curves and on the geometry of spacelike hypersurfaces are then given.

Metric and connection.

We are interested in a four-dimensional, connected, Hausdorff, paracompact manifold V_4. It is said to be Lorentzian if it is endowed with a metric of signature $(-, +, +, +)$:

$$ds^2 = g_{ab} dx^a \otimes dx^b$$

in local coordinates. We always use the convention of summation over repeated indices in different positions; Latin indices run from 0 to 3. In this section, V_4 is smooth and (g_{ab}) is of class C^3, but this can be relaxed slightly, see §6.2. The inverse of (g_{ab}) is denoted by (g^{ab}), and its determinant by g.

Unless otherwise specified, V_4 is assumed to be *oriented*. For pathologies on the global structure of space-time, see §6.4.

Minkowski space is $\mathbf{R}^4$ with the metric $(\eta_{ab}) = \text{diag}\,(-1, 1, 1, 1)$.

A vector field t^a is said to be *spacelike* (resp. *timelike*, *null*) if $g_{ab}t^a t^b > 0$ (resp. < 0, $= 0$). Unit timelike vectors satisfy $g_{ab}t^a t^b = -1$. While V_4 always admits Riemannian metrics without further assumptions, the situation is more complicated in the Lorentzian case. V_4 admits such a metric if and only if it admits a field of directions (*i.e.*, of pairs $(t, -t)$ of opposite, nonzero vectors), which can be thought of as the negative eigendirections of (g_{ab}) with respect to a reference Riemannian metric. One can show that a compact V_4 admits a Lorentzian metric if and only if its Euler-Poincaré characteristic vanishes (see Markus (1955)). We further assume, unless otherwise specified, the existence of oriented time-lines. The failure of this "time-orientability" is considered to be a serious pathology.

Latin indices will always be raised and lowered using g^{ab} and g_{ab}.

The connection. There is a unique symmetric connection ∇ such that the metric has vanishing covariant derivative. Its connection form is given locally by

$$\omega^a{}_b = \Gamma^a{}_{bc}\theta^c,$$

where $\theta^c = \partial/\partial x^c$ is the natural coframe, and the $\Gamma^a{}_{bc}$ are the *Christoffel symbols*

$$(6.2) \qquad \Gamma^a{}_{bc} = \frac{1}{2}g^{cm}[\partial_a g_{mb} + \partial_b g_{am} - \partial_m g_{ab}].$$

Covariant derivatives of tensors will be denoted by semi-colons and ordinary derivatives by commas: thus, if X_k is a covariant vector,

$$\nabla_j X_k = X_{k;j} = X_{k,j} - \Gamma^i_{jk} X_i.$$

The covariant derivative of tensors is defined by the usual rules of derivation; (6.2) follows easily from $\nabla_a g_{bc} = 0$. One computes easily that

$$\Gamma^c := g^{ab}\Gamma^c_{ab} = -(-g)^{-1/2}\partial_m(g^{cm}\sqrt{-g}).$$

It does not transform like a vector under coordinate transformations; one has $\Gamma^a = -g^{bc}\nabla_b \nabla_c x^a$, which is the Laplace-Beltrami operator applied to the coordinate function x^a. When it is zero, we say, that the coordinate system is *harmonic*.

Lie derivative. For any vector field X^a, the *Lie derivative* of any geometric object U is

$$\mathcal{L}_X U = \lim_{\varepsilon \to 0} \frac{U - \varphi^*(\varepsilon)U}{\varepsilon},$$

where $\varphi^*(\varepsilon)$ is the pull-back associated to the 1-parameter group $\{\varphi(\varepsilon)\}_{\varepsilon \in \mathbf{R}}$ generated by X^a. Thus,

$$\mathcal{L}_X v_a = X^b v_{a,b} + X^b_{,a} v_b; \quad \mathcal{L}_X v^a = X^b v^a_{,b} - X^A_{,b} v^b.$$

In these formulae, one may replace ordinary derivatives by covariant ones. X^a is a *Killing field* if $\mathcal{L}_X g_{ab} = 0$, which reduces to

$$\nabla_a X_b + \nabla_b X_a = 0.$$

A *conformal Killing field* satisfies $\mathcal{L}_X g_{ab} = \lambda(x) g_{ab}$.

Hodge duality. Let $\eta_{abcd} = \sqrt{-g}\, \varepsilon_{abcd}$, where ε_{abcd} is the sign of the permutation $(abcd)$ of $\{0,1,2,3\}$. We find $\eta^{abcd} = -(-g)^{-1/2} \varepsilon^{abcd}$, where ε^{abcd} has the same value as ε_{abcd}. Let F_{ab} be an antisymmetric tensor (which can be identified with a 2-form). Its *Hodge dual* is

$$F^*_{ab} = \frac{1}{2} \eta_{abcd} F^{cd}.$$

One finds $(F^*_{ab})^* = -F_{ab}$ (since we are in 4 dimensions). If F_{abcd} is antisymmetric in (a,b) and (c,d) ("double 2-form"), we let

$$F^*_{abcd} = \frac{1}{2} \eta_{cdrs} F_{ab}{}^{rs}; \quad {}^*F_{abcd} = \frac{1}{2} \eta_{abrs} F^{rs}{}_{cd}.$$

Curvature. The Riemann *curvature tensor* is defined by the commutation covariant derivatives, since the torsion is zero:

$$(\nabla_c \nabla_d - \nabla_d \nabla_c) X^a = R^a{}_{bcd} X^b;$$

this implies

$$R^a{}_{bcd} = \partial_c \Gamma^a_{bd} - \partial_d \Gamma^a_{bc} + \Gamma^m_{bd} \Gamma^a_{mc} - \Gamma^m_{bc} \Gamma^a_{md}.$$

An equivalent definition is

(6.3) $$d\omega^a{}_b + \omega^a{}_c \wedge \omega^c{}_b := \Omega^a{}_b = \frac{1}{2} R^a{}_{bcd}\, dx^c \wedge dx^d.$$

Even though $\Gamma^a{}_{bc}$ is not a tensor, $R^a{}_{bcd}$ is, and it has the following *algebraic properties*:

(6.4) $$R_{abcd} = -R_{abdc} = -R_{bacd},$$

and

(6.5) $$R^a{}_{[bcd]} = 0,$$

where

$$R^a{}_{[bcd]} := \frac{1}{3!}(R^a{}_{bcd} + R^a{}_{cdb} + R^a{}_{dbc} - R^a{}_{cbd} - R^a{}_{bdc} - R^a{}_{dcb}).$$

From these, one can derive (exercise 1)

$$R_{abcd} = R_{cdab}.$$

Taking these symmetry properties into account, one finds that R_{abcd} has 20 independent components $(n^2(n^2 - 1)/12$ in $V_n)$.

Furthermore, taking the exterior derivative of (6.3), we find that

(6.6) $$d\Omega^a{}_b + \omega^a{}_c \wedge \Omega^c{}_b - \omega^c{}_b \wedge \Omega^a{}_c = 0.$$

In terms of the curvature tensor, this means:

(6.7) $$R^a{}_{b[cd;e]} = 0.$$

Eqs. (6.5) and (6.7) are known as Bianchi's first and second identities. (Recall that semi-colons denote covariant derivatives.)

The *Ricci tensor* and the *scalar curvature* are defined by contraction from the curvature tensor:

$$R_{ab} = R^c{}_{acb}; \quad R = R_{ab}g^{ab}.$$

The Ricci tensor is symmetric. Contraction of (6.7) in (a, e) results in

(6.8) $$R^a{}_{bcd;a} - 2R_{b[c;d]} = 0,$$

where brackets denote antisymmetrization. Multiplying (6.8) by g^{bd}, we get

(6.9) $$(R^a{}_b - \frac{1}{2}R\delta^a{}_b)_{;a} = 0.$$

The Weyl tensor. The *conformal curvature* or *Weyl tensor* $C^a{}_{bcd}$ is defined by

(6.10) $$R^{ab}{}_{cd} = C^{ab}{}_{cd} + 2g^{[a}{}_{[c}R^{b]}{}_{d]} - \frac{1}{3}Rg^{[a}{}_{[c}g^{b]}{}_{d]}.$$

The Weyl tensor has the symmetries of the curvature tensor, and is invariant under conformal transformations $(g_{ab} \mapsto \Omega^2 g_{ab})$. It vanishes in conformally flat spaces (and identically in dimensions 1, 2 and 3).

The algebraic classification of the Weyl tensor plays an important role in General Relativity, because it has been connected with gravitational radiation.

It was first effected by the "matrix method" (Petrov (1954)): one views C_{abcd} as a 6×6 matrix on the space of bivectors, or alternatively, as a complex symmetric matrix on $\mathbf{C}^3$:

$$(6.11) \qquad Q_{ab} = E_{ab} + iH_{ab} = (C_{abcd} + iC^*{}_{abcd})u^c u^d,$$

where u^c is a unit timelike vector ($u^c u_c = -1$). The *Petrov types* are distinguished according to the number of distinct eigenvectors for (6.11):

Petrov type	*Eigenvalues of Q*
I	3 distinct
D	1 double, diagonalizable
II	2 distinct, non-diagonalizable
III	All zero, two independent eigenvectors ($Q^3 = 0$)
N	All zero, one eigenvector ($Q^2 = 0$)
O	$Q = 0$.

Type I is known as *algebraically general* and all the others as *algebraically special*.

There are two other methods of classification, based on null tetrads (Debever) and spinors (Penrose).

It is a non-trivial consequence of these methods that algebraically special solutions have at least one multiple null eigendirection, and that for principal null direction k^a of multiplicity 1, 2, 3, 4 respectively, one has

$$\begin{aligned}
k^b k^c k_{[e} C_{a]bc[d} k_{f]} &= 0; \\
k^b k^c C_{abc[d} k_{f]} &= 0; \\
k^c C_{abc[d} k_{f]} &= 0; \\
k^c C_{abcd} &= 0.
\end{aligned}$$

The Bel-Robinson tensor. It turns out that there is, in the vacuum case ($R_{ab} = 0$), a conservative 4-tensor which is quadratic in the curvature, and which has formal similarities with the energy-momentum tensor of electromagnetism. It is the *superenergy* or *Bel-Robinson* tensor:

$$(6.12) \qquad T_{abcd} = \frac{1}{2}(R^e{}_a{}^f{}_c R_{ebfd} + R^{*e}{}_a{}^f{}_c R^*{}_{ebfd}).$$

It is completely symmetric in $(abcd)$.

The derivation of the conservative character of this tensor is as follows: we let $B_{ab} = R_{ab} - \frac{1}{4}Rg_{ab}$. The *Bach tensor* B_{ab} vanishes if and only if $R_{ab} = \lambda g_{ab}$. It is convenient to rewrite (6.10) in terms of B_{ab}.

$$R_{abcd} = C_{abcd} + E_{abcd} + \frac{1}{12}Rg_{abcd},$$

where $g_{abcd} = 2g_{a[c}g_{d]b}$ and $E_{abcd} = -g_{abe[c}B^e{}_{d]}$, so that $E^a{}_{bad} = B_{bd}$, which has zero trace.

One then finds by direct calculation that

(6.13) $\qquad {}^*g^*{}_{abcd} = -g_{abcd}; \quad {}^*E^*{}_{abcd} = E_{abcd}; \quad {}^*C^*{}_{abcd} = -C_{abcd}.$

It follows that

$${}^*R^*{}_{abcd} + R_{abcd} = 2E_{abcd}.$$

Now, if R_{ab} is proportional to g_{ab}, $E_{abcd} = 0$. Taking the dual of this relation, we find that

$$-R^*{}_{abcd} + {}^*R_{abcd} = 0,$$

so that left and right Hodge duals coincide. Note also that in vacuum, $R_{abcd} = C_{abcd}$. Now, for any two tensors K_{abcd} and L_{abcd}, a calculation yields that

(6.14) $\qquad K^{am}{}_{pq}L_{bc}{}^{hq} = K^{**am}{}_{pq}L^{**}{}_{bc}{}^{hq}$

(6.15) $\qquad = \dfrac{1}{2}K^{am}{}_{rs}L_{bc}^{rs}\delta_p^h + K^{*amhs}L^*{}_{bcps}.$

For $K = L = C$, this gives $C^{abcd}C_{ebcd} - C^{*abcd}C^*{}_{ebcd} = \frac{1}{2}\delta_e^a C^{mhpq}C_{mhpq}$. But (6.13) implies $C^{*abcd}C^*{}_{ebcd} = C^{**abcd}C_{ebcd} = -C^{abcd}C_{ebcd}$; we therefore have

(6.16) $\qquad C^{abcd}C_{ebcd} = \dfrac{1}{4}C^{fbcd}C_{fbcd}\delta_e^a.$

To finish the proof, one computes first that

$$2T_{abcd} = C^e{}_a{}^f{}_c C_{ebfd} + C^e{}_a{}^f{}_d C_{ebfc} - \frac{1}{8}C^{hjkl}C_{hjkl}g_{ab}g_{cd}.$$

Using (6.8) in the vacuum case, we find $C^a{}_{bcd;a} = 0$. Using the antisymmetry of the Weyl tensor in its first two indices, and the Bianchi identities in the vacuum, we find

$$C^{abc}{}_d C_{aecf;b} = \frac{1}{2}C^{abc}{}_d C_{abcf;e}.$$

One can then verify that $T^a{}_{bcd;a} = 0$ expresses the vanishing of the covariant derivative of the difference between the sides of (6.16).

Curves and congruences. A curve $(\mu \mapsto \gamma^a(\mu))$ is said to be a geodesic if

$$\nabla_{\dot\gamma}\dot\gamma = f(\mu)\dot\gamma,$$

in other words, if $v^a = \dot\gamma(\mu) = d\gamma^a/d\mu$ satisfies

(6.17) $\qquad \dfrac{dv^a}{d\mu} + \Gamma^a{}_{bc}v^b v^c = fv^a.$

If we reparametrize the curve, we can arrange so that $f = 0$, in which case μ is called an *affine parameter*. Affine parameters are useful in Lorentzian spaces as a substitute for arclength on null curves. Geodesics can be obtained, as in the Riemannian case, by extremalizing the length

$$\int \sqrt{\pm g_{ab}\dot{\gamma}^a\dot{\gamma}^b}\, d\mu.$$

Eq. (6.17) with $f = 0$ can also be derived in the null case by variation of $\int g_{ab}\dot{\gamma}^a\dot{\gamma}^b\, d\mu$. This also parallels a well-known property in mechanics.

Let us now consider a *congruence* of curves, that is, a smooth 3-parameter family of curves: $x^a = x^a(v, \alpha^1, \alpha^2, \alpha^3)$, where v is a parameter along the curves of the congruence. We let $u^a = dx^a/dv$, and assume that (v, α^σ) can be taken as local coordinates on V_4. Thus, u^a can be thought of as a vector field. (Greek indices run from 1 to 3.) We note that for any scalar T, $dT/dv = u^a\nabla_a T := \dot{T}$. We will use quite generally the dot for the operator $u^a\nabla_a$. Let us now consider the effect of varying the α^σ for fixed v. This leads to the consideration of the *connecting vector*

$$\eta^a = \frac{\partial x^a}{\partial \alpha^\sigma}\delta\alpha^\sigma.$$

Since

$$u^b\partial_b\eta^a = \frac{\partial^2 x^a}{\partial v \partial \alpha^\sigma}\delta\alpha^\sigma = \frac{\partial}{\partial \alpha^\sigma}u^a\delta\alpha^\sigma = \eta^b\partial_b u^a,$$

we have

(6.18) $$\mathcal{L}_u\eta^a = u^b\nabla_b\eta^a - \eta^b\nabla_b u^a = 0.$$

Therefore,

$$\begin{aligned}
0 = \mathcal{L}_u\mathcal{L}_u\eta^a &= u^b\nabla_b(\mathcal{L}_u\eta^a) - (\mathcal{L}_u\eta^b)\nabla_b u^a \\
&= u^b\nabla_b(u^c\nabla_c\eta^a - \eta^c\nabla_c u^a) \\
&= \ddot{\eta}^a - u^b\eta^c\nabla_b\nabla_c u^a - u^b\nabla_b\eta^c\nabla_c u^a \\
&= \ddot{\eta}^a - u^b\eta^c\nabla_b\nabla_c u^a - u^c\nabla_c\eta^b\nabla_b u^a \\
&= \ddot{\eta}^a - u^b\eta^c\nabla_b\nabla_c u^a - \eta^c\nabla_c u^b\nabla_b u^a \\
&= \ddot{\eta}^a - u^b\eta^c(\nabla_b\nabla_c - \nabla_c\nabla_b)u^a - \eta^c\nabla_c(u^b\nabla_b u^a).
\end{aligned}$$

Therefore,

(6.19) $$\ddot{\eta}^a - R^a{}_{bcd}\, u^b\eta^c u^d = \eta^c\nabla_c\dot{u}^a,$$

where the r.h.s. equals zero for a geodesic congruence parametrized by an affine parameter. In that case, (6.19) is known as the *geodesic deviation equation*.

Let us now turn to a *timelike* congruence, which is not necessarily geodesic. We therefore have $u^a u_a = -1$. It is customary to split $\nabla_a u_b$ into four parts:

(6.20) $$\nabla_a u_b = \sigma_{ab} + \frac{1}{3}\theta h_{ab} + \omega_{ab} - u_a\dot{u}_b,$$

where $\omega_{ab} = u_{[a;b]} + \dot{u}_{[a}u_{b]}$ (twist), $\theta = u^a{}_{;a}$ (expansion), and $\sigma_{ab} = u_{(a;b)} + \dot{u}_{(a}u_{b)}$ (shear), with $h_{ab} = g_{ab} + u_au_b$. Note that h_{ab} is the projection on the 3-space orthogonal to u^a, and that the first three terms are all orthogonal to u^a. σ_{ab} is symmetric and traceless. Also, $\sigma_{ab}u^b = \sigma_{ab}h^{ab} = h_{ab}u^b = \omega_{ab}u^b = 0$.

Next, write the commutation of derivatives in the form

$$(\nabla_c\nabla_d - \nabla_d\nabla_c)u_a = R_{abcd}u^b,$$

and multiply by u^d. We find

$$\nabla_c\dot{u}_a - \nabla_cu^d\nabla_du_a - u^d\nabla_d(\nabla_cu_a) = R_{abcd}u^bu^d.$$

Inserting the decomposition (6.20), we find, after a relabeling of indices,

(6.21) $\quad (\sigma_{ab} + \dfrac{1}{3}\theta h_{ab} + \omega_{ab} - u_a\dot{u}_b)^{\cdot} - \nabla_a\dot{u}_b + R_{abcd}u^cu^d$

(6.22) $\quad\quad + (\sigma_a{}^c + \dfrac{1}{3}\theta h_a{}^c + \omega_a{}^c - u_a\dot{u}^c)(\sigma_{cb} + \dfrac{1}{3}\theta h_{cb} + \omega_{cb} - u_c\dot{u}_b) = 0.$

Mutliplying (6.22) by g^{ab}, we obtain the *Raychaudhuri equation*;

(6.23) $\qquad\qquad \dot{\theta} - \nabla_a\dot{u}^a + R_{ab}u^au^b + \sigma^{ac}\sigma_{ac} - \omega^{ac}\omega_{ac} + \dfrac{1}{3}\theta^2 = 0.$

Note that since σ_{ab} is symmetric and orthogonal to u^a, which is timelike, $\sigma^{ac}\sigma_{ac} \geq 0$.

As for the evolution of σ_{ab} and ω_{ab}, we find, by taking the antisymmetric part of (6.22),

(6.24) $\qquad \dot{\omega}_{ab} - \dfrac{1}{2}(\nabla_a\dot{u}_b - \nabla_b\dot{u}_a) + \dfrac{2}{3}\theta\omega_{ab} + \sigma_{cb}\omega_a{}^c - \sigma_{ca}\omega_b{}^c$

(6.25) $\qquad = \dfrac{1}{2}(u_a\ddot{u}_b - u_b\ddot{u}_a) + \dot{u}^cu_{[a}(\sigma_{c|b]} + \dfrac{1}{3}\theta h_{c|b]} + \omega_{c|b]}).$

The r.h.s. will vanish upon projection on the 3-space orthogonal to u^a; finally, the symmetric part yields

$$\dot{\sigma}_{ab} + \dfrac{1}{3}\theta(u_au_b)^{\cdot} - (u_{(a}\dot{u}_{b)})^{\cdot} - \nabla_{(a}\dot{u}_{b)} + \dfrac{2}{3}\theta\sigma_{ab} + \sigma_a{}^c\sigma_{cb} + \omega_{ac}\omega^c{}_b$$

$$+ (R_{acbd} - \dfrac{1}{3}R_{cd}h_{ab})u^cu^d + \dfrac{1}{3}h_{ab}[\nabla_a\dot{u}^a$$

(6.26) $\qquad\qquad\qquad - \dfrac{1}{3}\theta^2 - (\sigma^{cd}\sigma_{cd} - \omega^{cd}\omega_{cd})] = 0.$

See exercise 3 for analogous results for null congruences.

Hypersurfaces. We are interested in a spacelike hypersurface $S : \{x^0 = 0\}$ in V_4. We establish a few formulæ which are required in the discussion of the

Cauchy problem. Whenever we deal with hypersurfaces, we use the convention that *Greek indices* run from 1 to 3. The induced metric is a tensor on S defined by

$$\gamma_{\alpha\beta} = g_{\alpha\beta},$$

and $\gamma_{\alpha\beta}$ and its inverse $\gamma^{\alpha\beta}$ are used to lower and raise Greek indices. One also defines

$$h_{ij} = g_{ij} + n_i n_j.$$

This represents the projection on the space orthogonal to n^k.

The unit normal is

$$n_k = (-N, 0, 0, 0),$$

with $N^2 g^{00} = -1$, and $N > 0$. We have

$$n^k = (1/N, -N^\alpha/N),$$

where $N^\alpha = N^2 g^{\alpha 0}$. By expressing that $g_{ik}n^k = -\delta_{i0}N$, we find

$$g_{\alpha 0} = N_\alpha; \quad g_{00} = -N^2 + N^\alpha N_\alpha,$$

so that

$$ds^2 = g_{ij}dx^i \otimes dx^j = -(N\,dx^0)^2 + \gamma_{\alpha\beta}(dx^\alpha + N^\alpha\,dx^0) \otimes (dx^\beta + N^\beta\,dx^0).$$

$N\,dx^0$ is interpreted as a lapse of proper time, and $(dx^\alpha+N^\alpha\,dx^0)\otimes(dx^\beta+N^\beta\,dx^0)$ as a proper distance on S. For this reason, N and N^α are called the *lapse function* and *shift vector* respectively.

One verifies that

$$\gamma^{\alpha\beta} = g^{\alpha\beta} - g^{\alpha 0}g^{\beta 0}/g^{00}.$$

LEMMA $\gamma := \det(\gamma_{\alpha\beta}) = g/g^{00}$. In other words, $\sqrt{-g} = N\sqrt{\gamma}$.

Proof: Consider the matrix g^{ab} and perform the following elementary operations for $\alpha = 1, 2, 3$: multiply the first column by $g^{\alpha 0}/g^{00}$, and substract the result from the αth column. The result is

$$\begin{pmatrix} g^{00} & 0 & 0 & 0 \\ g^{10} & & & \\ g^{20} & & (\gamma^{\alpha\beta}) & \\ g^{30} & & & \end{pmatrix}.$$

The claim follows.

The second fundamental form is the (tangential part of) the covariant derivative of $-n$: let $e_i = \partial/\partial x^i$; we then have, since $(e_\alpha, n) = 0$,

$$K_{\alpha\beta} = -(e_\beta, \nabla_\alpha n) = (n, \nabla_\alpha e_\beta) = (-Ne_0, \Gamma^\gamma{}_{\alpha\beta}e_\gamma + \Gamma^0{}_{\alpha\beta}e_0).$$

Thus,

(6.27) $$K_{\alpha\beta} = -N\Gamma^0{}_{\alpha\beta},$$

and

(6.28) $$\nabla_\alpha e_\beta = \tilde{\Gamma}^\sigma{}_{\alpha\beta} e_\sigma - K_{\alpha\beta} n.$$

We denote by $\tilde{\Gamma}^\gamma{}_{\alpha\beta}$ and $\tilde{\nabla}$ the connection and covariant derivative associated with the induced metric. In particular, since $\Gamma_{\gamma|\alpha\beta} = \tilde{\Gamma}_{\gamma|\alpha\beta}$,[1]

$$
\begin{align}
(6.29) \quad K_{\alpha\beta} &= -N(g^{00}\Gamma_{0|\alpha\beta} + g^{0\gamma}\tilde{\Gamma}_{\gamma|\alpha\beta}), \\
(6.30) \quad &= \frac{1}{2N}(\tilde{\nabla}_\alpha N_\beta + \tilde{\nabla}_\beta N_\alpha - \partial_0 g_{\alpha\beta}),
\end{align}
$$

There are two other definitions of $K_{\alpha\beta}$. First, one checks directly form (6.27) that

$$K_{\alpha\beta} = \nabla_\alpha\nabla_\beta x^0.$$

Next, one can define a tensor on V_4 by

(6.31) $$K_{ij} = -\frac{1}{2}h_i{}^k h_j{}^l(\nabla_k n_l + \nabla_l n_k).$$

One checks directly that $K_{\alpha\beta}$ given by (6.31) agrees with (6.27). Using the relations $\mathcal{L}_n n^j = 0$, $\mathcal{L}_n n_j = \mathcal{L}_n(g_{ij}n^i) = (\nabla_i n_j + \nabla_j n_i)n^i = \dot{n}_j$, we find

(6.32) $$K_{ij} = -\frac{1}{2}[\mathcal{L}_n g_{ij} + n_i\dot{n}_j + \dot{n}_i n_j] = -\frac{1}{2}\mathcal{L}_n h_{ij}.$$

Note that since $n^k n_k = -1$, $\nabla_\alpha n$ has no normal component at all.

One can express the curvature of V_4 in terms of $K_{\alpha\beta}$ and the curvature tensor $^{(3)}R^\alpha{}_{\beta\gamma\delta}$:

THEOREM 6.1 *(Gauss-Codazzi)*

$$
\begin{align}
(6.33) \quad R^\alpha{}_{\beta\gamma\delta} &= {}^{(3)}R^\alpha{}_{\beta\gamma\delta} - (K_{\beta\gamma}K_\delta{}^\alpha - K_{\beta\delta}K_\gamma{}^\alpha) \\
(6.34) \quad n_k R^k{}_{\beta\gamma\delta} &= \tilde{\nabla}_\delta K_{\beta\gamma} - \tilde{\nabla}_\gamma K_{\beta\delta}.
\end{align}
$$

THEOREM 6.2 *(Gauss-Mainardi-Codazzi)*

(6.35) $$h_i{}^k h_j{}^l \mathcal{L}_n K_{kl} + K_{ik}K_j{}^k + h_i{}^k h_j{}^l\nabla_{(k}\dot{n}_{l)} + \dot{n}_i\dot{n}_j = n^k n^l R_{kilj}.$$

---

[1]We write $\Gamma_{\gamma|\alpha\beta}$ for $g_{\gamma m}\Gamma^m{}_{\alpha\beta}$.

Idea of Proofs: First define the frame (n, e_1, e_2, e_3) and the coframe

$$(dx^0, \omega^1, \omega^2, \omega^3),$$

where $\omega^\alpha = dx^\alpha + N^\alpha\, dx^0$. Then, pply the commutation formula

$$\mathcal{R}(\vec{u}, \vec{v})\vec{w} = ([\nabla_{\vec{u}}, \nabla_{\vec{v}}] - \nabla_{[\vec{u},\vec{v}]})\vec{w}$$

to $(\vec{u}, \vec{v}, \vec{w}) = (e_\gamma, e_\delta, e_\beta)$. We use (6.28) and multiply the result by the dual frame ω^α to get (6.33) and (6.34).

For (6.35), one computes $\mathcal{L}_n K_{ij}$, and uses (6.32) to express the result in terms of n_j. On needs to use the relations

$$\nabla_i n_j + K_{ij} + n_i \dot{n}_j = 0,$$

and

$$(6.36) \qquad\qquad \nabla_k n_i \nabla_j n^k = K_{ik} K^k{}_j.$$

Note in particular that

$$(6.37) \qquad\qquad \nabla_k n^k = -K := g^{ij} K_{ij} = -\mathrm{Tr}(K).$$

REMARK: One can compute the expression $R^k{}_{ikj} n^i n^j$ directly by expanding $n^h n^j R^k{}_{hjk} = n^j (\nabla_j \nabla_k - \nabla_k \nabla_j) n^k$, and by using (6.36) and (6.37). In particular,

$$(6.38) \qquad R = {}^{(3)}R + (\mathrm{Tr}\,(K^2) - (\mathrm{Tr}\,K)^2) - 2\nabla_k(\dot{n}^k + n^k \mathrm{Tr}\,K).$$

(Note that we abbreviated $K_\alpha{}^\alpha$ and $K_\alpha{}^\sigma K_\sigma{}^\alpha$ as $\mathrm{Tr}\,K$ and $\mathrm{Tr}(K^2)$ respectively; following a common abuse of notation, we will even use K for the former.)

Maximal hypersurfaces. If $K = 0$, we say that S is a maximal hypersurface. Indeed, this means that the mean curvature of S vanishes, and therefore, by the same calculation as in the Riemannian case, the first variation of the area integral $\int \sqrt{\gamma}\, dx^1 dx^2 dx^3$ vanishes. As in the Riemannian case, S may only be an extremum of the area integral. The Bernstein conjecture, solved by Calabi, states that any closed maximal hypersurface in Minkowski space M^4 is a plane. The n-dimensional version of this result is due to Cheng and Yau (1976). Maximal hypersurfaces are often used as initial hypersurfaces in the Cauchy problem, for reasons which will be given in the study of the constraint equations in §6.3.

6.2   EINSTEIN'S EQUATIONS

This section describes basic formal properties of Einstein's equations and discusses the structure of matter terms, boundary and asymptotic conditions, and variational formulations.

The basic equations of General Relativity are Einstein's equations:

$$G_{ab} := R_{ab} - \frac{1}{2}Rg_{ab} = \chi T_{ab}$$

where T_{ab} is a symmetric tensor, and χ is a constant. The energy-momentum tensor T_{ab} is part of the unknown; only its form or its algebraic type is prescribed. A fairly general fluid-mechanical T_{ab} is

$$T_{ab} = \mu u_a u_b + (u_a q_b + u_b q_a) + p\, h_{ab} + \pi_{ab} + \zeta\theta\, h_{ab},$$

where $u_a u^a = -1$, $h_{ab} = g_{ab} + u_a u_b$, $u_a q^a = 0$, $\pi_{ab} u^b = 0$ and π_{ab} is symmetric, $\theta = u^a{}_{;a}$. The second, fourth and fifth terms are related to heat flow, viscosity, and bulk viscosity respectively. A special case of importance is the *perfect-fluid case*:

$$(6.39) \qquad T_{ab} = (\rho + p)u_a u_b + p\, h_{ab},$$

often supplemented by an *equation of state* $\rho = f(p)$, to ensure that the number of unknowns is equal to the number of equations. In particular, the cases $p = 0$, $\rho = 3p$ and $\rho = p$ are known as "dust," "radiation," and "stiff matter." The latter is somewhat unphysical.

The T_{ab} may also contain an electromagnetic contribution:

$$(6.40) \qquad \tau_{ab} := F_{ac}F_b{}^c - \frac{1}{4}\delta_{ab}F_{rs}F^{rs},$$

where $F_{rs} = \partial_r A_s - \partial_s A_r$ is the electromagnetic field tensor. The case where T_{ab} reduces to (6.40) corresponds to the Einstein-Maxwell equations. (Recall that the equations $\nabla_a F^{ab} = 0$ and $\nabla_a F^{*ab} = 0$ must be adjoined to Einstein's equations when minimal coupling is assumed, and represent Maxwell electromagnetic field equations in the absence of sources.)

Rainich gave a set of conditions on the Ricci tensor which ensure by themselves that G_{ab} *must* have the form χT_{ab} with T_{ab} of the form (6.40), with F_{ab} obeying Maxwell's equations. In this already unified case, one replaces the Einstein-Maxwell equations by a set of equations on the metric alone.

Since we are particularly interested here in the difficulties created by the geometric nature of (6.1), we will often restrict ourselves to the *vacuum equations*, in which $T_{ab} = 0$.

Let us stress that T_{ab} is not a "source term" in the sense that it is part of the unknown. Furthermore, there is no "background metric," which means that the space-time is unknown *a priori*, even in its topological structure.

Structure of Einstein's equations. A first observation is that (6.9) implies

$$\nabla^a G_{ab} \equiv 0.$$

The 10 Einstein equations are therefore not independent, but are related by the above four *conservation identities*. Since there are 10 metric components to be determined, (6.1) is *underdetermined*. Note also that the relations $\nabla^a T_{ab} = 0$, which follow from Einstein's equations, lead in the case of fluids to the "hyperbolic conservation laws" of Ch. 3.

The appropriate initial-value problem for Einstein's equations is discussed in §6.3.

Boundary and asymptotic conditions must be considered next. If matter is confined to a finite and smoothly bounded region, as in the study of the vicinity of a star, one usually distinguishes the "exterior case" from the "interior case," the desired solution being obtained by matching solutions to these two sub-problems. The matching can be done by requiring that the induced metric and the second fundamental form agree on both sides of the boundary surface (Darmois). Other junction conditions proposed in the literature include (1) the continuity of all metric components and their first-order derivatives (Lichnerowicz); (2) the continuity of the g_{ij}, $\partial_0 g_{\alpha\beta}$ and $T_k{}^0$, if the boundary of the material region is given by $x^0 = 0$ (O'Brien-Synge).

Asymptotic conditions occur when representing ideally "isolated" systems, which would occupy a part of the space-time. Since space-time is not a non-dynamical background, there is already a difficulty in defining "going to infinity" in the first place. A commmon procedure to tackle this difficult problem is to consider a "physical" space-time $(\bar{V}, \bar{g}_{ab})$, representing the vicinity of some "isolated" massive object for instance, and to embed it *conformally* into an "unphysical" space-time (V, g_{ab}). We therefore have $g_{ab} = \Omega^2 \bar{g}_{ab}$. We then require that Ω vanish precisely on the boundary of $\bar{V}$ in V. This produces a boundary for $\bar{V}$, representing the points at infinity in $\bar{V}$. See problems 15 and 16 for an idea of the calculations involved in such a framework.

The simplest examples of such a conformal treatment of infinity is the conformal compactification of Minkowski space, used in Ch. 2.

The conformal boundary ∂V is not expected to be smooth. It is split into past and future null infinity $(\mathcal{I}^-, \mathcal{I}^+)$, and spatial infinity i_0 (usually represented by points), together with past and future timelike infinity (i_- and i_+). It is too restrictive to require that the nonphysical metric be regular at i_0, i_-, or i_+, except in very simple cases such as Minkowski.

At null infinity, the metric is asymptotically Minkowskian, with $g_{ab} = \eta_{ab} + O(1/v)$, $v \to \infty$, where $v = 2/\Omega$. At spatial infinity, the conditions for initial data sets $(S, h_{\alpha\beta}, K_{\alpha\beta})$ imply: $h_{\alpha\beta} = \delta_{\alpha\beta} + O(1/r)$, $K_{\alpha\beta} = O(1/r^2)$, $^{(3)}R_{\alpha\beta} = O(1/r^3)$ where r can be defined as the coordinate distance to a reference point.

Even though asymptotically flat space-times are approximately Minkowskian at infinity, the search for vector fields satisfying the Killing equations as one tends to infinity leads to a group that can be substantially larger than the Poincaré group. At null infinity, one finds an infinite-dimensional group, the Bondi-Metzner-Sachs (BMS) group, containing an infinite-dimensional abelian normal subgroup of "supertranslations." The quotient of the BMS group by this subgroup is the Lorentz group. At spatial infinity, one finds the *Spi* group, with a structure similar to that of the BMS group. Stronger fall-off conditions on the Weyl tensor can even lead to the Poincaré group itself.

Higher-order equations. It is sometimes helpful to derive Einstein's equations from a third-order system. More precisely, let $U_{ab} = \chi(T_{ab} - \frac{1}{2}T g_{ab})$, where $T = T_{ab}g^{ab}$. Then

THEOREM 6.3 *If the spacetime satisfies*

$$(6.41) \qquad R^a{}_{bcd;a} = U_{bd;c} - U_{bc;d},$$

then the contracted Bianchi identities (6.8) imply $\nabla^a(R_{ab} - U_{ab}) = 0$.

Proof: Eq. (6.8) imply

$$(6.42) \qquad R^a{}_{bcd;a} = R_{bd;c} - R_{bc;d}.$$

Let $M_{ab} = R_{ab} - U_{ab}$. Contracting (6.42) and (6.41) in (b, d), we find

$$(6.43) \qquad R^a{}_{c;a} = \nabla_c U - \nabla_b U^b{}_c + \nabla_c R - \nabla_b R^b{}_c.$$

Now $U/\chi = -T$, and $\nabla_b U^b{}_c = \chi \nabla^b(T_{bc} - \frac{1}{2}T g_{bc}) = -(\chi/2)T_{;b}\delta^b_c$, so that we find

$$\begin{cases} 2R^a{}_{c;a} - \nabla_c R &= 0 \\ R^a_{c;a} &= -\frac{\chi}{2}\nabla_c T, \end{cases}$$

hence $\nabla_c R = -\chi \nabla_c T$. Therefore,

$$(6.44) \qquad \begin{aligned} \nabla^c M_{bc} &= \nabla^c(R_{bc} - \chi T_{bc} + \frac{1}{2}\chi T g_{bc}) \\ &= -\frac{1}{2}\chi \nabla_b T + \frac{1}{2}\chi(\nabla^c T)g_{bc} = 0, \end{aligned}$$

QED.

The significance of this result is due to the fact that it can be used to treat Einstein's equations as initial conditions, *i.e.* to show that the equations $M_{ab} = 0$ on $x^0 = 0$ imply the same for all x^0. Let us show that in the analytic case. Let us first note that substracting (6.42) from (6.41) gives

$$(6.45) \qquad M_{bd;c} = M_{bc;d}.$$

Now, together with (6.44), these equations imply

$$\begin{cases} \nabla_0 M_{\alpha\beta} &= \nabla_\beta M_{\alpha 0} \\ \nabla_0 M_a{}^0 &= -\nabla_\alpha M_a{}^\alpha. \end{cases}$$

This is a first-order system, of Cauchy-Kowalewska type, for M_{ab}, and therefore $M_{ab} = 0$ for $x^0 = 0$ suffices to guarantee that Einstein's equations hold everywhere.

Variational structure. Einstein's equations can be derived from the *Hilbert Lagrangian*

$$(6.46) \qquad R\sqrt{-g}.$$

This Lagrangian, of second order, generates equations which are themselves of second order only; in fact, it is equivalent to a first-order Lagrangian:

$$(6.47) \qquad R\sqrt{-g} = L'\sqrt{-g} + \partial_j(g^{kj}\Gamma^r{}_{kr} - g^{kr}\Gamma^j{}_{kr}),$$

where

$$L' = \mathcal{G}^{ij}(\Gamma^r{}_{ij}\Gamma^s{}_{rs} - \Gamma^r{}_{is}\Gamma^s{}_{jr}),$$

with $\mathcal{G}^{ij} = \sqrt{-g}\,g^{ij}$. It is a quite general fact that to any translation-invariant Lagrangian, one can associate a divergence-free tensor called the canonical energy-momentum tensor:

$$\Theta_a{}^b = \frac{1}{2\chi}[L'\delta_a{}^b - \frac{\partial L'}{\partial g^{cd}{}_{,b}}\partial_a g^{cd}].$$

One can check that the Euler-Lagrange equations imply that $\nabla^a\Theta_a{}^b = 0$ (this follows from Noether's theorem).

Einstein's equations can be written

$$(6.48) \qquad \sqrt{-g}\,G_a{}^b = \chi[\partial_c \mathcal{U}_a{}^{bc} - \Theta_a{}^b],$$

where $\mathcal{U}_a{}^{bc} = -\mathcal{U}_a{}^{cb}$ (the "superpotential") is defined by

$$(6.49) \qquad 2\chi\sqrt{-g}\,\mathcal{U}_a{}^{bc} = g_{ak}\partial_h[\mathcal{G}^{kb}\mathcal{G}^{hc} - \mathcal{G}^{kc}\mathcal{G}^{hb}].$$

The advantage of this formulation is that it makes the origin of second-order derivative terms in Einstein's equations more transparent, and will also motivate the introduction of harmonic coordinates. In fact, writing $f(g,\partial g)$ for any first-order expression in g_{ab}, it follows from (6.48) and (6.49) that

$$(6.50) \qquad 2\sqrt{-g}\,G_a{}^b = 2\chi\partial_c \mathcal{U}_a{}^{bc} + f(g,\partial g)$$

$$(6.51) \qquad\qquad = g_{ak}[\frac{\mathcal{G}^{hc}}{\sqrt{-g}}\partial_{hc}\mathcal{G}^{kb} + \mathcal{G}^{kb}\partial_h F^h$$

$$(6.52) \qquad\qquad\quad - \mathcal{G}^{kc}\partial_c F^b - \mathcal{G}^{hb}\partial_h F^k] + f(g,\partial g),$$

where

$$F^h = -(-g)^{-1/2}\partial_a\mathcal{G}^{ah} = -\nabla^c\nabla_c x^h = \Gamma^h{}_{ab}g^{ab}.$$

This simplifies to

$$(6.53) \qquad G_{ab} = -\frac{1}{2}g^{hc}\partial_{hc}g_{ab} - \frac{1}{2}g_{ab}\partial_h F^h + \frac{1}{2}g_{bc}\partial_a F^c + \frac{1}{2}g_{ac}\partial_b F^c$$
$$+ f(g, \partial g) + \frac{1}{4}g_{ab}g^{kl}g^{hc}\partial_{hc}g_{kl}.$$

Contracting, we find

$$R = -\frac{1}{2}g^{kl}g^{hc}\partial_{hc}g_{kl} + \partial_h F^h,$$

hence

$$(6.54) \qquad R_{ab} = -\frac{1}{2}g^{cd}\partial_{cd}g_{ab} + H_{ab}(g, \partial g) + [g_{bc}\partial_a F^c + g_{ac}\partial_b F^c].$$

The field equations can also be written in terms of the induced metric and second fundamental form of a specified hypersurface $S : \{x^0 = 0\}$. This is not only useful to set the Cauchy problem, but also to derive a Hamiltonian form of Einstein's equations. In fact, in the notation of this section,

$$(6.55) \qquad n_k G^k{}_\alpha = (\tilde{\nabla}_\beta K_\alpha{}^\beta - \tilde{\nabla}_\alpha K_\beta{}^\beta),$$

and

$$(6.56) \qquad -G_0{}^0 = \frac{1}{2}{}^{(3)}R + \frac{1}{2}[(\mathrm{Tr}K)^2 - \mathrm{Tr}(K^2)],$$

while

$$(6.57) \qquad G_{\alpha\beta} = {}^{(3)}R_{\alpha\beta} - (K_{\alpha\sigma}K^\sigma{}_\beta - KK_{\alpha\beta}) - \mathcal{L}_n K_{\alpha\beta}$$
$$(6.58) \qquad \qquad - \nabla_{(\alpha}\dot{n}_{\beta)} - \dot{n}_\alpha\dot{n}_\beta - \frac{1}{2}R\gamma_{\alpha\beta}.$$

One can now take as conjugate variables $g_{\alpha\beta}$ and

$$\pi^{\alpha\beta} := \sqrt{\gamma}(\gamma^{\alpha\beta}K - K^{\alpha\beta})$$

(see the Notes for the Poisson structure).

The (Arnowitt-Deser-Misner) Hamiltonian is

$$(6.59) \qquad I_{ADM} = \int (\pi^{\alpha\beta}\partial_0\gamma_{\alpha\beta} - N\mathcal{H} - N_\alpha\mathcal{H}^\alpha)\, d^3x,$$

where

$$\mathcal{H} = \gamma^{-1/2}(\pi^{\alpha\beta}\pi_{\alpha\beta} - \frac{1}{2}(\pi^\rho{}_\rho)^2) - \sqrt{\gamma}\,{}^{(3)}R; \quad \mathcal{H}^\alpha = -\tilde{\nabla}_\beta\pi^{\alpha\beta}.$$

One usually adds to (6.59) a divergence term, representing a total energy, in the asymptotically flat case ("boundary term").

REMARK: It is sometimes useful to write Einstein's equations in terms of a background metric. One is thus led to considering the difference of two connections, which is a tensor. The resulting equations are known as the (Hawking-Ellis) *reduced equations*.

6.3   THE CAUCHY PROBLEM

We are interested in finding solutions of Einstein's equations near a hypersurface $S : \{x^0 = 0\}$. We would like to compute g from the values of g_{ij} and $\partial_0 g_{ij}$ on S. It turns out that these *Cauchy data* cannot be given arbitrarily, but must solve the *constraint equations* $G_a{}^0 = 0$. There are therefore two problems to be treated: the construction of a Cauchy data set by solving the constraint equations, and the evolution of g_{ij} from such a set. After a few further remarks on Einstein's equations, we discuss the evolution problem in the analytic and non-analytic cases, and then describe a few popular methods for solving the constraint equations. We limit ourselves exclusively to the vacuum equations. The characteristic initial-value problem is briefly discussed.

Preliminary remarks.

Einstein's equations being of second order, it would seem natural to hope that the metric is determined by the values of g_{ij} and $\partial_0 g_{ij}$ on S. In order to use the Cauchy-Kowalewska theorem, it would seem natural to try to compute $\partial_{00} g_{ij}$ in terms of other second-order derivatives, and lower-order terms. Let us therefore investigate the nature of second-derivative terms in the equations.

A first difficulty is that a coordinate change of the form

$$x'^i = x^i + \frac{1}{6}(x^0)^3[\varphi^i(x) + O(x^0)]$$

always preserves g_{ij} and $\partial_0 g_{ij}$ on S, as well as $\partial_{00} g_{\alpha\beta}$, but will, by choosing $\varphi^i(x)$ suitably, induce an arbitrary change in $\partial_{00} g_{0i}$. The above idea is therefore wrong, since some of the second-order derivatives can be modified arbitrarily without changing the Cauchy data.

To find a way out, we first isolate the difficulty by examining (6.48-6.49): second order time derivatives can arise only from the superpotential term, and in fact, are confined to the term

$$g_{ak} \partial_{00} [\mathcal{G}^{kb} \mathcal{G}^{00} - \mathcal{G}^{k0} \mathcal{G}^{b0}].$$

If $b = 0$, this expression vanishes, and $G_a{}^0$ *is expressible in terms of the Cauchy data alone*. The equations $G_a{}^0 = 0$ therefore represent constraints on the initial data.

This derivation of the constraints also points to a way out: assume that $F^c \equiv 0$, which means that the coordinate functions are *harmonic*.[2] Eq. (6.54) shows that Einstein's equations take the form

$$g^{cd}\partial_{cd}g_{ab} = F^{ab}(g, \partial g),$$

which is clearly a more appropriate form for an initial-value problem.

We will, in view of this form of the equations, say that S is characteristic if $g^{00} = 0$, non-characteristic otherwise. There are in fact other "trivial" characteristic directions for (6.1), see problem 9. In a more invariant way, if S has equation $f = 0$, S is characteristic if and only if

$$\Delta_1 f := g^{ab}\nabla_a f \nabla_b f = 0.$$

The characteristics of the above equation for characteristics, namely the bicharacteristics, are called the *rays* along which gravitational waves propagate. Across characteristic hypersurfaces, discontinuities of second order derivatives of g may occur, see problems 6 and 7.

Analytic Cauchy problem.

Assume we are given an initial data set satisfying the constraint equations $G_a{}^0 = 0$, and such that $g^{00} \neq 0$ on S. See exercise 9 for the symbol associated with Einstein's equations.

We find by inspection:

$$(6.60) \qquad R_{\alpha\beta} = -\frac{1}{2}g^{00}\partial_{00}g_{\alpha\beta} + F(\text{Cauchy data})$$

$$(6.61) \qquad R_{\alpha 0} = \frac{1}{2}g^{0\beta}\partial_{00}g_{\alpha\beta} + F(\text{Cauchy data})$$

$$(6.62) \qquad R_{00} = -\frac{1}{2}g^{\alpha\beta}\partial_{00}g_{\alpha\beta} + F(\text{Cauchy data}),$$

where $F(\text{Cauchy data})$ can be computed from the Cauchy data and their spatial derivatives. It follows that we can fix the g_{0a} on and off S in accordance with the given data, and then solve the six equations $R_{\alpha\beta} = 0$ by the Cauchy-Kowalewska theorem.

It remains to show that the equations $R_{0a} = 0$ also hold. We achieve this by showing that the constraint equations "propagate" off S: Note first that $R_{\alpha\beta} = 0$ implies that the R_{0a} can be expressed in terms of the $G_a{}^0$ and vice-versa. More

---

[2]Of course, these coordinates satisfy hyperbolic, rather than elliptic, equations in the Lorentzian case.

precisely,

$$(6.63) \qquad G_\alpha{}^0 \;=\; R_\alpha{}^0 = g^{0\beta} R_{\alpha\beta} + g^{00} R_{\alpha0} = g^{00} R_{\alpha0}$$

$$(6.64) \qquad G_0{}^0 \;=\; R_0{}^0 - \frac{1}{2} R$$

$$(6.65) \qquad\qquad =\; g^{0\alpha} R_{0\alpha} - \frac{1}{2} (g^{\beta\gamma} R_{\beta\gamma} + 2 g^{0\beta} R_{0\beta} + g^{00} R_{00})$$

$$(6.66) \qquad\qquad =\; \frac{1}{2} (g^{00} R_{00} - g^{\alpha\beta} R_{\alpha\beta}) = \frac{1}{2} g^{00} R_{00}$$

$$(6.67) \qquad G_\alpha{}^\beta \;=\; g^{\beta0} R_{0\alpha} - \frac{1}{2} \delta_\alpha^\beta (g^{00} R_{00} + 2 g^{0\beta} R_{0\beta}).$$

Thus,

$$(6.68) \qquad\qquad R_{\alpha0} \;=\; (g^{00})^{-1} G_\alpha{}^0;$$
$$(6.69) \qquad\qquad R_{00} \;=\; G_{00} = 2(g^{00})^{-1} G_0{}^0,$$

and all the $G_\alpha{}^\beta$ can be expressed linearly in terms of the $G_a{}^0$.

It follows that the contracted Bianchi identities $\nabla_c G_a{}^c = 0$ take the form

$$\begin{cases} g^{00} \partial_0 G_a{}^0 \;=\; \sum_{\rho,j} (A_\alpha^{\rho j} \partial_\rho G_j{}^0 + B_\alpha^j G_j{}^0) \\ \qquad\quad G_a{}^0 \;=\; \qquad 0 \quad \text{for} \quad x^0 = 0. \end{cases}$$

It follows finally that $G_a{}^0 \equiv 0$ hence $R_{ab} = 0$, QED.

Non-analytic Cauchy problem.

To solve the non-analytic Cauchy problem, we follow a different approach. We consider for definiteness the problem where g_{ab} is given on $S = \mathbf{R}^3$, together with its time derivative; we assume that g_{ab} is the sum of a tensor with components in H^s, and the Minkowski metric, although similar ideas apply to other situations, such as Compact Cauchy surfaces. Since we are in the vacuum, we must solve $R_{ab} = 0$. We will not have to separate the equations into a set of constraints and an evolution problem as in the analytic case. For simplicity, we assume $s \geq 4$, see however the references in the Notes.

The argument is as follows.

STEP 1: Using (6.54), we write

$$R_{ab} = R_{ab}^{(h)} + \frac{1}{2} [g_{bc} \partial_a F^c + g_{ac} \partial_b F^c].$$

We therefore begin by solving

$$(6.70) \qquad\qquad\qquad R_{ab}^{(h)} = 0$$

by writing it as a symmetric-hyperbolic system for the 10 unknowns g_{ab} (note that we solve 10 equations instead of 6 in the analytic case). Indeed, let $p_{ab} = \partial_0 g_{ab}$. We have

$$(6.71) \qquad \partial_0 g_{ab} = p_{ab}$$
$$(6.72) \qquad g^{\alpha\beta} \partial_0 g_{ab,\alpha} = g^{\alpha\beta} \partial_\alpha p_{ab}$$
$$(6.73) \qquad -g^{00} \partial_0 p_{ab} = 2 g^{0\alpha} \partial_\alpha p_{ab} + g^{\alpha\beta} \partial_\beta g_{ab,\alpha} - 2 H_{ab},$$

which we view as a first-order system in the 50-component unknown

$$(g_{ab}, g_{ab,i}, p_{ab})$$

Note that $g^{\alpha\beta}$ will be non-singular if $x^0 = 0$ is space-like. We obtain a local solution by applying Th. 2.3.

STEP 2: By solving $R_{ab}^{(h)} = 0$, we have constructed an H^s extension of our Cauchy data. We now show that if we knew that the Cauchy data satisfy $F^c = 0$ in addition to the constraints, then the coordinates are in fact globally harmonic, and $R_{ab} = 0$ as desired.

This will be achieved by showing that F^c solves a second-order hyperbolic system with zero Cauchy data.

Let us first prove that $\partial_0 F^c = 0$ as a consequence of the constraints. Taking $R_{ab}^{(h)} = 0$ into account, we find that

$$G_{ab} = \frac{1}{2}(g_{bm} \partial_a F^m + g_{am} \partial_b F^m - g_{ab} \partial_m F^m).$$

It follows that

$$G_a{}^0 = \frac{1}{2}(\partial_a F^o + g_{am} g^{0c} F^m - \delta_a^0 \partial_m F^m),$$

hence, since F^c (and its spatial derivatives) is initially zero, we find

$$g_{am} g^{00} \partial_0 F^m = 0,$$

or $\partial_0 F^m = 0$. Thus, F^m has zero Cauchy data.

We now turn to the equation satisfied by F^c. Let us define

$$L_{ab} = R_{ab} - R_{ab}^{(h)}; \quad L = L_{ab} g^{ab}.$$

Since $R_{ab}^{(h)} = 0$, we have

$$\nabla_a(L^{ab} - \frac{1}{2} L g^{ab}) = 0,$$

Writing this out, we find

$$g^{ac} \partial_{ac} F^b = \sum_{k,l} A_k^l \partial_l F^k,$$

so that F^c solves a linear homogeneous hyperbolic equations, and has zero data. It follows that $F^c \equiv 0$, QED.

STEP 3: One must finally prove that it is no restriction to assume that $F^c = 0$ on S. To see this, we start from the local development defined in Step 1, and define a new coordinate system by solving the wave equation

$$\Box x^a = 0,$$

with $x^a = (0, x^1, \dots)$ and $\partial_0 x^a = (1, 0, \dots)$ initially. The solutions are taken as new coordinates; this change of variables being of class H^{s+1}, the pulled-back metric is still H^s, and satisfies $F^c = 0$ by construction. A direct calculation shows that the constraint equations are still satisfied in this system. The reason for this property is that (6.55-6.56) show that the constraint equations do not involve $\partial_{00} g_{0a}$ or $\partial_0 g_{0a}$.

Now that we have obtained that the Cauchy data satisfy $F^c = 0$, we again solve the system $R_{ab}^{(h)} = 0$, and apply the argument of Step 2 to conclude that we have a solution of Einstein's equations.

STEP 4: We now turn to uniqueness. Given a solution of Einstein's equations, we note that one can always, as in Step 3, make a change of coordinates to ensure $F^c = 0$ on S. One will still have $R_{ab} = 0$ in the new coordinates, and therefore the harmonic condition propagates as in Step 2. We conclude that $R_{ab}^{(h)} = 0$, so that the metric is the unique solution of (6.71)-(6.73). (It is interesting to examine the regularity of the change of coordinates used here, see Fischer-Marsden (1972)).

There is a natural partial ordering on pairs (V_4, g) of space-times solving a given Cauchy problem ($V \subset V'$ and the inclusion is an isometry). It is not difficult to conclude, using Zorn's lemma, that there exists a *maximal development* of the Cauchy data on S, which is by definition a maximal element of this set of pairs.

Constraint equations.

One must now consider the issue of constructing initial data sets which satisfy the constraints. This problem is usually tackled by adding the restriction that $K = 0$ on the initial surface S. The reason is as follows: the constraint equations (6.55)-(6.56) consist of the *Hamiltonian constraint*

$$(6.74) \qquad {}^{(3)}R + \frac{1}{2}[(\mathrm{Tr}K)^2 - \mathrm{Tr}(K^2)] = 2\rho := -\chi T_0^0,$$

and the *momentum constraint*

$$(6.75) \qquad \tilde{\nabla}_\beta K_\alpha{}^\beta - \tilde{\nabla}_\alpha K_\beta^\beta = j_\alpha := \chi n_k T^k{}_\alpha.$$

If $K = 0$, these equations decouple into a scalar equation for γ and a system for $K_{\alpha\beta}$.

The momentum constraint is usually solved by reduction to an elliptic system. It is natural to distinguish the compact and non-compact cases.

If S is compact, and $\gamma \in H^{s+1}$, one seeks for every symmetric tensor in H^s, $s \geq 3$, an orthogonal decomposition

$$h_{\alpha\beta} = h^*_{\alpha\beta} + [l_\gamma(X)]_{\alpha\beta} + \frac{1}{3}\gamma_{\alpha\beta}\operatorname{Tr} h,$$

where $l_\gamma(X) := \mathcal{L}_X\gamma - (1/3)\gamma\operatorname{Tr}\mathcal{L}_X\gamma$, $\tilde{\nabla}^\alpha h^*_{\alpha\beta} = 0$ and $\operatorname{Tr} h^* = 0$. X is a field in H^{s+1}, such that

$$(\Delta^* X)_\beta := \tilde{\nabla}^\alpha (l_\gamma(X))_{\alpha\beta} = \tilde{\nabla}^\alpha (h_{\alpha\beta} - \frac{1}{2}h\gamma_{\alpha\beta}),$$

which turns out to be an elliptic system, the homogeneous solutions of which are the conformal Killing vectors. The star does not refer to adjunction here.

If $K = 0$, we get

$$K^{\alpha\beta} = K^{\alpha\beta}_* + [l_\gamma(W)]^{\alpha\beta}$$

for some W. We then find

$$\Delta^* W^\alpha = j^\alpha.$$

The $K^{\alpha\beta}_*$ are freely specifiable subject to the trace and divergence conditions.

Now, one can obtain more solutions by conformal scaling: let $\bar{\gamma}_{\alpha\beta} = \varphi^4\gamma_{\alpha\beta}$, $\bar{K}^{\alpha\beta} = \varphi^{-10}K^{\alpha\beta}$, so that a more general solution of the momentum constraint is

$$\bar{K}^{\alpha\beta} = \varphi^{-2}(K^{\alpha\beta} + [l_\gamma(W)]^{\alpha\beta}).$$

One then uses the Hamiltonian constraint to find φ. This leads to a nonlinear elliptic equation closely related to the Yamabe problem: (see Choquet-Bruhat and York for a classification of results).

In the non-compact case, we can apply splitting and conformal techniques with some modifications. First, one usually considers *asymptotically flat spacetimes*, which may be defined by the requirement

$$\gamma_{\alpha\beta} = e_{\alpha\beta} + \varphi_{\alpha\beta}; \quad \varphi_\alpha{}^\beta = O(1/r),$$

where e is the Euclidean metric, and r the Euclidean distance from the origin, supplemented by conditions of the form $\mathcal{L}_U\varphi = O(1/r^2)$, $\mathcal{L}_U\mathcal{L}_V\varphi = O(1/r^3)$ for e-preserving translational Killing fields U, V. Similarly, we require $K_{\alpha\beta} = O(1/r^2)$, $\mathcal{L}_U K_{\alpha\beta} = O(1/r^3)$. Also, the matter terms ρ and j^α are $O(1/r^4)$. These conditions ensure that the curvature tensor has components $O(1/r^3)$ in Gaussian coordinates, and that the total energy and total momentum

$$E = \lim_{r_0 \to \infty} \frac{1}{2}\int_{r=r_0} (\tilde{\nabla}^\beta K_\beta{}^\alpha - \tilde{\nabla}^\beta\operatorname{Tr} K)\, d^2 S_\beta;$$

$$P^\alpha = \lim_{r_0 \to \infty} \int_{r=r_0} (K^{\alpha\beta} - \gamma^{\alpha\beta}\operatorname{Tr} K)\, d^2 S_\beta$$

are well-defined. One also takes $N = 1 + O(1/r)$ and $N^\alpha = O(1/r^2)$, with $\mathcal{L}_U N = O(1/r^2)$; $\mathcal{L}_U \mathcal{L}_V N = O(1/r^3)$, and $\mathcal{L}_U N^\alpha = O(1/r^2)$. In intuitive terms, these conditions mean that the spacetime is close to Minkowski space at infinity. One can prove that E and P^α remain constant by virtue of the evolution equations.

One then applies splitting techniques in *weighted* Sobolev spaces, in which there are good invertibility properties for elliptic operators (due in particular to Cantor, Nirenberg-Walker). The main difference with the compact case is that there are no conformal Killing vectors which vanish at infinity.

The quantity E reduces to the mass parameter in the Schwarzschild and Kerr cases, and is therefore called the (ADM) mass. (There are other definitions of mass based on behavior at null infinity). The positive mass theorem (Schoen-Yau, Witten) says in particular that any flat spacetime with a maximal slice satisfies $E \geq 0$, with equality for flat space only.

Characteristic initial-value problem.

The case when initial data are prescribed on null hypersurfaces is well-adapted to :

(a) observational data gathered at a single point through light propagation;

(b) gravitational wave fronts propagating discontinuities of $\partial_{00} g_{\alpha\beta}$ across a surface $S : \{f = 0\}$ obeying $g^{ab} \partial_a f \, \partial_b f = 0$;

(c) numerical relativity.

Moreover, the use of null initial hypersurfaces gives a geometrically meningful way of avoiding constraints.

Initial data are given on a pair of intersecting hypersurfaces, one of which at least is null, and on their intersection V_2. Let us consider the case when both hypersurfaces are null, and are defined by $u = 0$ and $v = 0$ respectively. Let $k_a = u_{,a}$ define, in a small region of 4-space, a ray congruence such that $\nabla_a k^a \neq 0$, and complete k^a into a null tetrad[3] $(k, l, m, \bar{m})$, where l is tangent to the level sets $\{v = \text{const.}\}$ and the *complex* null vector m is tangent to V_2. Einstein's equations then split into:

1. Six main equations:

$$G_{ab} k^b = 0; \quad G_{ab} m^a m^b = 0.$$

The latter contains two equations by separating real and imaginary parts.

2. One "trivial" equation $G_{ab} m^a \bar{m}^b = 0$.

3. Three supplementary equations

$$G_{ab} l^a l^b = 0; G_{ab} l^a m^b = 0.$$

---

[3] a tetrad is a frame of four vectors.

The conservation identity $\nabla_a G^a{}_b = 0$ ensures that if the main equations hold everywhere, then (i) the trivial equation holds everywhere; (ii) the supplementary equations hold everywhere if they hold initially.

In coordinates adapted to those initial hypersurfaces, initial data should be of a considerably higher differentiability class than the solution is found to belong to; they generally have a direct geometrical interpretation.

In the analytic case, the characteristic IVP helps formulate the asymptotic behavior of the radiation field, and the notion of Bondi mass.

The procedure does not seem to be well compatible with the existence of a smooth structure at null infinity. However, Einstein's equations are sufficiently well-behaved under conformal changes of metric to admit of well-posed initial-value problems on initial surfaces which include part of null infinity. This leads to replacing the original Einstein's equations, which were written in a space $(\bar{V}, \bar{g}_{ab})$, by conformally regular vacuum field equations on an "unphysical" space $(V, g_{ab} = \Omega^2 \bar{g}_{ab})$ (see exercises 15 and 16). In that case, $\bar{V} = V \setminus I$. Note the similarity with the conformal method used in Ch. 2 for global existence.

6.4 LINEARIZATION STABILITY AND PERTURBATIONS

The problem of linearization stability is to decide whether or not every solution of the linearized Einstein equations (linearized at a given spacetime metric) arises by linearization of a family of exact solutions of the full equations. In other words, one asks whether the space of solutions has a tangent space spanned by the solutions of the linearized equations. It turns out that the answer is negative when the reference solution admits Killing fields. A few results on the Newtonian and Minkowskian limits are given, in relation to the problem of detection of gravitational waves.

Linearization stability.

The question of linearization stability can be asked about any nonlinear equation, written symbolically:

$$(6.76) \qquad F[x] = 0.$$

Let x_0 be a solution of (6.76) and φ be a solution of the linearized equation

$$(6.77) \qquad F'[x](\varphi) = 0.$$

The question is to find a family $\{x(\varepsilon)\}$ such that

$$F[x(\varepsilon)] = 0 \text{ for } |\varepsilon| \text{ small}, \quad x(0) = x_0; \quad \left.\frac{dx}{d\varepsilon}\right|_{\varepsilon=0} = \varphi.$$

Thus, x_0 is linearization stable if near $x_0 = 0$, every solution of the linearized equation is tangent to a curve of solutions of the exact problem.

Example 1: $F(x,y) = x^2 + y^2$, $(x,y) \in \mathbf{R}^2$. Then $(0,0)$ is *not* linearization stable because any (h,k) satisfies the linearized equations, but $F(x,y) = 0$ implies $x = y = 0$. Failure of linearization stability may be thought of as a failure of perturbation theory, and may be detected by second-order perturbation: let us substitute the expansion $(x,y) = \sum_{j \geq 1} \varepsilon^j (x_j, y_j)$ into $F = 0$. Identifying like powers of ε, we find no condition at order ε^1, but $x_1 = y_1 = 0$ at order ε^2. This discrepancy between first and second order perturbation is a symptom of failure of linearization stability. See also Problem 8.

Example 2: If $F(0,0) = 0$, $\partial_y F(0,0) \neq 0$, then $(0,0)$ is linearization stable. Indeed, $F = 0$ can be solved locally by the implicit function theorem, to give $y = f(y)$ near $x = y = 0$; $F'(0,0)(h,k) = 0$ if and only if $k = hf'(0)$. We take as family of solutions $(x(\varepsilon), y(\varepsilon)) = (\varepsilon h, f(\varepsilon h))$.

Example 3: Let X be the space of C^1 functions on $[0,T]$ with values in $\mathbf{R}^k$. Let $f \in C^\infty(\mathbf{R}^k \times \mathbf{R}; \mathbf{R})$, and define F by

$$
\begin{aligned}
F : X &\rightarrow C^0([0,T]; \mathbf{R}^k) \\
(t \mapsto x(t)) &\mapsto (t \mapsto \dot{x}(t) - f(x(t), t)).
\end{aligned}
$$

Any solution of the linearized equation near a reference solution $x_0(t)$ is obtained by solving

$$
\left\{
\begin{aligned}
\dot{y} &= f'_x(x_0(t))y \\
y(0) &= \varphi \in \mathbf{R}^k.
\end{aligned}
\right.
$$

We define $x(\varepsilon; t)$ by

$$
\left\{
\begin{aligned}
\dot{x} &= f(x(t), t) \\
x(0) &= x_0(0) + \varepsilon\varphi.
\end{aligned}
\right.
$$

It is clear that $\frac{dx}{d\varepsilon} = y$ for $\varepsilon = 0$. We therefore see that quite generally, solutions of well-posed initial-value problems should be linearization stable.

For General Relativity, linearization stability may fail because of the constraint equations. On the other hand, stability of the constraint equations implies, by the well-posedness of the Cauchy problem, linearization stability of the full equations.

Consider a smooth family $g_{ab}(\varepsilon)$ of solutions of Einstein's equations. Denote by $\varphi_{ab} := \delta g_{ab}$ the derivative of this family with respect to ε at $\varepsilon = 0$; the letter delta has the same meaning for other quantities in what follows.

THEOREM 6.4 *If X_a is a Killing field, the integral of $\nabla^k (X^i \delta^2 G_{ik}(\delta g, \delta g))$ on any spacelike hypersurface vanishes.*

Proof: The first step consists in establishing general expressions for the variations of the curvature tensor for arbitrary variations of the metric, using only the Bianchi identities. One then computes the second variation of Einstein's equations to conclude.

Let us therefore consider a general variation φ_{ab} of g_{ab}. We compute

$$\delta\Gamma^k{}_{ij} = \frac{1}{2}(\nabla_i\varphi^k{}_j + \nabla_j\varphi_i{}^k - \nabla^k\varphi_{ij}),$$

$$\delta R^h{}_{ijk} = \nabla_j\delta\Gamma^h{}_{ik} - \nabla_k\delta\Gamma^h{}_{ij},$$

$$\delta R_{ik} = \frac{1}{2}(\Delta\varphi_{ik} + \nabla_i k_k + \nabla_k k_i),$$

$$\delta R = \frac{1}{2}\Delta\varphi + \nabla_l k^l - R_{ij}\varphi^{ij},$$

where

$$k_i := \nabla_k[\varphi_{ik} - (1/2)\varphi\,\delta_i{}^k]; \quad \varphi = g^{ik}\varphi_{ik}.$$

We then find that if $G_{ij} = 0$ and φ_{ij} is an arbitrary variation (which does not necessarily solve the linearized equations), using the fact that the contracted Bianchi identities are satisfied for $g_{ij} + \varepsilon\varphi_{ij}$,

$$0 = \delta(-\nabla^k G_{ik}) = \varphi^{kl}\nabla_k G_{il} + G_{ih}g^{kl}\delta\Gamma^h{}_{ik} - \nabla^k\delta G_{ik}.$$

Therefore,

$$\nabla^k\delta G_{ik} = 0.$$

If X_i is a Killing field, we derive the conservation relation

$$\nabla^k[X^i\delta G_{ik}] = 0.$$

Since φ_{ij} is arbitrary, we may, given any domain bounded by a pair of compact hypersurfaces, modify it so that it vanishes on the outer one, and therefore we find that the integral of $X^i n^k \delta G_{ik}$ on such 3-surfaces vanishes.

Next, one assumes that φ_{ab} is tangent to a curve $g_{ab} + \varepsilon\varphi_{ab} + \varepsilon^2 k_{ab} + \ldots$ of solutions of Einstein's equations, which we write as a nonlinear operator equation $E[g] = 0$. Differentiating twice, and suppressing indices, we find

$$2E'[g](k) + E''[g](\varphi, \varphi) = 0.$$

Multiplying by X^i, we obtain a quantity quadratic in φ_{ab}, the integral of which will vanish on any hypersufrace, QED.

Other perturbation problems.

Many properties of the gravitational field have been inferred by perturbation of a background metric. We describe the outcome of these methods.

Perturbation of the Minkowski metric. Let

$$g_{ab} = \eta_{ab} + \varphi_{ab},$$

where η_{ab} is the Minkowski metric. The linearized Einstein equations become

$$G_{ab}^{(1)} = -\frac{1}{2}\partial^c\partial_c\varphi_{ab} - \frac{1}{2}\partial_{ab}\varphi + \partial^c\partial_{(b}\varphi_{a)c} - \frac{1}{2}(\partial^c\partial^d\varphi_{cd} - \partial^c\partial_c\varphi)\eta_{ab}.$$

The gauge freedom of General Relativity corresponding to the group of diffeomorphisms leads one to consider the perturbations φ_{ab} and

$$\varphi_{ab} + \mathcal{L}_X\eta_{ab} = \varphi_{ab} + \partial_a X_b + \partial_b X_a,$$

where X^a is the generator of a group of diffeomorphisms of Minkowski spacetime, as physically equivalent. Let $\bar\varphi_{ab} = \varphi_{ab} - \frac{1}{2}\varphi_{ab}$. Under this gauge transformation, it becomes

$$\bar\varphi_{ab} + \partial_a X_b + \partial_b X_a - \partial_c X^c\eta_{ab}.$$

By such a transformation, we may assume that

$$\partial^b\bar\varphi_{ab} = 0$$

(this expresses the linearized form of the harmonic coordinate condition). Note that we still have a "gauge freedom" which consists in taking transformations such that X^a satisfies

$$-\partial^a\partial_a X_b = 0.$$

It is easily checked that such restricted transformations preserve the linearized harmonicity condition.

We therefore find (with this gauge condition)

$$G_{ab}^{(1)} = \frac{1}{2}\partial^c\partial_c\bar\varphi_{ab},$$

which must be equated to χT_{ab}.

a) The *Newtonian limit* is obtained when gravity is weak, the relative motion of sources is slow ($v/c \ll 1$) and material stresses are much smaller than mass energy density:

$$T_{ab} \approx \rho u_a u_b, \quad u^a = \delta_0^a.$$

The slow variation of sources allows us to neglect time derivatives of h_{ab} so that the field equations $G_{ab}^{(1)} = \chi T_{ab}$ decompose into

$$\Delta\bar\varphi_{00} = -\chi\rho,$$

which recovers Poisson's equations, and

$$\Delta\bar\varphi_{0\alpha} = \Delta\bar\varphi_{\alpha\beta} = 0.$$

It follows that, with suitable asymptotic conditions, one can write $\bar{\varphi}_{0\alpha} = \bar{\varphi}_{\alpha\beta} = 0$ and finally

$$\varphi_{ab} = -(\eta_{ab} + 2u_a u_b)U,$$

where U solves Poisson's equation.

b) *Gravitational radiation.* Consider the vacuum equation

$$\partial_c \partial^c \bar{\varphi}_{ab} = 0$$

with $\partial^b \bar{\varphi}_{ab} = 0 = \partial^b \partial_b X_a$. We can choose the "radiation gauge:" $\varphi = \varphi_{0\alpha} = 0$ in a source-free region. We then get $\varphi_{00} = 0$ if no sources are present throughout the spacetime (see Wald (1984) for details). Plane-wave solutions of the source-free linearized equations with the form $\varphi_{ab} = A_{ab}\exp(ik_c x^c)$, A_{ab} constant, satisfy

$$k_a k^a = 0.$$

The radiation gauge conditions impose

$$A_{ab}k^b = 0; \quad A_{0\alpha} = 0; \quad A^c{}_c = 0.$$

Of these 9 equations, only 8 are independent; since there are 10 unknowns, this leaves two independent polarization states of plane gravitational waves.

The detection of those waves is an active experimental field of research, through measurement of the gravitational tidal force due to those waves, with the help of the geodesic deviation equation $\ddot{\eta}^a = R^a{}_{0b0}\eta^b$.

Remark 1. It has been observed that the solutions of well-posed initial-value problems do not, as a rule, depend smoothly on the data (there are old examples of Kato to this effect, see Ch. 2). This implies difficulties with the reliability of perturbation theory (Rendall (1990)). For instance, it is doubtful that perturbation theory will succeed if the domain of dependence of the reference metric is smaller than that of the exact solution. A possible justification of perturbation theory based on characteristic initial-value problems has been put forward by Damour and Schmidt (1990).

Remark 2. The Newtonian limit, as a limit of one-parameter families of solutions of Einstein's equations, appears to be singular, see Lottermoser (1992).

Remark 3. The parametrized post-Newtonian (PPN) expansion of Schwarzschild-type metrics has been used to study the Solar system (see Will (1981), Misner-Thorne-Wheeler (1973)).

6.5 SINGULARITIES AND COSMIC CENSORSHIP

The question of singularity formation takes a new aspect in General Relativity because of the absence of a background spacetime. After defining singularities in terms of geodesic incompleteness, sample singularity theorems are discussed. A few notions on the complementary issue of "absence of singularities," are then given, in relation to the various forms of the cosmic censorship problem.

The discussion of singularity formation in Einstein's equations creates special difficulties because singularities in metric or curvature components may be due to a poor choice of coordinates, and because pathological behavior may be associated with problems other than lack of smoothness, such as the failure of causality. Let us therefore review a few possible definitions of singular spacetimes, and show why geodesic incompleteness appears to be a satisfactory criterion for singularity.

What is a singularity?

Let us immediately dispose of singularities which by any definition must be considered as fictitious. A space such as $\mathbf{R}^4 \cap \{t > 0\}$, with metric

$$ds^2 = -t^{-2}dt^2 + dx^2 + dy^2 + dz^2,$$

is actually identical with Minkowski space, as can be seen by letting $\tau = \ln t$. Singularities of this type are called *coordinate singularities*. A famous example is the singularity $r = 2M$ of the Schwarschild metric (problem 14). The most common coordinate singularity occurs at the origin of polar coordinates.

One could try to define singularity by the property that *curvature scalars* such as R, $R_{ab}R^{ab}$ or $R_{abcd}R^{abcd}$ become infinite at some point on the boundary of a coordinate chart, but this is not satisfactory for three reasons.

First, there are spacetimes (of type N) which have a singular Weyl tensor, but for which all curvature scalars vanish.

Second, there are spacetimes with identically vanishing curvature tensor which one would still like to call singular: take the product of 2-dimensional Minkowski space by a cone in $\mathbf{R}^3$. Like any developable surface, the cone is locally isometric to a plane, and therefore, the curvature of this space vanishes identically. However, there is a "conical singularity," which can be reached by geodesics terminating on the tip of the cone.

Third, even if curvature scalars become infinite, they may do so only "at infinity" in terms of the spacetime metric, in which case there is no reason to call such a space singular.

In fact, a basic point is that singularities are *not* part of the spacetime, since spacetime itself is defined by a *regular* metric satisfying the field equations. We are therefore led to define singularities entirely in terms of properties of regular spacetimes. The consensus at the present time seems to be that one should define singularities by the existence of a timelike or null geodesic with bounded affine length; indeed, for a timelike geodesic, this means that some observer exists only for a finite extent of proper time, limited by its encounter with a "singularity." We then say that the spacetime is timelike or null incomplete. We will adopt this property as a definition of a singular spacetime.

Singularities and horizons.

Singular spacetimes are very common: the Schwarzschild, Kerr and Friedmann spacetimes are all singular; the latter in particular predicts a *big-bang* singularity at the "beginning" of the universe. It was however conceivable that this singular behavior was simply due to the high degree of symmetry of these solutions, and could be eliminated by small perturbations. That singularities are necessary under rather weak conditions on the energy-momentum tensor is the content of the singularity theorems initiated by Hawking and Penrose.

Before discussing them, let us introduce yet another way to get around singularities, which is to hide them behind a *horizon*. There are some differences in terminology in the literature, but two types of horizons are relevant here: event and Cauchy horizons.[4]

Let us introduce first some terminology which is frequently found in the literature.

One says that p is a *future endpoint* of the curve $\{\gamma(t)\}$ if $\forall U$ neighborhood of p, $\exists t_0 : \gamma(t) \in U$ for $t > t_0$. A curve is future *inextendible* if it has no future endpoint.

The (chronological) *future* of a set $S \subset M$ is the set of points that can be reached from S by a future-directed timelike curve. It is denoted by $I^+(S)$, and is open. The *causal future* $J^+(S)$ is defined similarly, with 'timelike' replaced by 'causal.' S is said to be *achronal* if $S \cap I^+(S)$ is empty.

M is *stably causal* if there is a differentiable function such that its gradient is past-directed timelike. One sometimes thinks of f as a "global time function." A weaker requirement (*strongly causal space-time*) is: $\forall p \in M$, $\forall U$ neighborhood of p, $\exists V$ neighborhood of p such that $V \subset U$ and no causal curve intersects V more than once.

A *Cauchy surface* is a closed achronal S set such that any point in M belongs to some inextendible causal curve which meets S. A spacetime which possesses a Cauchy surface is said to be *globally hyperbolic*. Such a space is in particular strongly causal. For two other equivalent definitions of a Cauchy surface, see problem 12.

The *Cauchy horizon* $H(S)$ of a set S is the boundary of the set of points which belong to some inextendible causal curve which meets S.

Let us now come back to horizons.

First, we may consider the boundary of the causal future of a given region. This would represent the boundary of the set of events accessible to observers

---

[4]Examples of Killing and particle horizons are given in §6.6, in the case of the Kerr and FLRW metrics respectively.

in that region. Such an event horizon might clothe singularities so that their existence would go unnoticed. On the other hand, *naked singularities* can be reached by timelike or null geodesics going "to infinity" in a prescribed way (related to conformal compactification).

Another procedure would be to solve the Cauchy data from a given spacelike hypersurface, and construct a maximal development from these data. One then considers the boundary of the domain of determinacy of the original surface. This Cauchy horizon represents the limit of how much of the spacetime can be predicted from the Cauchy data. Let us illustrate this point by two examples. First, if we consider the initial surface $t = -\sqrt{1 + x^2 + y^2 + z^2}$, we find that the backward light cone is a Cauchy horizon. Another example is the Misner spacetime $S^1 \times \mathbf{R}$, with metric

$$-\frac{1}{t}\, dt^2 + t\, d\vartheta^2,$$

where $0 \le \theta < 2\pi$, $t > 0$. Letting $\tau = t$ and $\psi_\pm = \theta \pm \ln t$, we find two extensions to $t < 0$ corresponding to the two choices of sign $(ds^2 = \mp 2d\psi_\pm d\tau + \tau\, d\psi^2)$. In both cases, $\tau > 0$ gives the original space. Note that the circles $\tau =$ const. turn timelike for $\tau < 0$, so that there are *closed timelike geodesics*. In this particular case, the two extensions are isometric to each other via

$$(\tau, \psi_\pm) \mapsto (\tau, 2\pi - \psi_\pm).$$

There are however examples with distinct extensions. These are therefore cases of non-uniqueness in the solution of the initial value problem.

One form of the Cosmic Censorship Conjecture states that "generally," solutions to the Cauchy problem from a spacelike hypersurface cannot be extended across Cauchy horizons. The formulation is usually left somewhat imprecise in the literature because the formulation of the conjecture is really part of the problem. This conjecture first arose (Penrose) in trying to see whether singularities due to gravitational collapse were not in fact hidden from observation, which would explain the lack of observational evidence for such singularities.[5] With the discovery of many possible mechanisms for the production of naked singularities, by imposing if necessary particular constraints on the matter terms, it became clear that naked singularities do occur in a variety of contexts. The Hawking-Penrose singularity theorems on the other hand predict *either* singularities *or* breakdown of causality. The issue is therefore to delineate precisely which possible singularities will be deemed relevant to realistic situations before settling their existence or nonexistence.

---

[5] "Does there exist a 'cosmic censor' who forbids the appearance of naked singularities, clothing each one in an absolute event horizon?" (Penrose (1969)).

Singularity theorems.

The singularity theorems reach their goal by showing that one timelike or null geodesic must have finite affine length. They proceed in two steps which will be outlined below, without striving for generality. Their usefulness is due to the fact that they use very weak conditions on the energy-momentum tensor, such as the *strong energy condition*:

$$(T_{ab} - \frac{1}{2}Tg_{ab})u^a u^b \geq 0$$

for all timelike (or null) u^a.

The first step is to show that if a timelike geodesic hypersurface-orthogonal congruence has uniformly negative expansion, its curves have finite affine length in one direction. Indeed, the Raychaudhuri equation shows that

$$R_{ab}u^a u^b = \omega_{ab}\omega^{ab} - \sigma_{ab}\sigma^{ab} - \dot{\theta} - \frac{1}{3}\theta^2.$$

Since u^a is hypersurface-orthogonal, we may rescale it and assume $u^a = \partial^a f$ for some function f, so $\omega_{ab} = 0$. Since the congruence is timelike, $\sigma_{ab}\sigma^{ab} \geq 0$. The energy condition and Einstein's equations imply $R_{ab}u^a u^b \geq 0$, so

$$\dot{\theta} + \frac{1}{3}\theta^2 \leq 0.$$

Therefore, if $\theta(0) < 0$, τ cannot become larger that $3/|\theta(0)|$. The congruence therefore develops singularities comparable to caustics; by themselves, they do not imply any pathology in the spacetime.

The second step consists in adding a geometric condition which guarantees that singularities must form. Here are examples:

(1) Assume the existence of a Cauchy surface S and consider its timelike normal congruence. Assume $\theta < -c_0 < 0$ on S and consider a timelike geodesic of affine length greater than $3/c_0$. We may assume that it maximizes length from its end p to S. One then sees that there is on this geodesic portion a point r conjugate to its intersection q with S. This means that beyond r, the geodesic is not maximal any more, hence a contradiction.

(2) Assume that spacetime is strongly causal and that there is a compact spacelike hypersurface S which has again $\theta < 0$ everywhere for the past normal congruence. Then one can see that the past Cauchy horizon $H^-(S)$ must be non-empty and compact. But then, it contains, from general properties of such horizons, a null geodesic, which, if it extends without limit, will have accumulation points. These almost closed geodesics cannot exist in a strongly causal spacetime.

Other possible assumptions leading to singularities include the existence of a trapped surface (such that geodesics from it in both directions converge), or "genericity," or the existence of non-trivial topology initially; see the Notes for details.

6.6 EXACT SOLUTIONS.

Exact solutions of Einstein's equations are by definition those which can be given in closed form. They provide, in the best of cases, an interior solution (with specified T_{ab}) matched to a vacuum solution. With the exception of a few, most of the known solutions are unphysical. They nevertheless have value as examples in the problem of singularities in particular. Techniques leading to those solutions are varied; we may single out three methods: (1) use of symmetries of T_{ab} and the resulting symmetries of the metric; (2) use of null tetrads and spinors to construct algebraically special spaces; (3) generation of new solutions form known ones through the use of Bäcklund transformations.

Spacetime symmetries.

The most widely used symmetry is the spacetime isometry, although other types (affine, conformal, etc ...) have been considered. A Lie group of isometries G_r (of dimension r) has generators X_A such that the Killing equations

$$\mathcal{L}_{X_A} g_{ab} := \nabla_a X_{Ab} + \nabla_b X_{Aa} = 0$$

hold ($A = 1, \ldots, r$). Such groups may be classified according to the number r, their Lie algebra structure, their invariant manifolds, their isotropy groups, etc ... We suppress the index A in what follows.

Differentiating the Killing equations and using the algebraic properties of the curvature tensor, we find

$$(6.78) \qquad\qquad \nabla_a \nabla_b X_c = R^d{}_{abc} X_d.$$

If we view the X_a as unknown, we see that they are constrained by 10 second-order equations. As in the Riemannian case, the maximum dimension of the isometry group is $n(n + 1) = 10$.

THEOREM 6.5 *In a space of dimension n, if the isometry group has maximum dimension, the spacetime has constant curvature:*

$$R^k{}_{abc} = \frac{R}{n(n-1)} (\delta^k_b g_{ac} - \delta^k_c g_{ab}).$$

Proof: We write first the integrability condition of (6.78):

$$
\begin{aligned}
(\nabla_d \nabla_a - \nabla_a \nabla_d) \nabla_b X_c &= (\nabla_d R^e{}_{abc} - \nabla_a R^e{}_{dbc}) X_e \\
&\quad - R^e{}_{abc} \nabla_d X_e + R^e{}_{dbc} \nabla_a X_e \\
&\quad - R^e{}_{bda} \nabla_e X_c - R^e{}_{cda} \nabla_b X_e.
\end{aligned}
$$

Since the isometry group has maximal dimension, the values of X_a and $\nabla_{(b}X_{c)}$ can be prescribed arbitrarily at any point. We may therefore set to zero separately their coefficients in this equation:

$$\nabla_d R^e{}_{abc} - \nabla_a R^e{}_{dbc} = 0;$$
$$\delta^h_a R^k{}_{dbc} - \delta^k_a R^h{}_{dbc} - \delta^h_d R^k{}_{abc} + \delta^k_d R^h{}_{abc} - \delta^k_e R^h{}_{bda} = 0.$$

Contracting in h and d in the second equation, we find

$$(n-1)R^k{}_{abc} = \delta^k_c R_{ba} - \delta^k_b R_{ca},$$
$$R^k{}_b = \frac{R}{n}\delta^k_b.$$

The result follows.

Important exact solutions.

Minkowski space. It is a space of maximum symmetry, of zero curvature, constituting the background spacetime of Special Relativity. It has in particular a nontrivial conformal boundary (which we used in Ch. 2) useful in the study of asymptotically flat spacetimes.

Spherical symmetry. The Schwarzschild vacuum solution is spherically symmetric (G_3, Bianchi type III, $X_1 = \cos\varphi\,\partial_\theta - \sin\varphi\cot\theta\,\partial_\varphi$, $X_2 = \partial/\partial_\varphi$, $X_3 = \sin\varphi\,\partial_\theta + \cos\varphi\cot\theta\,\partial_\varphi$), asymptotically flat and necessarily static, as a consequence of Birkhoff's theorem:

$$ds^2 = -(1 - \frac{2M}{r})\,dt^2 + (1 - \frac{2M}{r})^{-1}\,dr^2 + r^2(d\theta^2 + \sin^2\theta\,d\varphi^2).$$

It has singular behavior at $r = 2M$ due to the breakdown of coordinates, and at $r = 0$, where the curvature invariant $R_{abcd}R^{abcd}$ tends to infinity.

Its maximal extension (as a solution of the vacuum equations) is a *black hole* with a curious topology: two asymptotically flat spaces back to back, joined through a "Schwarzschild throat." It has two horizons: a Killing horizon (at $r = 2M$) where the timelike Killing vector becomes spacelike, and an event horizon delimiting those regions from which one cannot signal to infinity.

The classical tests of General Relativity are based on this solution:

(a) gravitational red-shift;

(b) advance of the perihelion of Mercury (or of any satellite orbiting a body exerting gravitational attraction);

(c) bending of light-rays when passing near massive bodies;

(d) time delay in the roundtrip delay of a radar signal which passes by the Sun or the Moon.

Static, spherically symmetric interior solutions for a perfect fluid can be obtained and used to study star sizes in relation to their mass.

The Lemaître-Bondi-Tolman solution has been used to study the gravitational collapse of a spherical mass to a black hole.

Cosmological models: Spatially isotropic and homogeneous models. The FLRW (Friedmann-Lemaître-Robertson-Walker) universe is locally isotropic everywhere and therefore spatially homogeneous with line element

$$ds^2 = a^2(t)[d\psi^2 + \Sigma^2(\psi, k)(d\theta^2 + \sin^2\theta \, d\varphi^2)] - dt^2,$$

where

$$\Sigma(\psi, k) = \begin{cases} \sin\psi & k = 1, \\ \psi & k = 0, \\ \sinh\psi & k = -1, \end{cases}$$

the matter content being that of a perfect fluid:

$$T_{ab} = \rho u_a u_b + p(g_{ab} + u_a u_b).$$

Einstein's equations for this metric are given in Problem 10. If $D = a\psi$ is the distance between two isotropic observers, we find

$$\dot{D} = \frac{\dot{a}}{a}D = HD,$$

where H is the *Hubble parameter*. We thus find that, since $\rho + 3p > 0$, will be always expanding or contracting except at the point where $\dot{a} = 0$. It follows that a photon of frequency ν_P emitted at a point P at instant t_1, is received, at a point Q at distance D_{PQ} from P, at time t_2, as a photon of frequency ν_Q such that

$$z := \frac{\nu_P}{\nu_Q} - 1 = \frac{a(t_1)}{a(t_2)} - 1 \approx \frac{\dot{a}}{a}(t_2 - t_1) \approx HD.$$

The redshift observed by Hubble leads to $\dot{a} > 0$, and hence to the *expansion of the universe*. On the other hand, since $\ddot{a} < 0$, the universe must have been expanding faster and faster as one goes backwards in time. We are thus led to postulate that at a time less than $T = a/\dot{a} = 1/H$ in the past, a was vanishingly small, and the universe was in a singular state, referred to as the "big bang," where the distance of all spatial points was zero, and the density of matter and the curvature were infinite.

This universe therefore starts at an initial singularity where from all matter content, along with space and time, originate; this view seems to be corroborated by the cosmic microwave blackbody radiation and the primaeval element abundances. However most of this matter is hidden from us by *particle horizons*, namely by the boundary of what a given observer can see.

The existence of nontrivial particle horizons can be easily seen in the case of the flat spatial geometry ($k = 0$):

$$ds^2 = -dt^2 + a^2(t)(dx^2 + dy^2 + dz^2).$$

By defining

$$\tau = \int \frac{dt}{a(t)},$$

we can write the line element as that of a conformally flat spacetime:

$$ds^2 = -a(\tau)^2(-d\tau^2 + dx^2 + dy^2 + dz^2),$$

so that we can join two events by a timelike or null curve in this metric, if and only if this can be done in the flat metric. Now, if the integral defining τ diverges as its lower limit tends to zero, τ will range down to $-\infty$, and an observer will be able to receive signals from all other observers. If the integral converges, the Robertson-Walker model will be conformally related only to the portion of Minkowski spacetime above some surface $\tau =$const. This is the case for instance if $a(t) = t^{2/3}$.

Another important issue is to assess the value of k, since (6.79) imply that an expanding universe will keep expanding forever if $k = 0$ of -1, but will eventually recontract if $k = 1$ (closed universe). Now, since $p \approx 0$ in the present state, define

$$\begin{aligned} \Omega &= 2q = -2\frac{a\ddot{a}}{\dot{a}^2} = \frac{8\pi\rho}{3H^2}; \\ H^2 &= \frac{8\pi\rho}{3} - \frac{k}{a^2}. \end{aligned}$$

The universe is closed if and only if $\Omega > 1$.

When a is constant, we obtain perfect fluid solutions for

$$G_{ab} + \Lambda g_{ab} = \chi T_{ab},$$

where Λ is the *cosmological constant*, which satisfies here

$$\Lambda = \frac{\chi}{2}(\rho + 3p) = \frac{3}{a^2} - \chi\mu,$$

with $k = 1$, $\Lambda > 0$. The metric is

$$ds^2 = a^2(d\psi^2 + \sin^2\psi(d\theta^2 + \sin^2\theta \, d\varphi^2)) - dt^2,$$

and admits a G_7. It is the *Einstein static universe* we encountered in the conformal method for global existence.

When $T_{ab} = 0$, we get the *de Sitter model*

$$ds^2 = \frac{dr^2}{1 - \frac{\Lambda}{3}r^2} + r^2(d\theta^2 + \sin^2\theta\, d\varphi^2) - (1 - \frac{\Lambda}{3}r^2)\, dt^2.$$

Cosmological models: Anisotropic or inhomogeneous models. Although isotropy and large scale homogeneity of spacetime are fairly well-supported, alternatives to the FLRW models have been studied because of the following theoretical considerations:

(i) Inhomogeneities at all stages of the universe seem to be necessary for galaxies to form by gravitational collapse;

(ii) Anisotropic singularities may have existed in the past;

(iii) Microwave radiation received from regions which never had causal communication suggests early "chaotic cosmology."

These models include:

(a) spatially homogeneous anisotropic models admitting a G_4 (Kantowski-Sachs metrics) or a G_3 (Bianchi metrics); they have in general singularities: at least one timelike geodesic is incomplete in any spatially homogeneous model in which $R_{ab}k^a k^b > 0$ for all timelike or null vectors (see Hawking-Ellis, p. 147). When matter terms or spatial curvature is important, the line element approximates the Kasner model (Bianchi I)

$$ds^2 = -dt^2 + t^{2p_1}\left(dx^1\right)^2 + t^{2p_2}\left(dx^2\right)^2 + t^{2p_3}\left(dx^3\right)^2$$

where $\sum_i p_i = \sum_i p_i^2 = 1$.

(b) inhomogeneous metrics permitting distorting and rotating universes. An example is given by those admitting a G_3:

$$ds^2 = -A^2 dt^2 + B dt\, dr + C^2 dr^2 + D^2(d\theta^2 + f^2(\theta)\, d\varphi^2),$$

where A, B, C, D are functions of (r, t) only, and $f(\theta) = \sin\theta$, θ, or $\sinh\theta$ respectively depending on the sign of the curvature of the metric $(d\theta^2 + f^2(\theta)\, d\varphi^2)$.

Stationary axisymmetric solutions. Their physical interest lies in the fact that they may describe equilibrium configurations of axisymmetric rotating bodies, or their exterior gravitational field. They possess two Killing vectors:

$$X = \partial/\partial t; \quad X^a X_a < 0, \quad \text{(stationarity)};$$
$$Y = \partial/\partial\varphi; \quad Y^a Y_a > 0, \quad \text{(axial symmetry)}.$$

The trajectories of Y are closed, compact curves; Y vanishes on the axis of rotation at all times. The group G_2 of isometries is abelian; this property is necessary when gravitational fields are asymptotically flat (Carter, 1970). It follows that $v_{ab} = 2X_{[a}Y_{b]}$ is surface-forming. The dual bivector

$$w_{ab} = \frac{1}{2}\eta_{abcd}v^{cd} = 2Z_{[a}\bar{Z}_{b]}$$

is surface-forming if and only if

(6.79) $v_{ab}\mathcal{L}_{\mathbf{Z}}Z^b = 0.$

The metric in coordinates adapted to Killing vectors and admitting two spaces orthogonal to each other can be written down explicitly (Lewis (1932), Papapetrou (1966)):

$$ds^2 = e^{-2U}(\gamma_{MN}dx^M\,dx^N + W^2\,d\varphi^2) - e^{2U}(dt + A\,d\varphi)^2,$$

where U, γ_{MN}, W and A depend only on the coordinates $x^M = (x^1, x^2)$. It becomes, in Weyl's canonical coordinates $x^1 = \rho$, $x^2 = z$, $\zeta = \rho + iz$

$$ds^2 = e^{-2U}(2e^{2K}\,d\zeta\,d\bar{\zeta} + \rho^2\,d\varphi^2) - e^{2U}(dt + A'd\varphi)^2.$$

K is an arbitrary function of (x^1, x^2). The case $U = U(\omega)$ generates the Papapetrou class of vacuum solutions, where ω is such that its gradient is the twist of the timelike Killing vector (see problem 4).

The family of Kerr metrics also belongs to this group; we will describe it briefly when dealing with algebraically special solutions.

Algebraically special solutions. These solutions have a Weyl tensor admitting at least one multiple principal null direction, see the discussion of Petrov types for criteria. Solutions considered in the literature have in common that the multiple null eigenvector is geodesic and shear-free. Let $\theta = \frac{1}{2}\nabla_a k^a$ (expansion) and $\omega = (\frac{1}{2}\nabla_{[b}k_{a]}\nabla^b k^a)^{1/2}$. Different classes of solutions can be described in terms of the values of ω and θ.

If $\omega = 0$, we get the Kundt class ($\theta = 0$) and the Robinson-Trautman class ($\theta \neq 0$).

In particular, one should mention the Kerr-Schild metrics:

$$g_{ab} = \eta_{ab} - 2Vk_ak_b,$$

with $g_{ab}k^ak^b = \eta_{ab}k^ak^b = 0$. They include in particular
(1) The charged Kerr solution

$$
\begin{aligned}
ds^2 \;=\; &-\left(\frac{\Delta - a^2\sin^2\theta}{\Sigma}\right)dt^2 - \frac{2a\sin^2\theta(r^2 + a^2 - \Delta)}{\Sigma}\,dt\,d\varphi \\
&+ \frac{(r^2 + a^2)^2 - \Delta a^2\sin^2\theta}{\Sigma}\sin^2\theta\,d\varphi^2 + \frac{\Sigma}{\Delta}\,dr^2 + \Sigma\,d\theta^2,
\end{aligned}
$$

where

$$\begin{aligned}\Sigma &= r^2 + a^2 \cos^2\theta \\ \Delta &= r^2 + a^2 + e^2 - 2Mr,\end{aligned}$$

with the electromagnetic potential

$$A_a = -\frac{er}{\Sigma}[\delta_a^0 - a\sin^2\theta\,\delta_a^r].$$

They are stationary axisymmetric and their Weyl tensor admits two double null eigenvectors. The parameters e, M and a are interpreted as the total electric charge of the spacetime, the total mass and the ratio J/M, where J is the total angular momentum.

It provides the current basic representation of the exterior field of a rotating black hole, although no satisfactory interior solution that matches it is available. It has a non-trivial global topology (when its maximal extension is considered). Unlike the Schwarzschild case, the Killing horizon and the event horizon do not coincide: there is an *ergosphere* between them. By processes carried within this ergosphere, one can in principle extract rotational energy from the black hole.

When $\theta = \omega = 0$, the vacuum Kerr-Schild field gives the plane-fronted gravitational wave, which belongs to the more general Kundt class.

Generation techniques. These methods work only if the spacetime admits at least one non null Killing vector. We summarize the method here and refer to Kramer *et al.* and Hoenselaers and Dietz for details and applications.

Let X^a be a non null Killing vector ($F := X^a X_a \neq 0$), and a 3-metric γ_{ab} defined by

$$\gamma_{ab} = |F|(g_{ab} - F^{-1}X_a X_b).$$

It turns out that for some fields sharing the spacetime symmetry, one can introduce scalar potentials f^A so that the field equations are exactly the Euler-Lagrange equations derived from

$$L(\gamma_{ab}, f^A) := \sqrt{\gamma}(\hat{R} + G_{AB}(f^C)\gamma^{ab}f^A{}_{,A}f^B{}_{,b}),$$

where $\hat{R}$ is the scalar curvature of γ_{ab}, and $\det|G_{AB}| \neq 0$.

Let

$$f^{A'} = f^{A'}(f^C); \quad \gamma'_{ab} = \gamma_{ab}$$

be a transformation of the potentials leaving L invariant; such a transformation then generates a new solution $(f^{A'}, \gamma_{ab})$.

On the other hand, those transformations may be obtained as symmetry transformations of the metric of the potential space:

$$dS^2 = G_{AB}(f^C)\,df^A\,df^B.$$

Generation of new vacuum Einstein-Maxwell solutions from a given vacuum or Einstein-Maxwell field has been the object of many studies in the case when the spacetime admits a non null Killing vector, or two commuting non null Killing vectors as in the stationary axisymmetric case. In particular, the counterpart of Bäcklund transformations have been found.

6.7 FURTHER RESULTS AND PROBLEMS

1. Show that $R_{abcd} = R_{cdab}$ follows from (6.4) and (6.5). [*Hint:* Let $2T_{abcd} = R_{abcd} + R_{cdab}$, and show that $R_{abcd} = T_{acbd} + T_{adcb}$.]

2 (Tetrad formalism) Let $\{e_a\}_{0 \le a \le 3}$ be a tetrad field (*i.e.*, a frame) and let $\{\theta^a\}$ be its dual. In local coordinates,

$$e_a = e_a{}^k \partial_k; \quad \theta^a = e_k{}^a \, dx^k,$$

where

$$e_a{}^k e_k{}^b = \delta_a{}^b; \quad e_i{}^a e_a{}^j = \delta_i{}^j.$$

We use a, b, ... for "tetrad indices" and i, j, ... for coordinate indices. The linear connection is determined by

$$\omega^a{}_b = \gamma^a{}_{bc} \theta^c,$$

where

$$\gamma^a{}_{bc} = e_i{}^a e_c{}^k \nabla_k e_b{}^i,$$

also known as the *Ricci rotation coefficients*.

The torsion of the connection, $\Sigma^a := d\theta^a + \omega^a{}_b \theta^b$, vanishes, and the curvature is given by $\Omega^a{}_b = d\omega^a{}_b + \omega_a{}^c \omega_c{}^b$.

(a) Show that the tetrad components of the curvature tensor are given by

$$R^a{}_{bcd} = \partial_c \gamma^a{}_{bd} - \partial_d \gamma^a{}_{bc} + \gamma^a{}_{ec} \gamma^e{}_{bd} - \gamma^a{}_{ed} \gamma^e{}_{bc} + \gamma^a{}_{be} b^e{}_{bc},$$

where $b^a{}_{bc}$ is defined by $d\theta^a = \frac{1}{2} b^a{}_{bc} \theta^b \theta^c = e_b{}^j e_c{}^k \partial_j e_k{}^a \, \theta^b \theta^c$.

(b) Derive (6.6) and (6.7) for the above $R^a{}_{bcd}$.

3. (Chern and Euler classes) Let Ω be the curvature form (omitting its indices). The ith Chern class is defined quite generally as the cohomology class of

$$c_i := \mathrm{tr} \left(\wedge^i \Omega \right) = \mathrm{tr} \left(\Omega \wedge \cdots \wedge \Omega \right).$$

Check that c_i is indeed closed, as a consequence of Bianchi's identities.

(a) Show that its integral on a closed (smooth) $2i$-cycle is independent of the choice of the connection.

Show that in a V_4,

$$c_2 = -R^{*ab}{}_{cd} R_{ab}{}^{cd} \, dV,$$

dV being the natural volume form on (V_4, g_{ab}).

(b) Show a similar independence property for the Euler form

$$\chi_n := \varepsilon^{a_1 a_2 \cdots a_{2n}} \Omega_{a_1 a_2} \wedge \cdots \wedge \Omega_{a_{2n-1} a_{2n}} / \beta(n),$$

where $\varepsilon^{a_1 a_2 \cdots a_{2n}}$ is the sign of the permutation

$$(a_1 a_2 \cdots a_{2n}),$$

in an even-dimensional manifold of dimension $2n$. $\beta(n)$ is a normalization constant, which ensures that the integral of this form on a compact manifold yields an integer.

One has, in four dimensions,

$$\chi_2 = {}^*R^{*ab}{}_{cd} R_{ab}{}^{cd}.$$

There are generalizations to other types of connections (connections on principal bundles), and topological invariants can be constructed from these classes (even though a differential structure seems necessary for their definition)

4. Let $X^a = F u^a$ be a timelike Killing field, with $u^a u_a = -1$.

(a) The Killing equations are equivalent to

$$\nabla_{[a} \dot{u}_{b]} = 0.$$

(b) Define its twist by

$$\omega^a = \eta^{abcd} X_b \nabla_c X_d.$$

Show that in a vacuum spacetime, one has $\nabla_{[a} \omega_{b]} = 0$, so that ω_a is locally a gradient. Show that $\mathcal{L}_X \omega_a = 0$.

5. Let $\mathcal{C}(k)$ be a congruence of null geodesics generated by a null vector field k^a, η^a being its connecting vector, and $f_a{}^b$ the projection operator on a space-like plane Π orthogonal to k^a and to l^a, where l^a is not proportional to k^a, $l^a l_a = 0$ and $l^a k_a = -1$.

(a) Show that l^a is uniquely determined and that

$$f_a{}^b = \delta_a{}^b + k_a l^b + k^b l_a.$$

We assume in the sequel that $k^c \nabla_c l^a = 0$.

(b) Prove that $k^c \nabla_c (k_a \eta^a) = 0$.

(c) Use a hat to denote the projected components of any tensor ($\hat{T}_{a \ldots} = f_a{}^m \cdots T_{m \ldots}$), and let $\theta_{ab} = \nabla_a k_b$. The trace ($\hat{\theta}$), trace-free symmetric part ($\hat{\sigma}_{ab}$) and antisymmetric part ($\hat{\omega}_{ab}$) of θ_{ab} are called respectively the expansion, shear, and twist of $\mathcal{C}(k^a)$.

(d) Show that $\hat{\theta}_{ab}k^b = \hat{\theta}_{ab}l^b = 0$ and that

$$k^c\nabla_c(\nabla_a k_b) = -\nabla_a k^c \nabla_c k_b - R_{abcd}k^c k^d.$$

(e) Projecting this equation on Π, and taking the trace, antisymmetric part, and trace-free symmetric part, derive

$$\begin{aligned}
k^c\nabla_c\hat{\theta} &= -\frac{1}{2}\hat{\theta}^2 - \hat{\sigma}_{ab}\hat{\sigma}^{ab} - \hat{\omega}_{ab}\hat{\omega}^{ab}; \\
k^c\nabla_c\hat{\omega}_{ab} &= -\hat{\theta}\hat{\omega}_{ab}; \\
k^c\nabla_c\hat{\sigma}_{ab} &= -\hat{\theta}\hat{\sigma}_{ab}f_a{}^c f_b{}^d C_{cmdn}k^m k^n.
\end{aligned}$$

6. Assume that g_{ab}, $g_{ab,c}$ are continuous across a smooth hypersurface S, while some of the derivatives $g_{ab,cd}$ have jumps. Recall (see Ch. 1) that if $S = \{f = 0\}$ and $k_a = f_{,a}$,

$$[g_{ab,cd}] = a_{ab}k_a k_b.$$

(a) Show that

(6.80) $$k_c[R_{abde}] + k_d[R_{abec}] + k_e[R_{abcd}] = 0$$

and

(6.81) $$[G_a{}^b]k_b = 0.$$

(b) Using the invariance of R_{abcd} under a change of coordinates tangent to the identity, show that $[R_{abcd}]$ does not change under a C^2 change of coordinates $x'^a = x'^a(x^b)$ such that

$$[x^a{}_{,b'c'd'}] = t^a k_{b'} k_{c'} k_{d'}.$$

Show that a_{ab} becomes $a_{ab} + t_a k_b + t_b k_a$. In particular, one can assume $a = a_{ab}g^{ab} = 0$, and this condition will be preserved by a change in which $t_a k^a = 0$.

7. In the situation of the preceding problem, assume that $R_{ab} = 0$.
(a) Show that

(6.82) $$k_c[R_{ab}{}^c{}_d] = 0; \quad (a_{ab} - \frac{a}{2}g_{ab})k^b.$$

Derive from (6.80) and (6.82) that $k_a k^a = 0$ and $k^c\nabla_c k_a = 0$.
(b) Assume $a = 0$. Show that

(6.83) $$(2k^m\nabla_m + \nabla_m k^m)[R_{abcd}] = 0.$$

(It will be first necessary to give a meaning to $k^m\nabla_m$ on S).

(c) Let $b_{ab} = -\frac{1}{2}a_{ab}$. Note that $b_{ab}k^b = 0$. Let $e = b_{ab}b^{ab}$. Show that $\nabla_a(ek^a) = 0$. Conclude, using $k^c\nabla_c k^a = 0$, that

$$\tau_{abcd} := \frac{1}{2}[R_a{}^m{}_c{}^p][R_{bmdp}] = e\,k_a k_b k_c k_d$$

is conservative. Compare this expression with the Bel-Robinson tensor.

8. Discuss the linearization stability of the trivial solution for $F(x) = (Ax, x)$, where A is a real symmetric matrix and $x \in \mathbf{R}^n$. Observe that the set of solutions has in general a conic singularity. How can one generalize to functions with a non-degenerate critical point? Investigate the relation to singularity theory.

9. Compute the symbol of the (linearization of) Einstein's equations. What is the significance of the non-null elements of the kernel of this symbol.

10. Show that Einstein's equations in FLRW space read

$$\frac{3\dot{a}^2}{a^2} = 8\pi\rho - \frac{3k}{a^2};$$
$$\frac{3\ddot{a}}{a} = -4\pi(\rho + 3p).$$

11. Prove (6.79).

12. The following are equivalent to global hyperbolicity:

(a) (Leray, 1951) M is strongly causal and $C(p,q)$ is compact for any two points p and q.

(b) (Hawking and Ellis, 1973) M is strongly causal and $J^+(p) \cap J^-(q)$ is compact for any two points p and q.

13. Check that Einstein's equations can be derived from the second-order Lagrangian $R\sqrt{-g}$ by varying independently the metric and the connection. How do you account for this?

14. The following describes an extension of Schwarzschild spacetime where the singularity at $r = 2M$ has been removed. It is useful to describe the causal structure of this spacetime.

(a) Show that radial null geodesics are defined by

$$t = \pm r_* + \text{ const.},$$

where

$$r_* = r + 2M\ln(\frac{r}{2M} - 1)$$

is the Regge-Wheeler tortoise coordinate.

(b) In the null coordinates $u = t - r_*$, $v = t + r_*$, the metric becomes

$$ds^2 = -(1 - \frac{2M}{r})\,du\,dv + r^2\,d\Omega^2,$$

where r is defined implicitly by (a), and $d\Omega^2 = d\theta^2 + \sin^2\theta \, d\varphi^2$. This may be written

$$-\frac{2M}{r}e^{-r/(2M)}e^{(v-u)/4M} \, du \, dv + r^2 \, d\Omega^2.$$

(c) Make the changes $U = e^{-u/(4M)}$, $V = e^{v/(4M)}$, $T = (U+V)/2$, $X = (U-V)/2$. Show that $(\frac{r}{2M} - 1)e^{r/(2M)} = X^2 - T^2$, $t/(2M) = 2\tanh^{-1}(T/X)$, and finally

$$ds^2 = \frac{32M^3e^{-r/(2M)}}{r}(-dT^2 + dX^2) + r^2 \, d\Omega^2.$$

(Kruskal (1960)).

15. Let $(\bar{V}, \bar{g}_{ab}) \to (V, g_{ab})$ be a conformal mapping:

$$g_{ab} = \Omega^2 \bar{g}_{ab}.$$

The pair (Ω, g_{ab}) is defined only upto a factor θ, since (Ω, g_{ab}) and $(\theta\Omega, \theta^2 g_{ab})$ correspond to the same $\bar{g}_{ab}$.

(a) Show that

$$
\begin{aligned}
\bar{R}^a{}_{bcd} &= R^a{}_{bcd} + 2\Omega^{-1}(\delta^a{}_{[c}\nabla_{d]}\Omega_b - g_{b[c}\nabla_{d]}\Omega^a) \\
&\quad - 2\Omega^{-2}\delta^a{}_{[c}g_{d]b}\Omega_m\Omega^m; \\
\bar{R}_{ab} &= R_{ab} + \Omega^{-1}(2\nabla_a\Omega_b + g_{ab}\Omega_m\Omega^m) \\
&\quad - 3\Omega^{-2}g_{ab}\Omega_m\Omega^m; \\
\bar{R} &= \Omega^2 R + 6\Omega\nabla_m\Omega^m - 12\Omega_m\Omega^m; \\
\bar{B}_{ab} &= B_{ab} + \Omega^{-1}(2\Omega_a\Omega_b - \frac{1}{2}g_{ab}\Omega_m\Omega^m),
\end{aligned}
$$

where $\Omega_a = \nabla_a\Omega$.

(b) Verify, using (6.10),

$$
\begin{aligned}
\bar{C}^a{}_{bcd} &= C^a{}_{bcd}; \\
S\Omega^{-1}\bar{\nabla}_e\bar{C}^a{}_{bcd} &= S\nabla_e(\Omega^{-1}C^a{}_{bcd}),
\end{aligned}
$$

where S stands for circular permutation over (c, d, e). Conclude that

(6.84) $$\Omega^{-1}\bar{\nabla}_a\bar{C}^a{}_{bcd} = S\nabla_a(\Omega^{-1}C^a{}_{bcd}).$$

Show that Bianchi's identities and (6.10) give

$$\nabla_a C^a{}_{bcd} = \nabla_{[c}R_{d]b} + \frac{1}{6}g_{b[c}\nabla_{d]}R.$$

16. With the set-up of the previous problem, assume now that

$$\bar{R}_{ab} = \Lambda \bar{g}_{ab}.$$

(a) Show that $\bar{B}_{ab} = 0$ implies

$$\nabla_a \Omega_b = -\frac{\Omega}{2} B_{ab} + \frac{1}{4} g_{ab} \nabla_m \Omega^m.$$

(b) Show that $\bar{\nabla}_a \bar{C}^a{}_{bcd} = 0$ implies

$$\nabla_{[c} R_{d]b} + \frac{1}{6} g_{b[c} \nabla_{d]} R - \Omega^{-1} C^a{}_{bcd} \Omega_a = 0.$$

(c) Show that $\Omega^{-1} \nabla_a \bar{R} + 12 \bar{B}_{ab} \Omega^b = 0$ implies

$$6 \nabla_a \nabla_b \Omega^b + \Omega \nabla_a R - R \Omega_a + 12 R_{ab} \Omega^b = 0.$$

This equation does not contain negative powers of Ω any more.

(d) $\bar{R} = 4\Lambda = \Omega^2 R + 6\Omega \nabla_m \Omega^m - 12 \Omega_m \Omega^m$.

(e) Show that $\nabla_a (\Omega^{-1} C^a{}_{bcd}) = 0$.

Comment: Two kinds of regular conformal vacuum field equations were considered in the literature, and the above formulae are useful in dealing with them. If we take as unknowns the quantities g_{ab}, Ω, $s := (1/4)\nabla_m \Omega^m$, $s_{ab} := (1/2)B_{ab}$, and $d^a{}_{bcd} := \Omega^{-1} C^a{}_{bcd}$, the equations established in the five questions of this problem constitute the system used by Friedrich (1983). If we take g_{ab}, Ω, $r_{ab} := R_{ab}$, $f := R$ and $d^a{}_{bcd}$ as unknowns, equation (6.84) with the results of (c), (d) and (e) form the system of Choquet-Bruhat and Novello (1987).

17. Let u^a be a timelike field and let $h_a{}^b$ be the orthogonal projection on the space orthogonal to u^a. We say that the congruence $\mathcal{C}(u)$ generated by u^a is *rigid* if

$$\mathcal{L}_u h_{ab} = 0.$$

Let us consider two points to be equivalent if they belong to the same curve of the congruence. Explain briefly why one can view the quotient S of spacetime by this relation as a manifold. A tensor field defines a tensor field on S if $T \perp u = 0$ and $\mathcal{L}_u T = 0$. We write

$$\perp T_{a...c}{}^{b...d} = h_a{}^m \dots h_c{}^n h_p{}^b \dots h_q{}^d T_{m...n}{}^{p...q}.$$

(a) Show that $\perp \nabla_a h_{bc} = 0$, so that h_{ab} may be considered as a metric on S.

(b) Show that $\mathcal{C}(u)$ is shear-free and without expansion.

(c) Show that for any v^a tangent to S,

$$\perp \nabla_a \perp \nabla_b v_c = \perp \nabla_a \nabla_b v_c + (h_b{}^m \omega_{ac} u^n + h_c{}^n \omega_{ab} u^m) \nabla_m v_n,$$

which leads to an expression for the curvature tensor of S:

$$R^a{}_{bcd}(h) = \perp \left[R^a{}_{bcd} + (\omega_a{}^d \omega_{bc} - \omega_b{}^d \omega_{ac}) - 2\omega_{ab}\omega_c{}^d \right],$$

or

$$R_{\alpha\beta\gamma\delta}(h) = R_{\alpha\beta\gamma\delta} + 3\omega_{\alpha\beta}\omega_{\gamma\delta}.$$

It follows in particular that

$$\perp \mathcal{L}_u(R_{abcd} + 3\omega_{ab}\omega_{cd}) = 0.$$

(c) Assume that V_4 is flat ($R^a{}_{bcd} = 0$). Show that

$$\mathcal{L}_u\omega_{ab} = 0,$$

which implies

$$\nabla_{[a}\dot{u}_{b]} = 0.$$

Prove the *Herglotz-Noether theorem*: in Minknowski spacetime, a rigid motion must be isometric. [*Hint:* Derive from the definition of rigidity the relations: $\nabla_a h_{bc} = 2\omega_{a(b}u_{c)} - 2u_a \dot{u}_{(b}u_{c)}$, $\nabla_a u_b = \omega_{ab} - u_a \dot{u}_b$, compute $(\nabla_a \mathcal{L} - \mathcal{L}\nabla_a)u_b$ where $\mathcal{L}$ is the Lie derivative in the direction of u^a, and use exercise 4.]

NOTES

The material in this chapter should give the reader a basis to read the specialized literature given in the references.

For background information on differential geometry, and in particular for the material in §6.1, see Kobayashi-Nomizu (1963) for the Riemannian case, Lichnerowicz (1967) (in Battelle Rencontres) and Misner-Thorne-Wheeler (1973) for the Lorentzian case; we generally followed the notation and sign conventions of the latter. Other standard texts include Weyl (1952), Pauli (1921), Fock (1959), Lichnerowicz (1955), Synge (1960), Hawking-Ellis (1973) and Wald (1984). The Petrov classification can be found in Petrov (1969); the Bel-Robinson tensor was introduced independently by Bel (1959) and Robinson (1959). See Marsden-Tipler (1980) for a review of results on maximal hypersurfaces, and Bartnik (1984) for more recent material.

Einstein's equations were proposed in 1915; Hilbert (1915) gave their variational derivation. The superpotential is due to Freud (1939). The early history of General Relativity is quite interesting and has recently received intense attention; see volumes 1, 3, 5 of the Einstein Studies (1989, 1992, 1993). The Hamiltonian formalism can be found in Arnowitt-Deser-Misner, and Fischer-Marsden, including the Poisson structure. For a discussion of Noether's theorem, see Olver (1986) and Kichenassamy (1994). The latter shows the equivalence of a method of computation of the canonical tensor, found in the physics literature, with a version of the prolongation method.

The Cauchy problem has evolved hand in hand with the theory of hyperbolic systems. After the solution of the analytic Cauchy problem, attention was devoted to showing in what sense Einstein's equations describe "waves." This was achieved on the one hand by the study of the propagation of discontinuities (see problem 6) and by the creation of an existence-uniqueness theory in the non-analytic case, in which the wave operator associated to the metric plays the prominent role. See Darmois (1927) Lichnerowicz (1955) for the early stages of the theory. The harmonic coordinate condition was introduced in linearized form by Einstein, and recognized in its present form by de Donder. It has been used in the non-analytic Cauchy problem by Choquet-Bruhat from 1952 onwards (see also the uniqueness results of Stellmacher (1937) and Choquet-Bruhat (1968)). The theory of symmetric-hyperbolic systems was applied to General Relativity by Fischer-Marsden and Hughes-Kato-Marsden, and it is this formulation that we mostly followed. See also Choquet-Bruhat, Christodoulou and Francaviglia for a further discussion of optimal regularity conditions. Note that Leray's theory of hyperbolic systems (which was the motivation for the Agmon-Douglis-Nirenberg definition of ellipticity for systems) is required to handle some non-vacuum situations such as perfect fluids. Recent surveys on the Cauchy problem include Choquet-Bruhat and York (1981) and the articles on cosmic censorship in the volume edited by Gotay *et al.* (1992). The former contains details on the construction of initial data sets.

Two different proofs of the positive mass conjecture are due to Schoen and Yau, and Witten.

Two further recent techniques for constructing solutions of Einstein's equations should be noted: H. Friedrich finds solutions that are global to the future of the hyperboloid $t = \sqrt{1 + |x|^2}$ in Minkowski space; data are assumed to have behavior at null infinity which ensures that conformal compactification is applicable. Christodoulou and Klainerman (1993) construct initial data sets leading to spaces which are complete (free of singularities) and are globally close to Minkowski space in a suitable sense; this circumvents the difficulty that harmonic coordinates are globally "unstable" (Choquet-Bruhat (1973)) by working in spacetimes with $g_{0a} = 0$ and data with $K = 0$. The construction rests on the properties of solutions of Bianchi's identities established in Christodoulou-Klainerman (1990), which parallel the sharp estimates for the wave equation obtained via the use of "invariant norms" (see Ch, 1). It uses in particular the conservation of the Bel-Robinson tensor to derive energy-type estimates.

The characteristic initial-value problem has been completely solved in the analytic case by Darmois (1927); its impact for gravitational radiation was studied by Bondi (1969) and Sachs (1962). The non-analytic case was considered by Müller zum Hagen and Seifert (1977). The conformal vacuum field equations (exercises 15 and 16) were solved by Friedrich (1983) and Choquet-Bruhat and

Novello (1987).

Our presentation of linearization stability is meant to be an introduction to Fischer-Marsden-Moncrief (1980) (and its references). The second-order conserved quantities are due to Taub. Our remarks on "reliability" of perturbation methods is based on Lottermoser (1992), Rendall (1989) and Damour-Schmidt (1990).

Exact solutions are studied in great detail in Kramer *et al.* and Hoenselaers and Dietz (1983).

The basic singularity theorems are due to Hawking and Penrose, see Hawking (1966) and Penrose. See also Marsden and Tipler, Clarke, and their references. The singularity theorems are so general that they do not give a model for the singularities. Two models may be mentioned: Lifschitz and Khalatnikov have proposed models of singular spacetimes with Kasner-like singular behavior. The Taub-NUT space can be perturbed into spacetimes with one Killing vector and a compact Cauchy horizon. In both cases, it is of interest to understand what the behavior of "generic" solutions close to those spaces is. The prevalent view prior to the singularity theorems was that singular spaces were always highly symmetric, and that one should therefore be able to "perturb away" singularities.

There are several attempts at constructing, given a singular spacetime, suitable boundaries which would help describe the singular behavior more precisely (b-boundaries, c-boundaries, TIPs and TIFs, etc ...), see Penrose-Rindler for a detailed discussion of conformal infinity and Clarke (1994) for a recent review of these issues, including some results on the extension of spacetimes beyond singularities.

These investigations have been a strong incentive to develop the detailed causal structure of spacetimes (Penrose, Geroch, Hawking, among others).

REFERENCES

ARNOWITT, R., DESER, S. AND MISNER, D. W. (1959) Dynamical structure and deformation of energy in General Relativity, *Phys. Rev., 116*: 1322–1330.

BARTNIK, R. (1984) Existence of maximal hypersurfaces in asymptotically flat spacetimes, *Comm. Math. Phys., 94*: 155–175.

BEL, L. (1959) Introduction d'un tenseur du 4ème ordre, *C. R. Acad. Sci. Paris, 247*: 1094.

BONNOR, W. B. AND VICKERS, P. A. (1981) Junction conditions in General Relativity, *Gen. Rel. and Grav., 13*: 29–36.

CHENG, S.-Y. AND YAU, S.-T. (1976) *Ann. Math., 104*: 407.

CHOQUET-BRUHAT, Y. (1972) Stabilité des solutions d'équations hyperboliques non linéaires, Applications à l'espace-temps de Minkowski en Relativité Générale, *C. R. Acad. Sci. Paris, 274A*: 843–846.

CHOQUET-BRUHAT, Y.CHRISTODOULOU, D. AND FRANCAVIGLIA, F. (1978) Cauchy data on a manifold, *Ann. Inst. H. Poincaré, 29*: 241–255.

CHOQUET-BRUHAT, Y. AND YORK, J. W., JR. (1980) The Cauchy Problem, in *General Relativity and Gravitation,* A. Held ed., Plenum.

CHOQUET-BRUHAT, Y. AND NOVELLO, M. (1987) Système conforme régulier pour les équations d'Einstein, *C. R. Acad. Sci. Paris, 305*: 155–160.

CHRISTODOULOU, D. AND KLAINERMAN, S. (1993) *The Global Nonlinear Stability of Minkowski Space,* Princeton.

CLARKE, C. J. S. (1993) *The analysis of space-time singularities,* Cambridge.

DAMOUR, T. AND SCHMIDT, B. (1990) Reliability of perturbation theory in General relativity, *J. Math. Phys., 31*: 2441–2453.

DARMOIS, G. (1927) Les équations de la gravitation einsteinienne, *Mém. des Sci. Math.,* fasc. 25, Gauthier-Villars, Paris.

EINSTEIN STUDIES (1989, 1992, 1993) vol. 1: *Einstein and the History of General Relativity,* D. Howard and J. Stachel Eds., vol. 3: *Studies in the History of General Relativity,* J. Eisenstaedt and A. Kox Eds., vol. 5: *The Attraction of Gravitation,* J. Earman, M. Janssen and J. D. Norton Eds., Birkhäuser, Boston.

FISCHER, A. E. AND MARSDEN, J. E. (1972) *Comm. Math. Phys., 28*: 1–38.

FISCHER, A. E., MARSDEN, J. E. AND MONCRIEF, V. The structure of the space of solutions of Einstein's equations 1: One Killing field, *Ann. Inst. H. Poincaré, 33*: 147–194.

FOCK, V. (1959) *The Theory of Space, Time and Gravitation,* Pergamon, New York.

FOURÈS-BRUHAT, Y. (1952) Théorème d'existence pour certains systèmes d'equations aux dérivees partielles non-linéaires, *Acta Math., 88*: 141–225.

FRIEDRICH, H. (1983) Cauchy problems for the conformal vacuum field equations in General Relativity, *Comm. Math. Phys., 91*: 445–472.

FRIEDRICH, H. (1987) On the existence of *n*-geodesically complete or future complete solutions of Einstein's field equations with smooth asymptotic structure, *Comm. Math. Phys., 107*: 587–609.

GANNON, D. (1976) On the topology of spacelike hypersurfaces, singularities, and black holes, *Gen. Rel. and Grav., 7*: 219–232. See also *J. Math. Phys., 16*: 2364–2367 (1975).

GOTAY, M. J., MARSDEN, J. E., AND MONCRIEF, V. (1992) Mathematical Aspects of Classical Field Theory, Proc. AMS-IMS-SIAM Joint Summer Research Conference, July 20–26, 1991, AMS Providence.

HAWKING, S. W. (1966) The occurrence of singularities in General Relativity, I: *Proc. Roy. Soc., 294A*: 511; II: *Proc. Roy. Soc., 295A*: 490; III: *Proc. Roy. Soc., 294A*: 187 (1967).

HAWKING, S. W. AND ELLIS, G. F. R. (1973) *The Large-Scale Structure of Space-Time,* Cambridge.

HAWKING, S. W., AND ISRAEL, W. (Eds.) (1979) General Relativity, an Einstein Centenary Survey, Cambridge.

HAWKING, S. W., AND PENROSE, R. (1970) The singularities of gravitational collapse and cosmology, *Proc. Roy. Soc. London, A314*: 529–548.

HOENSELAERS, C., AND DIETZ, W. (1983) *Solutions of Einstein's Equations: Techniques and Results,* Proceedings, Retzbach, Germany, *Lect. Notes in Physics, 205,* Springer, Berlin.

HUGHES, T., KATO, T. AND MARSDEN, J. E. (1976) Well-posed quasi-linear second-order second order hyperbolic systems with applications to nonlinear elastodynamics and General Relativity, *Arch. Rat. Mech. Anal., 63*: 273–294.

KICHENASSAMY, S. (1994) The prolongation formula for tensor fields, *J. Phys., A: Math. Gen., 27*: 7857–7874.

KOBAYASHI, S. AND NOMIZU, K. (1963) *Foundations of Differential Geometry,* Interscience, N. Y.

KRAMER, D., STEPHANI, H., HERTL, E., AND MACCALLUM, M.; (E. SCHMUTZER Ed.) (1980) *Exact Solutions of Einstein's Field Equations,* Cambridge University Press.

KRAMER, D. AND NEUGEBAUER, G. (1983) Bäcklund Transformations in General Relativity, in *Lect. Notes in Physics, 205.*

KUNDT, W., AND TRÜMPER (1962) Beiträge zur Theorie der Gravitations-strahlungsfelder, *Akad. Wiss. Univ. Mainz, Abhandl. Math.-Nat. Kl., 12*.

LERAY, J. (1951) *Hyperbolic Differential Equations,* Lecture Notes, Inst. for Adv. Study, Princeton.

LICHNEROWICZ, A. (1955) *Theéories Relativistes de la Gravitation et de l'Electromagnétisme,* Masson, Paris.

LICHNEROWICZ, A. (1960) Ondes et radiations électromagnétiques et gravitationnelles en Relativité Générale, *Ann. Mat. Pura Appl., 150*: 1–95.

LICHNEROWICZ, A. (1990) Relativity and Mathematical Physics, in *Albert Einstein, 1879–1979,* d'Inverno ed.

LICHNEROWICZ, A. (1961) Propagateurs et Commutateurs en Relativité Générale, *Publ. Sci. IHES,* Presses Universitaires de France.

LOTTERMOSER, M. (1992) A convergent post-Newtonian approximation for the constraint equations in general relativity, *Ann. Inst. H. Poincaré, 57 (3)*: 279–317.

MARSDEN, J. E. AND TIPLER, F. (1980) Maximal hypersurfaces and foliations of constant mean curvature in general relativity, *Phys. Rep., 66*: 109–139.

MARKUS, L. (1955) Line element fields and Lorentz structures on differentiable manifolds, *Ann. Math. 62*: 411–417.

MISNER, C. W., THORNE, K. S. AND WHEELER, J. A. (1973) *Gravitation,* Freeman, San Francisco.

MÜLLER ZUM HAGEN, H. AND SEIFERT, H.-J. (1977) On characteristic initial value and mixed problems, *Gen. Rel. and Grav., 8*: 259–301.

OLVER, P. J. (1986) *Application of Lie Groups to Differential Equations,* Springer, New York, 1st edition, 1986; 2nd edition, 1993.

PAULI, W. (1921) *Relativitätstheorie,* Teubner, Leipzig (translation reprinted by Dover (1981)).

PENROSE, R. (1972) Techniques of differential topology in relativity, SIAM, Philadelphia.

PENROSE, R. AND RINDLER, W. (1984–1988) *Spinors and Spacetimes,* Cambridge (2 volumes).

PETROV, A. Z. (1969) *Einstein spaces,* Oxford.

RENDALL, A. D. (1990) Convergent and divergent perturbation series and the post-Newtonian approximation scheme, *Class. Quantum Grav., 7*: 803–812.

ROBINSON, I. (1959) unpublished communication at the 'Colloque International sur les Théories Relativistes de la Gravitation' held in Royaumont, France.

SACHS, R. K. (1962) On the characteristic initial value problem in gravitational theory, *J. Math. Phys., 3*: 908–914.

SCHOEN, R. AND YAU, S.-T. (1979) On the proof of the positive mass conjecture in General Relativity, *Comm. Math. Phys., 65*: 45–76.

STELLMACHER, K. (1937) *Math. Ann., 115*: 136–152.

SYNGE, J. L. (1960) *Relativity: The General Theory,* North-Holland, Amsterdam.

WALD, R. M. (1984) *General Relativity,* U. of Chicago Press.

WEYL, H. (1952) *Space-time-matter,* Dover.

WILL, C. M. (1981) *Theory and Experiment in Gravitational Physics,* Cambridge.

WITTEN, E. (1981) A new proof of the positive energy theorem, *Comm. Math. Phys., 80*: 381–402.

Index of Notation

$\| \ \|_p$	L^p norm (except in§5.2)
$\| \ \|_s$	H^s norm (except in§5.2)
$\| \ \|_{s,p}$	$L^{s,p}$ norm
$\|u; X\|$	norm of u in the space X.
$\square$	$\partial_t^2 - \Delta$
Δ	Laplacian
D_j	$-i\partial_j$
$|D|^\mu$	$\mathcal{F}^{-1}|\xi|^\mu \mathcal{F}$
$\mathcal{F}u, \hat{u}$	Fourier transform of u
$\text{meas}(E)$	Lebesgue measure of the set E
$\text{Res}_z f$	residue of f at z
$u_a = u_{,a} = \partial_a u$	partial derivatives

Indices are raised and lowered using the underlying metric.

$u_{;a} = \nabla_a u$	covariant derivatives
$u_{[ab]}$	$(1/2)(u_{ab} + u_{ba})$
$u_{(ab)}$	$(1/2)(u_{ab} - u_{ba})$

more generally, $[ab\dots m]$ indicates averaging over all permutations,
and $(ab\dots m)$ indicates averaging weighted by the signature of each permutation.

$u_{[a|\dots|b]}$	permutations of a and b only.
$WF(u); WF_s(u)$	C^∞- and H^s- wave-front sets.

273

Index

Printed and bound by CPI Group (UK) Ltd, Croydon, CR0 4YY

17/10/2024

01775700-0010